EMIS

Serie

Properties of
III-V Quantum Wells
and Superlattices

ELECTRONIC MATERIALS INFORMATION SERVICE
Other books in the EMIS Data Reviews Series from INSPEC:

Details of books in the EMIS Data Reviews Series are available at www.theiet.org/publications/booksandconferences

Properties of
III-V Quantum Wells
and Superlattices

Edited by Pallab Bhattacharya
University of Michigan, USA

The Institution of Engineering and Technology

Published by The Institution of Engineering and Technology, London, United Kingdom

First edition © 1996 The Institution of Electrical Engineers
Paperback edition © 2007 The Institution of Engineering and Technology

First published 1996 (0 85296 881 7)
Paperback edition 2007

The Institution of Engineering and Technology
Michael Faraday House
Six Hills Way, Stevenage
Herts, SG1 2AY, United Kingdom

www.theiet.org

British Library Cataloguing in Publication Data
A catalogue record for this product is available from the British Library

ISBN 978-0-86341-778-8

Printed in the UK by Lightning Source UK Ltd, Milton Keynes

Contents

Introduction

Ever since the first proposal of a semiconductor superlattice by Esaki and Tsu in 1969-1970, the development of superlattices and quantum wells has been one of the most exciting and useful in modern semiconductor electronics. The first experimental superlattice was synthesised by Blakeslee and Aliotta using chemical vapour deposition in 1970 with the $GaAs/GaAs_{1-x}P_x$ system. At almost the same time A. Cho (1971) and J. Woodall (1972) demonstrated similar structures with GaAs/GaAlAs by the molecular beam epitaxy (MBE) and liquid phase epitaxy (LPE) growth techniques, respectively. Pioneering work on the understanding of the optical and electronic properties of superlattices and quantum wells and effects related to size quantisation has been done by many researchers. Notable amongst this work are the demonstration of resonant tunnelling phenomena by L. Esaki and co-workers, and the demonstration of the quantised Hall effect by K. von Klitzing and co-workers.

It is perhaps instructive at this point to distinguish between quantum wells and superlattices, and the different types of these structures. The type of superlattice originally proposed by Esaki is a compositionally modulated structure. These synthetically modulated structures are classified as Type I, Type I′ and Type II, depending on the band line-up of the constituent semiconductors. The most commonly used form is the Type I, in which the quantum wells for electrons and holes overlap spatially. In the other two types of superlattices, the wells for electrons and holes are staggered, so that there is no overlap. In a superlattice the constituent quantum wells are so close to each other, that their carrier wavefunctions overlap and form minibands. The composite material acquires its own band structure and bandgap. In a multiple quantum well structure, the individual quantum wells are separated by thick enough barriers, so that they are quantum mechanically decoupled. Both are synthesised and used for fundamental studies and device applications. It is also fair to state that the Type I structures have been most intensively studied and applied. Another type of staggered superlattice structure is produced by heavily doped n- and p-type alternating layers. These doping superlattices were studied and demonstrated by G. Dohler and co-workers starting in 1972.

The development of crystal growth techniques such as MBE and metal-organic vapour phase epitaxy (MOVPE) ushered in a new era in which semiconductor heterostructures, including quantum wells and superlattices, can be grown with unprecedented control over thickness, composition and doping. The ability to tailor the electronic and optoelectronic properties of multilayered materials by altering thickness, composition and band line-up of the constituent semiconductors is commonly termed 'bandgap engineering', and this is put to good use in enhancing the performance of electronic and optoelectronic devices and in providing building blocks for new devices. It is no accident that quantum well lasers outperform any other type of semiconductor laser. The step-like density of states function in quantum wells offers advantages in terms of laser performance. Similarly, use of quantum wells and superlattices enables the tailoring of the impact ionisation coefficients in avalanche photodiodes.

In the early days of development and understanding, most of the work and applications centred around the lattice matched GaAs/GaAlAs quantum wells and the InP-based GaInAs/InP or $Ga_{0.47}In_{0.53}As/Al_{0.48}In_{0.52}As$ quantum wells. This imposed some restrictions on tailorability of material properties. Another degree of freedom became available with the use of strained layer superlattices. Use of strained layers not only enabled additional tailoring of bandgaps, band offsets and miniband parameters, additionally the material properties such as band structure and effective mass are also altered by the biaxial strain imposed during epitaxy. Such strained heterostructures are most useful in the pseudomorphic form, where the composite structure is coherently strained. Strained heterostructures have become the mainstay of device physics and design.

Needless to say, quantum wells and superlattices have established themselves as important materials for enabling technologies in microelectronics and optoelectronics. Although the enormous amount of research done with these materials has been disseminated in journal articles, monographs and textbooks, it is befitting at this time to dedicate a volume of the EMIS series to the progress in the understanding and applications of these tailored materials. This volume is titled 'Properties of III-V Quantum Wells and Superlattices'; however, most of the Datareviews describe properties or applications of GaAs- and InP-based structures, since these have been investigated most extensively. Nonetheless, a large Datareview is dedicated to antimony-based structures. After an initial chapter providing a historical perspective the following topics are reviewed: theoretical aspects; epitaxial growth and doping; structure; electronic properties; optical properties; modulation doping; and device applications. The individual Datareviews in each chapter have been contributed by renowned experts in the field, whose work has helped in the understanding and application of these important materials, so the volume should prove to be a valuable resource for students, researchers and practising engineers.

Before concluding, I would like to thank John L. Sears, Managing Editor of the EMIS Datareviews Series, for his support and patience, and the authors and reviewers for their contributions.

Pallab Bhattacharya
Department of Electrical Engineering and Computer Science
University of Michigan
Ann Arbor, Michigan 48109-2122
USA

July 1996

Contributing Authors

K.K. Bajaj	Emory University, Department of Physics, 1501 N. Clifton Road, Atlanta, GA 30322, USA	2.4, 4.4
R. Bhat	Bell Communications Research Inc., 331 Newman Springs Road, Room 3z-203, Red Bank, NJ 07701, USA	3.3
P.K. Bhattacharya	University of Michigan, Department of Electrical Engineering and Computer Science, Ann Arbor, MI 48109-2122, USA	3.4, 6.5
A.S. Brown	Georgia Institute of Technology, Department of Electrical and Computer Engineering, Atlanta, GA 30322, USA	3.2
A.C. Bryce	University of Glasgow, Department of Electronic and Electrical Engineering, Glasgow, G12 8QQ, Scotland	4.5
C.C. Button	University of Oxford, Department of Engineering Science, Parks Road, Oxford, OX1 3PJ, UK	8.2
J.C. Campbell	University of Texas at Austin, Microelectronics Research Center, Building ENS 439, Austin, TX 78712, USA	8.3
Y.C. Chang	University of Illinois at Urbana-Champaign, Department of Physics, 1110 West Green Street, Urbana, IL 61801, USA	2.1, 2.2
J.E. Cunningham	AT&T Bell Laboratories, Room 4B-513, Crawfords Corner Road, Holmdel, NJ 07733, USA	6.4
G. Dohler	University of California, c/o P. Petroff, Engineering Materials Department, Santa Barbara, CA 93106, USA	1.2
M. Dutta	US Army Research Laboratory, Physical Division, Physical Science Directorate, Fort Monmouth, NJ 07703-5601, USA	5.3, 8.1
H.O. Everitt	US Army Research Office, Electronics Division, PO Box 1211, Research Triangle Park, NC 27709-2211, USA	5.3

D. Gammon   University of Michigan, Physics Department, 6.3
       3035-C Randall, Ann Arbor, MI 48109-2122, USA

G.J. Iafrate   US Army Research Office, Electronics Division, 5.3
       PO Box 1211, Research Triangle Park,
       NC 27709-2211, USA

R.I. Killey   University of Oxford, Department of Engineering 8.2
       Science, Parks Road, Oxford, OX1 3PJ, UK

K.W. Kim   North Carolina State University, Department of 5.3
       Electrical and Computer Engineering, Raleigh,
       NC 27695-7911, USA

B.F. Levine   AT&T Bell Laboratories, Room 7A-215, 6.9
       600 Mountain Avenue, Murray Hill, NJ 07974,
       USA

M.A. Littlejohn   North Carolina State University, Department of 5.3
       Electrical and Computer Engineering, Raleigh,
       NC 27695-7911, USA

J. Loehr   Wright-Patterson AFB, WL/ELDM, Building 620, 2.5
       2241 Avionics Circle, OH 45433, USA

S. Luryi   State University of New York at Stony Brook, 8.4
       Department of Electrical Engineering, Stony Brook,
       NY 11794-2350, USA

J.H. Marsh   University of Glasgow, Department of Electronic 4.5
       and Electrical Engineering, Glasgow, G12 8QQ,
       Scotland

E.E. Mendez   State University of New York at Stony Brook, 1.1
       Department of Physics, Stony Brook, NY 11794-3800,
       USA

R. Merlin   University of Michigan, Physics Department, 6.3
       3035-C Randall, Ann Arbor, MI 48109-2122, USA

H. Morkoc   University of Illinois at Urbana-Champaign, 3.1, 7.2
       104 S. Goodwin Avenue, Urbana, IL 61801, USA

B.R. Nag   University of Calcutta, Institute of Radio Physics 2.3
       and Electronics, 92 Acharya P C Ray Road,
       Calcutta 700 009, W B India

L.D. Nguyen   Hughes Research Laboratories, 3001 Malibu 7.3
       Canyon Road, Malibu, CA 90265, USA

R.J. Nicholas	University of Oxford, Clarendon Laboratory, Parks Road, Oxford, OX1 3PU, UK	2.6
S. Noor Mohammad	University of Illinois at Urbana-Champaign, 104 S. Goodwin Avenue, Urbana, IL 61801, USA	7.2
T.B. Norris	University of Michigan, Center for Ultrafast Optical Science, M1 IST Building 2099, Ann Arbor, MI 48109, USA	5.2
H. Okamoto	Chiba University, Department of Materials Science, 1-33, Yayoicho, Inageku, Chiba 263, Japan	4.1
G.C. Osbourn	Sandia National Laboratories, Mail Stop 0350, Albuquerque, NM 87185, USA	1.3
G. Parry	University of Oxford, Department of Engineering Science, Parks Road, Oxford, OX1 3PJ, UK	8.2
F. Pollak	Brooklyn College, Physics Department, 2900 Bedford Avenue, Brooklyn, NY 11210, USA	6.2
D.C. Reynolds	Wright State University, University Research Center, Dayton, OH 45435, USA	6.1
H. Rhan	University of Munich, Department of Physics at HASYLAB/DESY, Notkestrasse 85, D-22603 Hamburg, Germany	4.3
A. Salvador	University of Illinois at Urbana-Champaign, 104 S. Goodwin Avenue, Urbana, IL 61801, USA	3.1
A.C. Schaefer	University of Michigan, Department of Electrical Engineering and Computer Science, Ann Arbor, MI 48109-2122, USA	6.6
M. Shayegan	Princeton University, Department of Electrical Engineering, Princeton, NJ 08544, USA	7.1
J. Singh	University of Michigan, Department of Electrical Engineering and Computer Science, Ann Arbor, MI 48109-2122, USA	2.7
Y. Sirenko	North Carolina State University, Department of Electrical and Computer Engineering, Raleigh, NC 27695-7911, USA	5.3

P. Snyder	University of Nebraska, Department of Electrical Engineering, Lincoln, NE 68688-0511, USA	6.7
P.N. Stavrinou	University of Oxford, Department of Engineering Science, Parks Road, Oxford, OX1 3PJ, UK	8.2
D.G. Steel	University of Michigan, Department of Electrical Engineering and Computer Science, Ann Arbor, MI 48109-2122, USA	6.6
R.A. Stradling	Blackett Laboratory, Imperial College, Prince Consort Road, London, SW7 2BZ, UK	8.5
G. Strasser	Institut Festkoerperelektronik, Technische Universitaet Wien, Floragasse 7, 1040 Wien, Austria	5.4
M.A. Stroscio	US Army Research Office, Electronics Division, PO Box 1211, Research Triangle Park, NC 27709-2211, USA	5.3
I. Suemune	Hokkaido University, Research Institute of Electronic Science, Kita-12, Nishi-6, Kita-ku, Sapporo 060, Japan	6.8
K. Suzue	University of Illinois at Urbana-Champaign, 104 S. Goodwin Avenue, Urbana, IL 61801, USA	7.2
Y. Suzuki	Chiba University, Department of Materials Science, 1-33, Yayoicho, Inageku, Chiba 263, Japan	4.1
C.F. Tang	University of Oxford, Department of Engineering Science, Parks Road, Oxford, OX1 3PJ, UK	8.2
L. Tapfer	Scientific Director - Div. III, PASTIS - CNRSM, SS7 Appia km 712, I-72100 Brindisi, Italy	4.2
B. Vinter	Thomson-CSF, Laboratoire Central de Recherches, Domain de Corbeville, F-91404 Orsay Cedex, France	5.1
M. Whitehead	University of Oxford, Department of Engineering Science, Parks Road, Oxford, OX1 3PJ, UK	8.2
S. Yu	University of Oxford, Department of Engineering Science, Parks Road, Oxford, OX1 3PJ, UK	5.3
A. Zaslavsky	State University of New York at Stony Brook, Department of Electrical Engineering, Stony Brook, NY 11794-2350, USA	8.4

Acknowledgements

It is a pleasure to acknowledge the work both of the contributing authors named on the previous pages and of the following experts in the field.

K. Alavi	L.P. Leburton
S. Alterovitz	F. Madarasz
P. Asbeck	T. Masselink
N. Chand	M. Melloch
P.-C. Chao	F. Menendez
Y.-C. Chen	E. O'Reilley
K.Y. Cheng	A.K. Ramdas
S. Datta	J. Schulman
J.P.R. David	A. Seabaugh
L. Davis	J. Shah
D. Deppe	D. Streit
M. Dutta	R. Tober
S.J. Eglash	C.W. Tu
R. Evrard	I. Vurgaftman
S. Goodnick	B.L. Weiss
A. Gossard	C.R. Wie
G. Hasnain	M. Wojtowicz
J. Hinckley	S. Yalisove
T. Kagawa	

Abbreviations

The following abbreviations are used in this book.

2DEG	two-dimensional electron gas
2DES	two-dimensional electron system
2DHS	two-dimensional hole system
A-HHFE	antisymmetric heavy-hole free exciton
A-LHFE	antisymmetric light-hole free exciton
AC	alternating current
AFM	atomic force microscopy
ALE	atomic layer epitaxy
APD	avalanche photodiode
BH	buried heterostructure
CAT	composition analysis by thickness-fringe
CB	centre of the barrier
CBE	chemical beam epitaxy
CBED	convergent beam electron diffraction
CDE	charge-density excitation
CER	contactless electroreflectance
CIP	conduction in the plane
CMOS	complementary metal oxide semiconductor
CPP	conduction perpendicular to the plane
CR	cyclotron resonance
CSBC	clamped-surface boundary condition
CVD	chemical vapour deposition
CW	centre of the well
CW	continuous wave
D-SEED	diode based self electro-optic device
DBR	distributed Bragg reflector
DBS	double barrier structure
DC	direct current
DFB	distributed feedback
DH	double heterostructure
DOS	density of states
DQW	double quantum well
EBOM	effective bond orbital method
EL	electroluminescence
EM	electromodulation
EMA	effective mass approximation
ESM-MBE	epitaxial shadow mask molecular beam epitaxy
EW	edge of the well

F-SEED	FET biased self electro-optic device
FET	field effect transistor
FKO	Franz-Keldysh oscillation
FSBC	free-standing boundary condition
FWHM	full width at half maximum
FWM	four-wave-mixing
GID	grazing incidence diffraction
GRIN	graded-index
HBT	heterostructure bipolar transistor
HEMT	high electron mobility transistor
HFET	heterostructure field effect transistor
HH	heavy hole
HRTEM	high resolution transmission electron microscopy
HWHM	half width at half maximum
IB	interband
IC	integrated circuit
IF	interface
IFVD	impurity-free vacancy disordering
IID	impurity induced disordering
IR	image rejection
IR	infrared
LA	longitudinal acoustic
LH	light hole
LID	laser induced disordering
LL	Landau level
LNA	low noise amplifier
LO	longitudinal optic
LPE	liquid phase epitaxy
MBE	molecular beam epitaxy
MEE	migration enhanced epitaxy
MESFET	metal-semiconductor field effect transistor
MIC	microwave integrated circuit
MMIC	monolithic microwave integrated circuit
MOCVD	metalorganic chemical vapour deposition
MODFET	modulation doped field effect transistor
MOMBE	metalorganic molecular beam epitaxy
MOS	metal oxide semiconductor
MOVPE	metalorganic vapour phase epitaxy
MQW	multiple quantum well
MSEO	modified single-effective oscillator
NDR	negative differential resistance
NE(Δ)	noise equivalent temperature difference
NMR	nuclear magnetic resonance

OEIC	optoelectronic integrated circuit
OMCVD	organometallic chemical vapour deposition
OMVPE	organometallic vapour phase epitaxy
PAE	power added efficiency
PAID	photo-absorption induced disordering
PHEMT	pseudomorphic high electron mobility transistor
PIC	photonic integrated circuit
PIN	p-type intrinsic n-type
PL	photoluminescence
PLE	photoluminescence excitation spectroscopy
PMODFET	pseudomorphic modulation doped field effect transistor
PR	photoreflectance
QCL	quantum cascade layer
QCSE	quantum confined Stark effect
QHE	quantum Hall effect
QW	quantum well
QWI	quantum well intermixing
QWIP	quantum well infrared photodetector
R-SEED	resistor biased self electro-optic device
REM	reflection electron microscope
RF	radio frequency
RHEED	reflection high energy electron diffraction
RHET	resonant hot-electron transistor
RISOPS	resonant inter-subband optical phonon scattering
RRS	resonant Raman scattering
RS	Raman scattering
RTA	rapid thermal annealing
RTD	resonant tunnelling diode
RTP	rapid thermal processing
RTS	resonant tunnelling structure
S-HHFE	symmetric heavy-hole free exciton
S-LHFE	symmetric light-hole free exciton
S-SEED	symmetric self electro-optic device
S-TEM	scanning transmission electron microscopy
SAM	separate absorption and multiplication
SCFL	source coupled field effect transistor logic
SCH	separate confinement heterostructure
SDE	spin-density excitation
SE	spectroscopic ellipsometry
SEED	self electro-optic device
SET	single-electron transistor
SIMS	secondary ion mass spectrometry
SISA	selective intermixing in selected areas
SL	superlattice
SLS	strained-layer superlattice

SO	surface optic
SQW	single quantum well
STM	scanning tunnelling microscopy
SUC	superlattice unit cell
TA	transverse acoustic
TE	transverse electric
TED	transmission electron diffraction
TEM	transmission electron microscopy
TM	transverse magnetic
TO	transverse optic
UV	ultraviolet
VCSEL	vertical cavity surface emitting laser
VLSI	very large scale integration
VPE	vapour phase epitaxy
VWIR	very long wave infrared (detectors)
XRD	X-ray diffraction

CHAPTER 1

HISTORICAL PERSPECTIVE

1.1 Compositionally modulated structures: a historical perspective

E.E. Mendez

April 1996

A INTRODUCTION

The term modulated structure refers here to a crystalline material whose physical properties are controlled by spatial variations in its composition. In semiconductors, these variations are achieved by epitaxial deposition of multilayer films on a crystalline substrate. Broadly speaking, composition can be changed by varying either the material, say, from GaAs to AlAs, or the doping in a given semiconductor, for instance, in InGaAs from n-type to p-type. In the former case, we speak of compositionally modulated structures, which is the subject of this Datareview. Modulation by periodic doping is discussed in Datareview 1.2 on doping superlattices. In either case, to ensure the perfection of a modulated structure, the lattice constant of its material constituents should be the same as that of the substrate on which it is grown. However, almost unavoidably there is a mismatch between those lattice parameters, which can strain some of the layers considerably. As discussed in detail in the articles on strained-layer superlattices, it is possible to take advantage of that built-in strain to modify, and improve, the properties of modulated structures.

Incipient suggestions of modulated structures can be traced back to a visionary address by R. Feynman at a meeting of the American Physical Society in 1959, but it was a more concrete proposal by L. Esaki and R. Tsu in 1969 that gave birth to the area of semiconductor modulated structures. The research explosion that followed is illustrated in FIGURE 1, showing the number of papers on modulated structures presented at the biannual International Conference on the Physics of Semiconductors (ICPS) through the years. FIGURE 1 does not include the large body of work presented in a satellite conference devoted to Superlattices, Microstructures and Microdevices, dealing primarily with compositionally modulated structures. Other regular meetings on the subject include the International Conference on Modulated Semiconductor Structures, which in recent editions has had more than two hundred contributions.

The rapid development of this field is the result of a combination of novel physics ideas, new techniques for epitaxial growth to implement them, and applications in electronics and optoelectronics based on those concepts. The size of this interdisciplinary area makes impossible a complete review of its historical evolution within the limits of this Datareview. We will therefore restrict ourselves to highlighting the conceptual milestones and their consequences, referring only to some seminal papers, and to monographs and books for detailed reading on the various topics. We start by presenting in loose chronological order the key concepts underlying the plethora of phenomena observed and configurations implemented in modulated semiconductor structures [1-3]. Then we introduce selected applications based on them and finally we offer our view of the direction of the field. The organisation of this Datareview makes it quite general regarding materials, but, where the need for specificity exists, the focus is on III-V compounds.

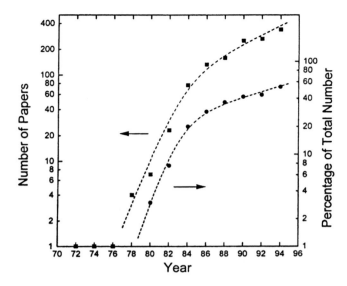

FIGURE 1. Evolution with time of the number of papers on Quantum Heterostructures (and its percentage relative to the total number) presented at the biannual International Conference on the Physics of Semiconductors. In the 1994 Vancouver conference there were more contributions on that subject than on all the others combined.

B   MILESTONES

Possibly the most basic concept for modulated structures is that of restricted dimensionality. The properties of bulk semiconductors are three-dimensional (although sometimes anisotropic), as, for example, the electronic density of states of the conduction band. In contrast, in a compositionally modulated semiconductor the corresponding density of states can be three-dimensional but highly anisotropic (superlattice), two-dimensional (quantum well), one-dimensional (quantum wire) or even of zero dimensionality (quantum dot). The first two situations can be achieved simply by varying the configuration of the structure or by an external parameter like an electric field. The formation of wires and dots very often requires additional lithographic techniques, and will be discussed here only briefly.

The superlattice concept was introduced in 1969, searching for ways of using semiconductor heterostructures to induce 'negative conductance (that) could lead to ultra-high-speed devices' [4]. The idea was to superimpose on the atomic periodicity a longer one-dimensional periodic potential (superlattice) by a periodic variation of alloy composition or impurity density during epitaxial growth. Along the superlattice direction a new Brillouin zone would be created that, compared with the other two dimensions, would have a reduced wavevector range and narrower energy bands. An electron in one of these minibands would in principle be excited much more easily to energies beyond the inflection point of the E-k dispersion curve and even to the top of the miniband, so that not only negative conductance but also Bloch oscillations would be possible.

In compositionally modulated semiconductors, the superlattice potential is created by the discontinuities in the conduction-band (or valence-band) profile induced by the periodic alternation of two semiconductors. For the effects of this potential to be appreciable, the mean free path of electrons along the superlattice direction must be much larger than the period of the potential. Since in a superlattice an electron must travel through several interfaces between two different materials, their quality is of paramount importance. It should not be surprising then that although metallurgically a superlattice was realised in 1973, it took several years before improvements in materials purity and interface smoothness made possible the demonstration of a superlattice in the 'electronic' sense.

When the electron mean free path is only slightly larger than one period, it is more appropriate to speak of multi-quantum wells than of a superlattice, in which case the electronic density is two-dimensional and materials periodicity is unnecessary. A single quantum well is formed when a semiconductor thin film is clad on each end by a thick layer of a second semiconductor whose conduction-band edge is higher in energy than that of the film, confining the electronic wavefunctions and quantising the energy levels.

Another basic quantum-mechanics concept in which the field of modulated structures is rooted is that of resonant tunnelling. The probability for an electron to tunnel through a potential barrier decreases exponentially with the barrier width. But if the energy of the electron is the same as any of the quantised energies of a well confined by two consecutive barriers, then there is a resonance between the two energies and the tunnelling probability is dramatically enhanced, even being equal to one when the barriers are identical.

It was through a resonant-tunnelling experiment that the formation of discrete energy states in semiconductor quantum wells was first demonstrated. Since changing the energy of an electron incident on a double-barrier structure would not be easy, in experiments involving resonant tunnelling the heterostructure is usually terminated on both ends by heavily doped electrodes that act as reservoirs of electrons with energies from zero up to the Fermi energy. The resonance is induced by applying a voltage between the electrodes, which 'aligns' the discrete states in the well with the energy of the electrons. Thus the current between the electrodes increases sharply for voltages at which the alignment occurs and almost vanishes for all other voltages. Every maximum in the current corresponds to the alignment of a discrete state with the bottom of the conduction band of the 'emitter' electrode, as first shown experimentally by Chang, Tsu and Esaki in 1974 [5].

Almost simultaneously, optical-absorption experiments also demonstrated the formation of discrete states. In a semiconductor, the optical absorption increases monotonically with energy, beyond a threshold that corresponds to the energy gap between the valence and conduction band. It is quite frequent that when a quantum well is formed in the conduction band of a heterostructure a similar well is formed in the valence band. Discrete states are then created in both bands, so that the optical absorption consists of steps corresponding to the two-dimensional joint density of states for the conduction and valence bands. Experiments by Dingle, Wiegmann and Henry showed those steps in the absorption spectra and confirmed the formation of quantum states in wells with widths ranging from 140 Å to 4000 Å [6].

The reduced dimensionality of discrete states in quantum wells was directly revealed in high-field magnetotransport experiments by Chang and co-workers in 1977 [7]. The energy quantisation imposed by a strong magnetic field on electrons in semiconductors leads to pronounced Shubnikov-de Haas oscillations of the electric resistance with increasing field. The period of those oscillations is inversely proportional to the cross section of the electrons' Fermi surface perpendicular to the field, so that a study of the dependence of the oscillations on the orientation of the magnetic field relative to the crystallographic directions of the semiconductor provides information about the size and shape of the Fermi surface. Magnetotransport experiments showed the highly anisotropic nature of a superlattice's Fermi surface, which in the quantum-well limit became cylindrical, corresponding to a true two-dimensional system.

In these early transport experiments, conduction-band electrons, although localised in quantum wells, were the result of uniform doping throughout the heterostructure. Consequently, electron-impurity scattering limited the mobility of charge carriers, especially at low temperature. A major breakthrough was the demonstration by Dingle et al in 1978 of the spatial separation between electrons and parent impurities [8]. By selectively doping with silicon (which acts as a donor) the GaAlAs regions of a GaAlAs/GaAs/GaAlAs heterostructure, they were able to show a dramatic enhancement in the electron low-temperature mobility, from about 3×10^3 cm^2/V s in bulk GaAs to 1.2×10^4 cm^2/V s in a selectively doped structure.

When Si is introduced in GaAlAs as an impurity, electrons are transferred to GaAs and confined to this region by the potential profile of GaAlAs/GaAs, in which the former acts as a barrier and the latter as a well. The reduced electron-impurity scattering was responsible for the original increase of the electronic mobility and for subsequent enhancements obtained by the introduction of an 'undoped' spacer between electrons and donor impurities. Combined with significant improvements in the elimination of residual impurities, this technique, frequently called modulation doping, has made it possible to achieve mobilities over 1×10^7 cm^2/V s at temperatures of 2 K and has opened new directions in the physics of low-dimensionality systems.

Although first observed in metal-oxide-semiconductor Si heterostructures, the quantum Hall effect (QHE) in two-dimensional electrons, discovered in 1980 by von Klitzing, Dorda, and Pepper [9], was soon confirmed in modulation-doped GaAlAs/GaAs structures. Since then, most studies on the QHE have been carried out in this system (and other III-V compounds) because of the relative ease of preparing two-dimensional electron and hole gases with high mobility. It was this capability that led in 1982 to the discovery by Tsui, Stormer, and Gossard of the 1/3 fractional quantisation of the Hall effect (as opposed to integer quantisation in the 1980 discovery), when the highest electron mobility was about 10^5 cm^2/V s [10]. The subsequent increase of this record number by two orders of magnitude has permitted the observation of many new fractional states, from which a very complete picture of the novel physics of two-dimensional electrons at high magnetic fields is emerging [11].

When such high mobilities are achieved, the electronic mean free path can easily become larger than the sample's size, especially if this is in the mesoscopic regime, a loosely defined term that frequently refers to dimensions in the 10 nm to 1000 nm range. In that regime, the motion of electrons becomes ballistic when, on the average, the only scattering that they suffer is with the sample boundaries. Many ballistic-transport phenomena, among them the quantisation of the zero magnetic field conductance and coherent electron focusing by a magnetic field, are best observed using quantum point contacts, that is, constrictions in a two-dimensional gas of the order of the Fermi wavelength (typically 50 to 100 nm). These phenomena are similar to those in electron transport in vacuum and have analogies in optics as well: thus the term solid state electron optics is often used to refer to ballistic transport in two-dimensional electron gases [12].

Much as in atomic physics, the eigenstates of quantum wells can be modified by an external electric field. In 1982 Mendez and co-workers proposed and demonstrated the use of a field perpendicular to the well planes to change both the energy and wavefunction of the ground state of the quantum well [13]. The effect, called the quantum-confined Stark effect, or Stark effect for short, has been extensively studied since then and is the basis for electro-optical modulators and self electro-optic devices (SEED). A related effect is that of an electric field on a semiconductor superlattice, by which the quasi continuum of states that constitutes the superlattice minibands is split into a series of discrete states called the Wannier-Stark ladder, effectively inducing a transition from a three-dimensional to a two-dimensional density of states. Although predicted long ago for solids in general, the Wannier-Stark ladder was not observed until 1988, when semiconductor superlattices with a periodicity ten times larger than that of bulk semiconductors were used [14]. The demonstration of the Wannier-Stark ladder is at the core of recent reports on the observation of Bloch oscillations.

Crucial to the implementation of these physical ideas has been the dramatic development of materials preparation techniques, in particular thin-film epitaxial deposition of semiconductors by metallo-organic vapour phase deposition and molecular beam epitaxy, summarised in Section C of this Datareview. The former has advantages for production of large batches, such as those found in development laboratories, while the flexibility of the latter makes it ideal as a research tool. In both techniques growth processes are far from equilibrium, which allows low growth rates and control of desirable abrupt interfaces between different materials.

Molecular beam epitaxy, in a primitive form, was already being used in the 1960s to prepare CdS and GaAs, but it was the advances in the understanding of growth kinetics in the late 1960s that led to the deposition of the first high quality films in the early 1970s [15], undoubtedly spurred by the just-announced superlattice proposal. The initial 'home-made' MBE systems soon gave way to commercial units with better control of the growth environment (vacuum, Knudsen cells, etc.) and purer source materials. As at many other times in history, the confluence of a novel concept and a new technique gave rise to a new field, in which science and technology fed each other.

C   APPLICATIONS

Many of the numerous applications that semiconductor modulated structures are finding in electronic and optoelectronic devices benefit from the interaction between those two elements: reduced dimensionality (physics) and sharp interfaces (technology). Probably the best example of the former is the quantum-well laser, which has its precedent in the double-heterostructure laser invented in 1970. This device heralded the practical use of the semiconductor laser by achieving low threshold currents through good optical and electron confinements provided by the different indices of refraction and band discontinuities in a double heterostructure.

The quantum-well laser represented a further improvement in that direction due to the lower dimensionality of the density of states in a quantum well [16]. Two-dimensional charge carriers, with a step-function density of states, contribute more efficiently to the peak gain than three-dimensional carriers and consequently lower the threshold required for lasing action. By using multiple quantum wells and special configurations (e.g. graded index of refraction and separate confinement), threshold currents as low as 200 A cm^{-2} have been achieved.

The ability to grow epitaxially a large array of alternating layers has also made possible the realisation of vertical quantum-well lasers, in which the light is emitted perpendicular to the multilayers instead of parallel to them as in conventional semiconductor lasers. In vertical lasers, the optical cavity is created by the region between two 'dielectric' mirrors, grown epitaxially under and above the active region that contains the quantum well [17]. The mirrors consist of alternating layers of two semiconductors with different indices of refraction (quarter-wave stack Bragg reflectors), with reflectivities of up to 99.9%. Vertical lasers are desirable because of the flexibility they offer for direct chip-to-chip communications and for two-dimensional optical imaging arrays.

Modulated semiconductor structures have also been very successful for devices based on electronic transport. The numerous structures proposed and realised experimentally fall into the broad categories of parallel-transport devices, in which the electronic transport is in the plane of the heterostructure, and vertical-transport devices, where electronic motion is perpendicular to the structure interfaces. Among the former, the high-mobility field-effect transistor (HFET) deserves special mention. A natural consequence of the advantages of modulation doping for high-speed operation, the HFET, invented in 1980 by groups in Japan and France, consisted initially of a doped AlGaAs layer deposited on an undoped GaAs region [18]. The conduction in the two-dimensional electron gas at the GaAs/AlGaAs interface was modulated and controlled by means of a top metal gate. Since then many variations have been introduced, both on the details of the structure and on the III-V semiconductors employed, that have enhanced the performance of the initial devices by, for example, increasing band discontinuities (as in pseudomorphic InGaAs/AlGaAs and InGaAs/InAlAs). Nowadays HFETs are in commercial use as high-frequency, low-noise amplifiers and are actively researched for specialised digital applications.

Vertical-transport devices are for the most part either heterostructure bipolar transistors (HBT) or tunnelling (or resonant-tunnelling) diodes and transistors. The design and operation of a GaAs/GaAlAs HBT is similar to that of an n-p-n GaAs homojunction, except that the emitter consists of AlGaAs, so that its wider bandgap reduces the base current from hole injection. In this way, the base can be heavily doped and its resistance correspondingly reduced, thus enhancing the maximum operating frequency. Although the underlying concept for the HBT goes back to the early 1950s, integrable HBTs with high current gain and high speed have become a reality only with the development of modern epitaxial-deposition techniques [19].

A second important category of vertical-transport devices is based on the concept of resonant tunnelling, discussed above. The current-voltage characteristics of resonant-tunnelling diodes show negative differential resistance regions, which give rise to high-frequency oscillations when the diodes are placed in a resonant cavity. Although similar in operation to the p-n tunnel diode, the resonant tunnelling diode can generate higher power at higher frequencies, due in part to the lower capacitance inherent to lower doping and correspondingly longer depletion lengths. Resonant tunnelling in semiconductor heterostructures was demonstrated in 1974, but it was the observations of fast intrinsic response times (1983) and of room-temperature operation (1985) of resonant-tunnelling devices that spurred the interest for practical applications [20]. In 1991, oscillation frequencies of up to 700 GHz with an output power of 0.3 μW had already been reached using the InAs/AlSb materials system.

Oscillators are the most mature devices based on resonant tunnelling, but are by no means the only ones being explored. For instance, resonant-tunnelling hot-electron and bipolar transistors have also been demonstrated. The former can act as a frequency multiplier or as an exclusive-NOR gate, which allows the implementation of many logic circuits with fewer transistors than in more conventional MESFET technology. Recently, room-temperature operation of a 17-transistor full adder with coexisting resonant-tunnelling and heterojunction bipolar transistors has been shown.

In addition to the devices highlighted above, either already commercialised or approaching commercialisation, a myriad of compositionally modulated structures have been proposed and demonstrated with electronic or optoelectronic applications in mind, many of which have shown great potential. Examples of these include heterojunctions, Gunn and IMPATT diodes, the quantum-cascade laser, and various optical modulators and photodetectors, described in detail in numerous monographs and reviews.

D   OUTLOOK

Although predicting the future is risky at the very least, based on present trends it is possible to foresee certain directions in the field of compositionally modulated structures for the next few years. The recent advances in epitaxial growth of wide-gap III-V semiconductors such as GaN and related ternary compounds for blue light sources will continue to attract a large number of applied scientists interested in the ultimate goal of a robust blue laser operating continually at room temperature. Eventually we could see blue lasers both in the 'side' and vertical configurations.

A significant effort will be devoted to quantum wires and dots, this both as a result of a receding frontier in two-dimensional physics and as a recognition of the potential of very low dimensionality structures, for physics and applications. In spite of many of the inherent advantages of low-dimensionality devices over conventional ones, the difficulty in displacing such mainstream technologies as silicon MOS integrated circuits may become in some cases an insurmountable barrier to their widespread use. On the other hand, low-dimensionality structures are quite promising for discrete devices, normally seen in optoelectronic applications. An example of these is the quantum-dot laser, which could lead to unprecedentedly low threshold currents if the limitations of a recently recognised 'phonon bottleneck' and of a small active volume for output intensity are overcome. An incipient area that will see significant growth is that of semiconductor-superconductor heterostructures, in which the advantages of high-mobility two-dimensional electron gases and superconductivity will be combined to study new phenomena or to implement certain devices for specialised applications that tolerate the temperature limitations (at least for the time being) of such structures.

The availability of new experimental tools will offer ways of probing semiconductor modulated structures in new regimes. A recent example of these tools is the free-electron laser, which is allowing the study of ultra-high frequency emission and absorption in quantum wells and superlattices, in search of a better understanding of the conditions under which Bloch oscillations can occur. The current interest in this topic summarises well the goal that has driven the field of compositionally modulated structures from its inception: the search for novel physical phenomena that may one day lead to important applications.

REFERENCES

[1]   G. Bastard [*Wave Mechanics Applied to Semiconductor Heterostructures* (Wiley, 1989)]

[2]   C. Weisbuch, B. Vinter [*Quantum Semiconductor Structures* (Academic Press, 1991)]

[3]   M.J. Kelly [*Low-Dimensional Semiconductors* (Oxford University Press, 1995)]

[4]   L. Esaki, R. Tsu [*IBM J. Res. Dev. (USA)* vol.14 (1970) p.61-5]

[5]   L.L. Chang, L. Esaki, R. Tsu [*Appl. Phys. Lett. (USA)* vol.24 (1974) p.593-5]

[6]   R. Dingle, W. Wiegmann, C.H. Henry [*Phys. Rev. Lett. (USA)* vol.33 (1974) p.827-30]

[7]   L.L. Chang, H. Sakaki, C.-A. Chang, L. Esaki [*Phys. Rev. Lett. (USA)* vol.45 (1977) p.1489-92]

[8]   R. Dingle, H.L. Stormer, A.C. Gossard, W. Wiegmann [*Appl. Phys. Lett. (USA)* vol.33 (1974) p.665-7]

[9]   K. von Klitzing, G. Dorda, M. Pepper [*Phys. Rev. Lett. (USA)* vol.45 (1980) p.494-7]

[10]   D.C. Tsui, H. Stormer, A.C. Gossard [*Phys. Rev. Lett. (USA)* vol.48 (1982) p.1559-62]

[11]   T. Chakraborty, P. Pietilainen [*The Quantum Hall Effects - Fractional and Integer* (Springer Verlag, 1995)]

[12]   C.W.J. Beenakker, H. van Houten [*Solid State Physics* vol.44 (Academic Press, 1991) p.1-228]

[13]   E.E. Mendez, G. Bastard, L.L. Chang, L. Esaki, H. Morkoc, R. Fischer [*Phys. Rev. B (USA)* vol.26 (1982) p.7101-4]

[14]   E.E. Mendez, F. Agullo-Rueda, J.M. Hong [*Phys. Rev. Lett. (USA)* vol.60 (1988) p.2426-9]

[15]   A. Cho [*Molecular Beam Epitaxy* (AIP Press, 1994)]

[16]   N. Holonyak Jr., K. Hess [in *Synthetic Modulated Structures* Eds L.L. Chang, B.C. Giessen (Academic, 1985) p.257-310]

[17]   K. Iga, F. Koyama, S. Kinoshita [*IEEE J. Quantum Electron. (USA)* vol.24 (1988) p.1845-55]

[18]   N.T. Linh [*Semicond. Semimet. (USA)* vol.24 (1987) p.203-48]

[19]   P.M. Asbeck, M.-C.F. Chang, K.C. Wang, D.L. Miller [in *Semiconductor Technology: GaAs and Related Compounds* Ed. C.T. Wang (Wiley, 1990) ch.4 p.170-230]

[20]   H.C. Liu, T.C.L.G. Sollner [*Semicond. Semimet. (USA)* vol.41 (1994) p.359-419]

1.2 Doping superlattices - historical overview

G.H. Dohler

December 1995

A INTRODUCTION

In this brief review on doping superlattices in III-V compounds only the most significant developments in a wide field, which is covered by hundreds of publications, can be summarised. Although a historical perspective is used, at the same time an attempt has been made to structure the material presented in a way suitable for reference to a specific topic.

B PREHISTORIC PERIOD (1970-1980)

In 1971 the author of this Datareview, working on the kinetics of donor acceptor pair luminescence, realised that dramatic effects are to be expected, if donors and acceptors were not randomly distributed, but placed periodically on donor (n-doped) and acceptor (p-doped) layers, separated by a distance d_i in an otherwise intrinsic (i-) host semiconductor crystal. In the first two papers on these 'n-i-p-i' (or, in modern terminology, 'δ-doped n-i-p-i') superlattices many of the peculiarities of the electronic structure [1] and the electrical and optical properties [2] were outlined. The electrical field of the ionised dopants causes a periodic space charge potential, which is superposed on the crystal potential. The motion of carriers in the direction normal to the layers (the growth direction for real structures) becomes quantised. The energy levels corresponding to the bottom of the various subbands were calculated for a wide range of sheet doping densities, assuming a homogeneous distribution of the impurity space charge over the layer. Also the first calculations of lifetimes for phonon assisted intersubband transitions of free carriers were presented.

Independently several other authors had also considered the use of periodic n- and p-doping for the generation of a periodic confining potential at about the same time [3-6]. None of them, however, had anticipated the unique feature of n-i-p-i structures, the possibility of 'dynamically tuning' the electronic properties.

In fact, the most important observation reported in [1,2] concerns the specific electronic structure resulting from the spatial separation between electron and hole states in these systems (FIGURE 1(a)). Theoretically calculated electron-hole recombination lifetimes are increased by factors ranging from slightly larger than unity up to arbitrarily large values depending on the design parameters (n- and p-doping density and layer thickness) due to the (strongly) decreased overlap between electron and hole subband wave functions in these systems with an 'indirect gap in real space'. (This situation is similar to the case of a 'type-II' compositionally modulated superlattice (see Datareview 1.1) formed by two semiconductors with 'staggered' conduction and valence band line-up; the effects on the electronic properties, however, are much more drastic.)

The long recombination lifetimes enable the accumulation of large non-equilibrium free electron and hole space charge densities in the n- and p-layers. With increasing electron and hole densities the donor and acceptor space charge fields become increasingly screened. As a consequence the effective bandgap, E_g^{eff}, defined as the energy difference between the lowest conduction and uppermost valence subband, as well as the separation between the quasi Fermi levels of the electrons, Φ_n, and the holes, Φ_p, increase. Simultaneously, the subband spacing decreases and the electron-hole recombination lifetimes drop even on an exponential scale, due to an increasing overlap of the wave functions (FIGURE 1(b)).

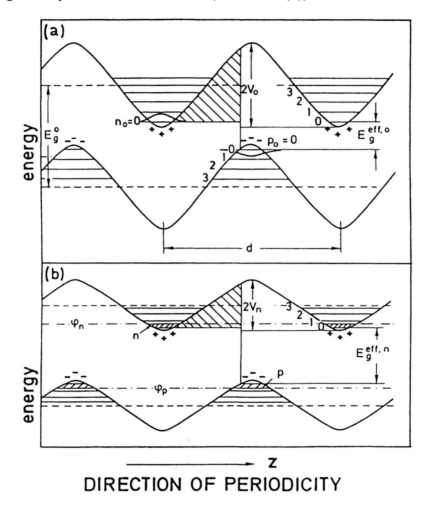

FIGURE 1. Schematic band diagram of an n-i-p-i doping superlattice for the ground state (a) and an excited state (b). The superlattice potential which modulates the conduction and valence band edges E_c and E_v is due to the space charge potential of the ionised donors and acceptors. Optical or electrical excitation leads to a partial screening of the superlattice potential by the space charge of the excess carriers. The shaded areas depict the barriers for the electron-hole recombination.

In [2] simple estimates were presented showing that the luminescence resulting from the recombination of the spatially separated electrons and holes in n-i-p-i structures should be tunable within a wide spectral range, for instance, by varying the excitation density. The relaxation times of photo-induced electrons and holes within the conduction and valence subbands, respectively, are many orders of magnitude faster than the recombination lifetimes. The steady state electron and hole densities accumulated in the n- and p-doped layers and, hence, the effective bandgap, E_g^{eff}, are expected to increase with increasing excitation density.

An approximately logarithmic relation between excitation density and spectral shift of the luminescence is expected because of the above mentioned exponential relation between effective bandgap and recombination lifetime.

The effect of an external field on the transport and the re-emission of 'stored' optical excitation energy was also discussed in [2]. During this prehistoric period the first self-consistent calculations of the electronic subband structure as a function of the dynamically tunable sheet carrier density were also carried out [7].

The modulation of the electronic structure by the injection or extraction of electrons and holes via an external bias U_{pn} applied between 'selective n- and p-contacts', providing ohmic contacts to the corresponding n-i-p-i layers, but forming blocking p-n-junctions with respect to the layers of opposite doping type, was first discussed in [8]. In this paper the possibility of using this modulation scheme for tunable electroluminescence with $\hbar\omega \approx eU_{pn}$ or tunable electroabsorption was also outlined.

C   ANTIQUITY AND MIDDLE AGES (1980-1986)

C1   Tunable Photoluminescence and Subbands

Once the first high quality GaAs n-i-p-i structures had been grown by MBE by Ploog and co-workers around 1980 [9] the experimental verification of many of the theoretically predicted basic properties was a matter of a relatively short period of time.

In [10] the dynamical tuning of the effective bandgap was demonstrated by photoluminescence experiments in GaAs doping superlattices consisting of uniformly doped n- and p-layers (thickness $d_n = d_p = 40$ nm, doping density $n_D = n_A = 10^{18}$ cm^{-3}) with no intrinsic layers. By varying the excitation density between 2 and 10^4 W/cm^2 the peak energy of the luminescence spectra, which is expected to lie between the effective bandgap and the quasi Fermi level difference, could be shifted by about 200 meV from 1.30 to 1.50 eV with this specific sample design.

At the same time the existence of 2-D subbands in these purely space charge induced quantum wells could be demonstrated by resonant electronic Raman scattering experiments using laser energies corresponding to the gap between the split-off valence and the conduction band of bulk GaAs [10]. The separation between the peaks in the Raman spectra decreased as a function of increasing excitation density. In fact, the observed Raman shifts turned out to agree very well with the self-consistently calculated excitation dependent intersubband spacings. Before, it had not been clear whether the observation of subbands in purely space charge induced quantum wells would not be completely obscured by a strong broadening due to the potential fluctuations resulting from the random spatial distribution of the dopants.

A quantitative comparison between experiment and theory was made possible by the fact that the tunable effective bandgap E_g^{eff} (as observed by the luminescence spectra), the tunable free electron density and the tunable subband structure (as deduced from the Raman shift) are directly related to each other. Later on more results on Raman studies of n-i-p-i structures with higher doping levels, exhibiting larger subband spacings, were reported [11,12].

At cryogenic temperatures, where all of the early luminescence studies were performed, electron-hole recombination by tunnelling represents the only radiative process. A simple qualitative argument developed by Döhler in 1982 [13] indicated that these processes, which are the basis for the tunability of the luminescence, should not only persist, but dominate over the thermally activated tunnelling and over the thermally activated vertical recombination in the barriers up to room temperature if the doping density exceeds a certain level. This prediction however was verified by temperature dependent luminescence experiments only in 1986 [14,15].

Although the radiative recombination lifetimes are increased by many orders of magnitude compared with those of bulk semiconductors it was found that the luminescence efficiency was nevertheless quite high. This observation, in fact, confirmed the expectation that the non-radiative recombination across the indirect gap in real space should be impeded by a factor of comparable magnitude. In these early PL investigations the relation between excitation density, PL peak position and luminescence intensity was also studied, which bears most of the information about radiative and nonradiative lifetimes [16]. As expected, the lifetimes exhibited huge variations, depending on the design parameters of the doping superlattice and, in particular, on the degree of excitation. These findings were corroborated by time resolved PL studies [17] which provided total lifetimes ranging from 1 to 10^4 μs depending on the excitation level and the design of the samples. The doping levels ranged from 0.5 to 2 x 10^{18} cm^{-3}.

C2   Stimulated Emission

Large photo-excited carrier densities up to population inversion can be achieved at much lower optical pumping level in suitably designed n-i-p-i structures compared to uniform bulk semiconductors due to the strongly enhanced lifetimes for spontaneous recombination. Therefore, a low threshold pumping power for the onset of stimulated emission had been predicted [8]. In fact, in the first study of stimulated emission in n-i-p-i structures a tunable gain spectrum with maximum gain values of 300 cm^{-1} at 1.35 eV at an excitation density of 20 W/cm^2 and of 400 cm^{-1} at 1.40 eV at 500 W/cm^2, though at 2 K, was reported [18].

C3   Tunable Absorption

Whereas the previously described samples were particularly suitable for studies of the tunable luminescence owing to the relatively short superlattice period, first investigations of the tunable absorption coefficient were performed in an n-i-p-i structure with a large superlattice period of 380 nm [19]. The calculated height of the space charge potential in the ground state was about equal to the GaAs bulk bandgap. Therefore, very long recombination lifetimes and large photo-induced changes of the electronic structure, in this case specifically of the built-in electric fields, were expected. A careful analysis of the observed transient photoconductivity changes at different photon energies allowed the determination of the excitation dependent changes of the absorption spectra. The tunable absorption spectra agreed well with the results of a simple model which was based on a spatial average of the Franz-Keldysh absorption associated with the local space charge fields [19].

Also extremely long lifetimes could be demonstrated for these structures. At 4 K values exceeding 10^3 s were determined from the transient absorption studies. However, even these large values turned out to be still orders of magnitude smaller than theoretically expected values for band to band tunnelling recombination, probably because of the presence of deep traps.

C4   Electrical Modulation of n-i-p-i Structures

Although it was clear from the very beginning that electrical modulation of the carrier density, of the electric fields, of the optical transition probabilities and of many other properties, by the injection or extraction of electrons and holes, would be of high interest for both fundamental investigations and device applications, there were only a few reports on electrical investigations of n-i-p-i systems during these first years. This was due to the lack of a reliable and satisfactory technology for the fabrication of 'selective' n- and p-type contacts, that would provide good ohmic contacts with low contact resistance to all the doping layers of one kind but simultaneously form high quality p-n junctions to the layers of opposite doping sign. The most frequently used technique consisted in depositing small Sn and Sn/Zn balls and alloying them to form selective n- and p-type contacts, respectively [20]. These contacts usually showed large leakage currents, in particular when the doping densities were high. In addition there was the problem that the dopants would often not reliably diffuse through the whole structure down to the bottom layers. Nevertheless, it could be demonstrated for n-i-p-i structures with not too high a doping level ($<\sim 10^{18}$ cm^{-3}), that both the n- and p-layer conductances (measured between two n- or two p-contacts) could be modulated by the application of a bias between an n- and a p-contact for n-i-p-i structures grown by MBE [20] and LPE [21]. If the product of doping density and layer thicknesses exceeds certain values complete depletion of the layers occurs under reverse bias ('negative effective bandgap').

For the observation of tunable luminescence significantly below the bandgap energy of the host material large doping densities are required [13]. Satisfactory selective alloyed contacts could not be obtained for such highly doped ($>10^{18}$ cm^{-3}) n-i-p-i structures. However, electroluminescence with peak energies tunable between 1.18 and 1.48 eV could be demonstrated at 4 K [22,23]. The applied forward bias and currents, however, exceeded 10 V and 400 mA, respectively, because of the large contact and layer series resistances. Also, the light emission was mostly confined to the area around the p-contacts.

C5   Hetero n-i-p-i Structures

Shortly after the advantages of 'modulation' doping for achieving high mobilities had been demonstrated, the concept of hetero n-i-p-i structures was introduced [24]. In 'type-I hetero n-i-p-i structures' (FIGURE 2) the centre of the n- and/or the p-layer consists of undoped quantum wells of a lower bandgap material, separated by an undoped spacer layer from the n- and p-doped regions, thus leading to reduced impurity scattering. The feasibility of the concept was demonstrated in [25] although only a relatively modest increase in the electron and hole mobilities was observed in this specific case.

The concept of 'type-II hetero n-i-p-i' structures (see FIGURE 3) was also proposed [26]. In these systems the i-layers consist of a material of lower bandgap. The potential fluctuations are expected to be strongly reduced due to the spatial separation of the electrons and holes from their parent dopants.

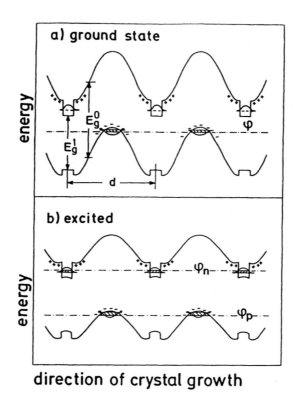

FIGURE 2. Schematic real space picture of the band diagram of a type-I hetero n-i-p-i structure for the ground state and an excited state.

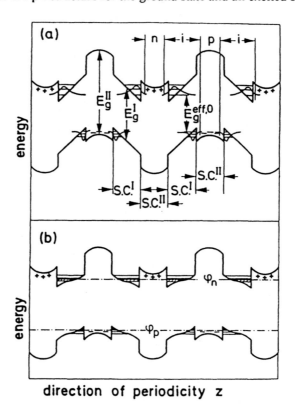

FIGURE 3. Schematic real space picture of the band diagram of a type-II hetero n-i-p-i structure for the ground state and an excited state.

This expectation was confirmed by PL experiments on an $Al_{0.3}Ga_{0.7}As$/GaAs type-II hetero n-i-p-i structure with 30 nm wide i-GaAs layers [27,28]. The luminescence was also tunable, but the spectra were much narrower compared with those observed in homo n-i-p-i crystals. Several lines were resolved which could be consistently attributed to transitions originating from different electron and hole subbands.

C6 Theoretical Work: Electronic Structure, Elementary Excitations, Optical Nonlinearities

The experimental verification of most of the predicted properties of the n-i-p-i doping superlattices stimulated a significant amount of further more fundamental theoretical studies.

The electronic structure of homo and hetero n-i-p-i structures was calculated self-consistently including exchange and correlation effects in the local density approximation [29]. Electronic excitations including spin-density and charge-density excitations and their coupling to the LO phonons, as well as plasmons were studied theoretically in [30,31]. In particular it was found that the 'depolarisation shift' to be observed in the charge density excitations should compensate the decreasing subband spacing in the parabolic quantum wells of uniformly n- and p-doped n-i-p-i structures when the carrier density is increasing. It was also predicted that the excitation of 'backfolded' acoustic phonons (turning into 'optical' phonons!) with wavelengths corresponding to integer fractions of the superlattice period by microwaves in the >100 GHz range should become possible due to the excess charge of the ionised impurity atoms [32].

The fact that due to the long recombination lifetime large densities of photogenerated carriers can be induced by low optical power leads to large optical nonlinearities. This phenomenon was first studied in [33]. The contribution of both bandfilling and changes of the internal fields was discussed. It was shown that changes of the absorption coefficient and the refractive index could be induced by optical power levels which are orders of magnitude smaller than those required in bulk semiconductors or other semiconductor structures. The absolute and relative changes of the absorption coefficient were predicted to be extremely large but also the expected changes of the refractive index turned out to be large in comparison with other semiconductor structures. In [33] a proposal was made that the disadvantage of the long (intrinsic) decay times of the non-linear response could be overcome by selective n- and p-contacts, providing an external recombination channel and resulting in adjustable ('RC') recombination lifetimes which depend on the capacitance of the n-i-p-i structure and the series resistance of the doping layers and the external circuit. It was also realised that the extremely non-linear intensity dependence of the absorption and refractive index changes which results from the degradation of the recombination lifetimes at high excitation levels can be overcome by the use of selective contacts.

D MODERN TIMES (>1986)

The discussion of the results reviewed in Section C has made clear that the strongly excitation dependent properties of n-i-p-i structures are quite interesting and also appealing for device applications. However, it also became obvious that both the study of the properties as well as their application, in particular for fast devices, are strongly hampered by the complicated relation between the excitation level and the electron-hole generation and recombination dynamics. This is particularly true as the photo-induced electron-hole generation rate as well as the radiative and the nonradiative recombination rate depend in a strongly non-linear manner on the excitation level. This problem can be overcome completely if the excitation level can be controlled electrically. If electrons and holes can be injected and extracted via selective n- and p-type contacts, the quasi Fermi level splitting, $\Delta\Phi_{np}$, is related to the applied voltage U_{pn} by the simple relation $\Delta\Phi_{np} = eU_{pn}$, if the voltage drop within the n- and p-layers due to their finite series resistance can be neglected. In this case the excitation level is uniquely defined by the applied voltage.

In particular, it can be varied very fast in structures of sufficiently small lateral dimensions. In this case the relevant time constant for fast modulation is the RC-time, determined by the series resistance of the n- and p-doping layers and the capacitance associated with the system of interdigitated charged n- and p-layers. The minimum values for these 'external' lifetimes depend quadratically on the lateral contact separation. Simple estimates [34] show that sub-ns time constants should be achievable.

On the other hand the dark current flowing between the n- and the p-contacts contains the full information about the electron-hole recombination (or, in the case of reverse bias, the generation) rate. Moreover, the photo current measured under illumination provides the full information about the optical generation rate at the given excitation level.

In view of the generic importance of high quality selective n- and p-contacts for both fundamental research and device applications, the moment when they became available marks the turning point for the transition from the middle ages to modern times in the history of n-i-p-i crystals.

D1   Selective n- and p-Type Contacts

Alloyed contacts, though conceptually neat and simple, could not be developed to the necessary stage to provide satisfactory results. Therefore, alternative solutions were required. The method which has finally turned out most suitable to overcome all contact related problems is based on lateral in situ structuring of the doping profile by the use of a shadow mask during MBE growth. In its original version [35] windows were first opened photolithographically in a Si-wafer using an anisotropic etch. For the MBE growth this mask was clipped, upside down, onto the GaAs wafer. Taking advantage of the different angle of incidence the shadow effect on the Si-donor and Be-acceptor beams resulted in the growth of structures with a central n-i-p-i region laterally flanked by n-i-n-i and p-i-p-i structures. These could easily be contacted by high quality ohmic n- and p-contacts, respectively. The problem of contact induced generation or recombination leakage currents could be avoided completely by the lateral separation of the contacts from the doping layers of opposite sign. This technique turned out to represent the breakthrough needed for more sophisticated investigations of n-i-p-i structures and for device applications. Due to the low contact resistances and the absence of leakage currents at the contacts tunable conductivity [35], tunable electro-absorption [36] and even tunable room temperature electroluminescence [37], which requires particularly high doping densities exceeding about 3 x 10^{18} cm^{-3} [13], could be demonstrated immediately.

In spite of this success there was still need for improvement. In particular a miniaturisation of the mask pattern was required in order to minimise the voltage drop in the n- and p-layers due to their series resistances, in particular for high carrier injection levels in the case of electroluminescence, but also to minimise the RC time constant, which increases quadratically with the lateral dimensions. The first successful step towards this goal consisted in the growth of an n-i-p-i structure within a window etched about 10 μm deep with undercut into the GaAs substrate [38]. In this case the substrate was also forming the mask. A further improvement was finally achieved by the introduction of a two-layer epitaxial shadow mask, consisting of a relatively thin (typically 3 μm) GaAs layer on a thicker (typically 9 μm) AlGaAs layer. By a combination of anisotropic non-selective etching through the GaAs layer and highly selective etching through the AlGaAs a window with well defined width and distance from the substrate can be achieved [39]. Finally, this 'epitaxial shadow mask' MBE technique (ESM-MBE) turned out to become even more powerful and flexible, if the operation of the shutters for the doping cells (and actually also of the group V effusion cells) becomes correlated with the sample rotation [40,41].

Alternative approaches to providing selective contacts to n-i-p-i structures consist in dopant implantation or diffusion as well as lateral epitaxial overgrowth of mesa structures with highly doped material [42-44]. We are only aware of reasonably successful work using the diffusion process.

D2   Luminescence, Experiment and Theory

After the introduction of the shadow mask technique the emphasis with regard to experimental luminescence studies of n-i-p-i systems has shifted from PL to electroluminescence (EL). High efficiency (external quantum efficiencies comparable to those of commercial double heterostructure light emitting devices) and a wide tunability range (0.7 eV $< \hbar\omega < 1.4$ eV) were reported [37,45]. Using the epitaxial shadow mask technique narrow structures exhibiting low voltage drops in the doping layers even at high lateral injection currents were successfully grown [46,47]. In these structures the position of the luminescence peak followed very closely the relation $\hbar\omega \sim eU_{pn}$. The voltage drop in the doping layers increased significantly only when the luminescence peak was approaching the bulk bandgap, but did not exceed 0.5 V.

Whereas straightforward self-consistent calculations of the luminescence spectra provided satisfactory results for the subband spacing [10,12,48], the relation between effective bandgap and lifetime [17,29] and the high energy tail of the spectra at elevated temperatures [47,49], a non-acceptable discrepancy was observed with regard to the low energy side of the spectra. It was clear, that an extended low energy tail ought to be expected due to strong potential fluctuations, resulting from the random distribution of the dopants within the doping layers. In fact, excellent agreement between experiment and theory has recently been achieved by including the effects of these potential fluctuations and the screening of the free carriers. Both a semiclassical treatment based on the Kane model [45] and a quantum mechanical model [47] reproduce the observed spectra within an intensity range of three orders of magnitude without using an adjustable parameter. In particular, they indicate that the disorder effects are so strong that to observe a signature of transitions from different occupied subbands appears unlikely. In this light the only report on the observation of the subband structure in n-i-p-i luminescence spectra [50] appears quite remarkable.

D3   Electro-absorption, Bandfilling and Optical Modulators

Quite a few studies of optically induced changes of the absorption coefficient in various kinds of n-i-p-i structures have been reported [43,51-55]. Most of these investigations were not dealing with the non-linear optical behaviour, as usually the modulation was performed by 'pump' light with photon energies above bandgap, and the absorption and refractive index changes were probed with light in the near-bandgap-and-below range.

Simpson et al [51] have reported the first observation of significant light induced absorption changes in an n-i-p-i structure. The GaAs n-i-p-i structure consisted of 50 nm thick n- and p-doped layers with $n_D = n_A = 2 \times 10^{18}$ cm^{-3}. The maximum transmission change observed in the 1 μm thick superlattice for photon energies just below the bandgap at T = 14 K was 10%. Kost and Garmire [52] used a 'multiple quantum well (MQW) hetero n-i-p-i structure' in order to take advantage of the larger maximum absorption changes for a given field change. Ando et al [53] and Law et al [54] reported even larger changes in nearly identical structures. A different phenomenon, namely bandfilling, was used by Larson et al [55] in a type-I heterostructure with pseudomorphic strained InGaAs QWs.

In all these studies it turned out that unusually large changes of the absorption coefficient were achieved by optical power densities which were orders of magnitude smaller than in bulk semiconductors or other microstructures. However, the changes were found to depend basically logarithmically on the optical power density. This is not surprising, if the exponential dependence between recombination lifetime on one hand and effective bandgap, internal fields and free carrier densities on the other hand is considered. These 'opto-optical' modulators are of limited interest not only by themselves but also due to the very long recovery times.

The whole scenario for tunable absorption changes once electrical modulation becomes feasible. The advantages listed in the introduction of Section D yield the following specific features for the electrical modulation of the absorption:

(a)   In n-i-p-i semimetals very high internal electric fields, exceeding breakdown fields of bulk semiconductors, can be achieved under reverse bias U_{pn}. They yield correspondingly large absorption changes within a wide photon energy range. Thus, electric fields exceeding 900 kV/cm were observed and a switching contrast exceeding 6:1 was observed in a 3.9 μm thick n-i-p-i structure grown by the epitaxial shadow mask technique when the voltage U_{pn} was swept between -6 and 1.0 V [40,56].

(b)   The unusually large field changes are associated with unusually large changes of the carrier density per n-i-p-i period. Thus, very large changes of the averaged free carrier density, resulting in unusually large changes of the refractive index, can be electrically induced. The changes of the free carrier densities achieved in [56] and [57] correspond to values of about 10^{18} cm^{-3} and 5×10^{17} cm^{-3}, respectively. The induced refractive index changes were used in an electro-optical Mach-Zehnder interferometer [43].

(c)   Hetero n-i-p-i crystals offer the unique possibility of utilising electric field and bandfilling effects simultaneously by constructive superposition, if appropriately designed with regard to the bandgaps in the intrinsic and the doped layers. First results on electrically modulated structures of this kind were reported [57-59].

(d)   A detailed study of electroabsorption and bandfilling phenomena as a function of electric field and carrier density, respectively, becomes easily possible. Bandfilling induced changes of the absorption coefficient exceeding 1 μm, referred to quantum wells of 8 nm width, were recently reported [57].

(e)   The internal recombination (or generation) times can be studied most easily from current vs. voltage and capacitance measurements. These lifetimes can be extremely high under reverse bias [60].

(f)   The modulation speed is determined by the RC time constants, which can be made small in structures of reduced lateral dimensions. Recently a turn-on and turn-off time of about 1 ns was reported for a 5 μm wide structure whose design, however, was still far from being optimised [61].

D4   Photoresponse and n-i-p-i Photo Detectors

The photoresponse of n-i-p-i structures differs from the bulk response by the following characteristics [34]:

(a)   Due to the built-in fields there is electroabsorption for photon energies smaller than the bandgap of the host material of the n-i-p-i structure and, hence, also a more or less pronounced photo response. Moreover, the response spectrum can be strongly dependent on the optical power, including spectral ranges with falling photo-current vs. voltage characteristics.

(b)   The extremely long electron-hole recombination times can be used to achieve an extremely high photoconductive response with gain exceeding values of 10^6 even by orders of magnitude.

(c)   Best performance is achieved under reverse bias in 'n-i-p-i semimetals' (i.e. under 'negative effective bandgap' conditions). In this regime the photoresponse is linear over many orders of magnitude. The shot noise due to thermal or field-induced electron-hole generation can be kept very low and the thermal noise (Johnson noise) can be minimised by reducing the dark conductivity to a value close to zero by working near the threshold voltage.

(d)   The photoconductive gain is adjustable within a range of many orders of magnitude by variation of the external load resistor in the U_{pn} circuit. More sophisticated versions, like bistable or threshold switches, or quenching of the photoconductive signal are also possible [60,62,63].

The first results on n-i-p-i based photodetectors were reported for structures with a selective n- and p-contact obtained by alloying [64]. This structure represented basically a pin photo diode, though with an unusual topography. Unfortunately, the reported results were of a more qualitative than quantitative character. Therefore, no information about responsivity, detectivity or linearity of these devices is available.

In [60] a detailed theoretical study of n-i-p-i photoconductive detectors was presented together with experimental results obtained from a 'single-period GaAs n-i-p-i', consisting of a selectively contacted pnp structure grown by the shadow mask technique using a Si-mask. An adjustable gain, reaching values up to 2.2×10^5, and very good linearity over 8 orders of magnitude were demonstrated. Values of the adjustable lifetime ranging from 1.5 μs to 1.5 ms were demonstrated using external resistors between 10 kΩ and 10 GΩ.

A major fraction of the n-i-p-i detector activity was devoted to small bandgap semiconductors with the goal of achieving superior performance for detectors in the long wavelength spectral range [65-68].

D5 Anomalous Ambipolar Diffusion in Doping Superlattices

A very remarkable property of n-i-p-i doping superlattices is their 'giant ambipolar diffusion coefficient' for diffusion parallel to the layers. The spatial separation of the electron and hole plasma results in internal in-plane fields which yield huge enhancement factors for the diffusion coefficient. Depending on the design parameters and the degree of excitation typical values exceed those of uniform bulk semiconductors by 2 to 4 orders of magnitude. Values of 5000 cm^2/s corresponding to an enhancement factor of 300 were reported [69]. In the same letter it was also demonstrated that recombination at distances in the cm-range was negligible, even at room temperature, due to the long recombination lifetimes.

These properties are very appealing for high speed optoelectronic applications. The fast ambipolar diffusion has, in fact, very recently been used in an n-i-p-, i.e. a '1/2-period n-i-p-i', detector structure [70]. In this device the fast ambipolar diffusion and the low capacitance associated with the depleted n-channel were used simultaneously.

E PERSPECTIVES FOR THE FUTURE

Within one and a half decades since the first doping superlattice structures were grown it has been demonstrated by a large amount of experimental and theoretical work that this class of superlattices represents indeed a fascinating new material with remarkable novel features. Some of these properties have already proved suitable for optoelectronic device applications. At this point it appears likely that future work on this material will be focused on the following issues:

(a)   Exploitation of its potential for the fabrication of devices with improved performance, in particular for modulators, detectors, non-linear optical switches, tunable light sources and devices using the - compared to other semiconductors - large voltage induced changes of the refractive index, or the anomalously fast ambipolar diffusion.

(b)   Further research on properties resulting from the indirect bandgap in real space. Recently, investigations of transport and luminescence under the influence of an electric field applied in the growth direction were started [71]. Also first results on lateral n-i-p-i structures were reported [72-74]. Drastic effects on the luminescence due to wave function tuning by in-plane magnetic fields are expected [75] and first results were reported in [76].

(c)   n-i-p-i systems represent particularly suitable model systems for the study of a wide variety of problems or phenomena whose relevance is quite general. For instance, many-body phenomena can be studied as a function of the variable carrier density in systems with disorder (e.g. in δ-doped layers) or without disorder (undoped quantum wells in large period hetero n-i-p-i crystals). Another example is the recent demonstration of weakly δ-doped n-i-p-i systems as an ideal model system for the study of the density of states of impurity bands [77,78]. As a last example, the interesting aspects for the study of fs hot carrier dynamics associated with simultaneous relaxation in energy and real space should be mentioned [79].

REFERENCES

[1]   G.H. Döhler [*Phys. Status Solidi B (Germany)* vol.52 (1972) p.79-92]

[2]   G.H. Döhler [*Phys. Status Solidi B (Germany)* vol.52 (1972) p.533-45]

[3]   L. Esaki, R. Tsu [*IBM J. Res. Dev. (USA)* vol.14 (1970) p.61]

[4]   R.F. Kazarinov, R.A. Suris [*Sov. Phys.-Semicond. (USA)* vol.5 (1971) p.707]; R.F. Kazarinov, Yu.V. Shmarsev [*Sov. Phys.-Semicond. (USA)* vol.5 (1971) p.710]

[5]   V.I. Stafeev [*Sov. Phys.-Semicond. (USA)* vol.5 (1971) p.359]

[6]   M.I. Ovsyannikov, Yu.A. Romanov, V.N. Shabanov, R.G. Loginova [*Sov. Phys.-Semicond. (USA)* vol.4 (1971) p.1919]; M.I. Ovsyannikov, Yu.A. Romanov, V.N. Shabanov [*Proc. Int. Conf. on the Physics and Chemistry of Semiconductor Heterojunctions and Layer Structures VI*, Budapest (1971) p.205]

[7]   G.H. Döhler [*Surf. Sci. (Netherlands)* vol.73 (1978) p.97-105]

[8]   G.H. Döhler [*J. Vac. Sci. Technol. (USA)* vol.16 (1979) p.851-6]

[9]   K. Ploog, A. Fischer, G.H. Döhler, H. Künzel [*Inst. Phys. Conf. Ser. (UK)* no.59 (1981) p.721]

[10]   G.H. Döhler et al [*Phys. Rev. Lett. (USA)* vol.47 (1981) p.864-7]

[11]   Ch. Zeller, G. Abstreiter, K. Ploog [*Surf. Sci. (Netherlands)* vol.142 (1984) p.456-9]

[12]   G. Fasol, P. Ruden, K. Ploog [*J. Phys. C, Solid State Phys. (UK)* vol.17 no.8 (20 March 1984) p.1395-403]

[13]   G.H. Döhler [*J. Vac. Sci. Technol. B (USA)* vol.1 (1983) p.278-84]

[14]   G.H. Döhler, G. Fasol, T.S. Low, J.N. Miller [*Solid State Commun. (USA)* vol.57 (1986) p.563-6]

[15]   K. Köhler, G.H. Döhler, J.N. Miller, K. Ploog [*Superlattices Microstruct. (UK)* vol.2 (1986) p.339-43]

[16]   H. Jung, G.H. Döhler, H. Künzel, K. Ploog, P. Ruden, H.J. Stolz [*Solid State Commun. (USA)* vol.43 (1982) p.291-4]

[17]   W. Rehm, P. Ruden, G.H. Döhler, K. Ploog [*Phys. Rev. B (USA)* vol.28 (1983) p.5937-42]

[18]   H. Jung, G.H. Döhler, E.O. Göbel, K. Ploog [*Appl. Phys. Lett. (USA)* vol.43 (1983) p.40-2]

[19]   G.H. Döhler, H. Künzel, K. Ploog [*Phys. Rev. B (USA)* vol.25 (1982) p.2616-26]

[20]   K. Ploog, H. Künzel, J. Knecht, A. Fischer, G.H. Döhler [*Appl. Phys. Lett. (USA)* vol.38 (1981) p.870-3]

[21]   P. Zwicknagl, W. Rehm, E. Bauser [*J. Electron. Mater. (USA)* vol.13 (1984) p.545]; P. Zwicknagl [*J. Appl. Phys. (USA)* vol.55 (1984) p.1513]

[22]   H. Künzel, G.H. Döhler, P. Ruden, K. Ploog [*Appl. Phys. Lett. (USA)* vol.41 (1982) p.852-4]

[23]   G. Abstreiter, H. Kirchstaetter, K. Ploog [*J. Vac. Sci. Technol. B (USA)* vol.3 (1985) p.623]

[24]   P. Ruden, G.H. Döhler [*Surf. Sci. (Netherlands)* vol.132 (1983) p.540]

[25]   H. Künzel, A. Fischer, J. Knecht, K. Ploog [*Appl. Phys. A (Germany)* vol.30 (1983) p.73-81]

[26]   G.H. Döhler, P. Ruden [*Surf. Sci. (Netherlands)* vol.142 (1984) p.474-85]

[27]   R.A. Street, G.H. Döhler, J.N. Miller, P.P. Ruden [*Phys. Rev. B (USA)* vol.33 (1986) p.7043-6]

[28]   R.A. Street, G.H. Döhler, J.N. Miller, R.D. Burnham, P.P. Ruden [*Proc. 18th Int. Conf. on the Physics of Semiconductors* Ed. O. Engström (World Scientific, Singapore, 1987) p.215]

[29]   P. Ruden, G.H. Döhler [*Phys. Rev. B (USA)* vol.27 (1983) p.3538-46]

[30]   P. Ruden, G.H. Döhler [*Phys. Rev. B (USA)* vol.27 (1983) p.3547-53]

[31]   P. Ruden [*J. Vac. Sci. Technol. B (USA)* vol.1 (1983) p.285]

[32]   P. Ruden, G.H. Döhler [*Solid State Commun. (USA)* vol.45 (1983) p.23-5]

[33]   P. Ruden, G.H. Döhler [*Proc 17th Int. Conf. on the Physics of Semiconductors* Eds W.H. Harrison, J.D. Chadi (Springer, New York, 1985) p.535-8]

[34]   G.H. Döhler [*CRC Crit. Rev. Solid State Mater. Sci. (USA)* vol.13 no.2 (1987) p.97-141]

[35]   G.H. Döhler, G. Hasnain, J.N. Miller [*Appl. Phys. Lett. (USA)* vol.49 (1986) p.704-6]

[36]   C.J. Chang-Hasnain et al [*Appl. Phys. Lett. (USA)* vol.50 (1987) p.915-7]

[37]   G. Hasnain, G.H. Döhler, J.R. Whinnery, J.N. Miller, A. Dienes [*Appl. Phys. Lett. (USA)* vol.49 (1986) p.1357-9]

[38]   G. Hasnain, D. Mars, G.H. Döhler, M. Ogura, J.S. Smith [*Appl. Phys. Lett. (USA)* vol.51 (1987) p.831-3]

[39]   X. Wu et al [*Appl. Phys. Lett. (USA)* vol.62 (1993) p.152-3]

[40]   K.H. Gulden et al [*Appl. Phys. Lett. (USA)* vol.62 (1993) p.3180-2]

[41]   S. Malzer et al [accepted for publication in *J. Vac. Sci. Technol. B (USA)*]

[42]   D.E. Ackley, J. Mantz, H. Lee, N. Nouri, C.-L. Shieh [*Appl. Phys. Lett. (USA)* vol.53 (1988) p.125]

[43]   G.W. Yoffe, J. Brübach, F. Karouta, W.C. van der Vleuten, L.M. Kaufmann, J.H. Wolter [*Appl. Phys. Lett. (USA)* vol.63 (1993) p.1456]

[44]   S.D. Koehler, E.M. Garmire, A.R. Kost, D. Yap, D.P. Docter, T.C. Hasenberg [*IEEE Photonics Technol. Lett. (USA)* vol.7 (1993) p.878]

[45]   M. Renn, G.H. Döhler [*Phys. Rev. B (USA)* vol.48 (1993) p.11220-7]

[46]   K. Schrüfer, C. Metzner, U. Wieser, M. Kneissl, G.H. Döhler [*Superlattices Microstruct. (UK)* vol.15 (1994) p.413-20]

[47]   C. Metzner, K. Schrüfer, U. Wieser, M. Luber, M. Kneissl, G.H. Döhler [*Phys. Rev. B (USA)* vol.51 no.8 (1995) p.5106-15]

[48]   Ch. Zeller, B. Vinter, G. Abstreiter, K. Ploog [*Physica B (Netherlands)* vol.117 (1983) p.729]

[49]   H.J. Beyer, C. Metzner, J. Heitzer, G.H. Döhler [*Superlattices Microstruct. (UK)* vol.6 (1989) p.351-6]

[50]   E.F. Schubert, J.E. Cunningham, W.T. Tsang, G.L. Timp [*Appl. Phys. Lett. (USA)* vol.51 (1987) p.15]

[51]   T.B. Simpson, C.A. Pennise, B.E. Gordon, J.E. Anthony, T.R. AuCoin [*Appl. Phys. Lett. (USA)* vol.49 (1986) p.590]

[52]   A. Kost, E. Garmire, A. Danner, P.D. Dapkus [*Appl. Phys. Lett. (USA)* vol.54 (1989) p.301]; A. Kost, M. Kawase, E. Garmire, A. Danner, H.C. Lee, P.D. Dapkus [*Proc. SPIE (USA)* vol.943 (1988) p.114]

[53]   H. Ando, H. Iwamura, H. Oohashi, H. Kanbe [*IEEE J. Quantum Electron. (USA)* vol.25 (1989) p.2135]

[54]   K.K. Law, J. Maserjan, R.J. Simes, L.A. Coldren, A.C. Gossard, J.L. Merz [*Opt. Lett. (USA)* vol.14 (1989) p.230]

[55]   A. Larson, J. Maserjan [*Appl. Phys. Lett. (USA)* vol.59 (1991) p.1946]

[56]   P. Kiesel et al [*Superlattices Microstruct. (UK)* vol.13 (1993) p.21-4]

[57]   J. Schultz et al [*Solid-State Electron. (UK)* vol.40 (1996) p.683-6]

[58]   S. Malzer et al [*Phys. Status Solidi B (Germany)* vol.173 (1992) p.459-72]

[59]   M. Kneissl, K.H. Gulden, P. Kiesel, A. Luczak, S. Malzer, G.H. Döhler [*Solid-State Electron. (UK)* vol.37 (1994) p.1251-3]

[60]   P. Kiesel, P. Riel, H. Lin, N. Linder, J.N. Miller, G.H. Döhler [*Superlattices Microstruct. (UK)* vol.6 (1989) p.363-8]

[61]   U. Pfeiffer et al [*Appl. Phys. Lett. (USA)* vol.68 no.13 (1996)]

[62]   P. Riel et al [*Proc. SPIE (USA)* vol.1675 (1992) p.242-54]

[63]   A. Höfler et al [*Appl. Phys. Lett. (USA)* vol.62 (1993) p.3399-401]

[64]   Y. Horikoshi, A. Fischer, K. Ploog [*Appl. Phys. Lett. (USA)* vol.45 (1984) p.919]

[65]   J. Oswald, M. Pipan [*Semicond. Sci. Technol. (UK)* vol.8 (1992) p.435-42]

[66]   R.A. Stradling [*Semicond. Sci. Technol. (UK)* vol.6 (1991) p.52]

[67]   R.A. Stradling [*Proc. SPIE (USA)* vol.1361 (1991) p.630]

[68]   V.B. Kulikov, G.H. Avetisyan, I.D. Zalevsky, P.V. Bulaev [*Proc. SPIE (USA)* vol.2694 (1996) paper no.26 in press]

[69]   K.H. Gulden, H. Lin, P. Kiesel, P. Riel, G.H. Döhler, K.J. Ebeling [*Phys. Rev. Lett. (USA)* vol.66 (1991) p.373-6]

[70]   P. Riel et al [*Appl. Phys. Lett. (USA)* vol.66 (1995) p.1367-9]

[71]   H. Böhner, S. Malzer, G.H. Döhler, A. Förster, H. Lüth [*Verh. Dtsch. Phys. Ges. (Germany)* vol.29 (1994) p.1165]

[72]   F. Hirler et al [*Surf. Sci. (Netherlands)* vol.263 (1992) p.536-40]

[73]   A. Huber, H. Lorenz, J.P. Kotthaus, S. Bakker [*Phys. Rev. B (USA)* vol.51 (1995) p.5028-32]

[74]   M. Fritze, A. Nurmikko, P. Hawrylak [*Surf. Sci. (Netherlands)* vol.305 (1994) p.580-4]

[75]   G.H. Döhler [*Springer Ser. Solid-State Sci. (Germany)* vol.87 (1989) p.174-84]

[76]   M. Forkel et al [Magnetic tuning of spatially indirect interband transitions in parabolic and rectangular quantum wells in *High magnetic fields in the physics of semiconductors* Ed. D. Heiman (World Scientific, Singapore, 1995)]

[77]   G.H. Döhler [*NATO ASI Ser. B (USA)* vol.183 (1988) p.159]

[78]   J. Schönhut et al [*Solid-State Electron. (UK)* vol.40 (1996) p.701-5]

[79]   N. Moritz, H. Hauenstein, A. Seilmeier, G. Bickel, G.H. Döhler [*Proc. 22nd Int. Conf. on the Physics of Semiconductors* Ed. D.J. Lockwood (World Scientific, Singapore, 1995) vol.1 p.859-62]

1.3   Strained layer superlattices - a historical perspective

G.C. Osbourn

January 1996

A   INTRODUCTION

Lattice-mismatched strained-layer heterostructures have become a common element of many high performance device structures and the theme of much current research. A newcomer to the field might expect that the introduction of lattice-mismatched materials into quantum well and superlattice structures was an obvious evolutionary extension of work on the lattice-matched AlGaAs/GaAs system, and that developments in this area lagged behind analogous lattice-matched work perhaps because of limitations in crystal growth technology. In fact, the realisation that strained-layer semiconductor structures could exhibit both useful material quality and advantageous optoelectronic properties was a major departure from conventional wisdom. Indeed, early experimental demonstrations of these properties encountered widespread scepticism. This article will briefly recount the interesting history of the research and development of large-mismatch III-V strained-layer heterostructures, and will describe the paradigm shift which ultimately produced new materials/heterostructures of choice for certain semiconductor lasers and low noise, high frequency amplifiers. The discussions will emphasise interesting events which controlled the progress in this field but which are difficult to learn from the literature alone.

B   MISMATCHED HETEROSTRUCTURES circa 1982

The conventional wisdom before 1982 was that significant lattice constant mismatch, e.g. 1%, at heterostructure interfaces lowered the material quality and the long-term stability of these structures, making these heterostructures undesirable for most device applications. This view was based on extensive research on III-V heterostructures and devices grown primarily by liquid phase epitaxy (LPE). Such structures typically contained layers that were quite thick by quantum well (QW) and superlattice (SL) standards, and any large lattice mismatches in such structures produced misfit dislocations. Dislocations are defects which degrade carrier mobilities and provide undesirable non-radiative recombination sites for electron-hole pairs. Thus, mismatched interfaces and their associated misfit dislocations were undesirable for use in the active regions of high quality devices.

Many III-V alloys with interesting optical and transport properties, e.g. GaAs-rich InGaAs, cannot be grown on available binary substrates without introducing lattice mismatch. In some cases, this lattice mismatch could be contained in graded alloy buffer layers grown between the substrate and the final alloy layer. This approach could contain many, but not all, of the dislocations in passive regions of the structure buried below the active region. However, even physically separated dislocations proved to be problematic in high power light emitting devices. Dislocations could climb into the active regions of light-emitting diodes and lasers during operation, producing a network of dark line defects which ultimately produced device failure. These observations provided further motivation for avoiding lattice mismatch in any part of III-V devices.

It was also recognised that lattice mismatch could be accommodated entirely by elastic strain, i.e. without any dislocations, if the mismatched layer thicknesses were less than a critical value. The physics underlying this was well studied since the work of Frank and van der Merwe in 1949 [1]. Theoretical calculations indicated that the energy required to elastically accommodate the mismatch for sufficiently thin layers can be less than the energy required to create a combination of dislocations and residual strain for mismatch accommodation [2]. The critical thickness (hc) for elastic strain accommodation increases as the lattice mismatch decreases, and early III-V experimental studies focused primarily on the need for very small lattice mismatches for typical device layer thicknesses in the micrometre range. However, Matthews and Blakeslee [3-5] examined the critical thicknesses that corresponded to larger mismatches and showed that mismatches exceeding one percent could be accommodated elastically with layers in the range from atomic thicknesses to tens of nm.

Heterostructures with such thin layers became of interest at the time of Blakeslee and Matthew's work due to Esaki's prediction of novel transport properties in thin-layered SLs [6]. Also, such thin layers were becoming achievable by the emerging techniques of molecular beam epitaxy (MBE) and chemical vapour deposition (CVD). It might seem in retrospect that the interest in the physics of thin-layer QWs and SLs would have led to the development of SLs with large lattice mismatches, i.e. strained-layer superlattices (SLSs). This was not the case. The first attempts to verify Esaki's prediction at IBM were with mismatched GaAs/GaAsP superlattice structures grown by Blakeslee [7]. These structures did not exhibit the predicted transport effects, and in fact contained many dislocations. Attention shifted to the lattice matched GaAs/AlGaAs system, and novel quantum confinement effects were observed in this system. Blakeslee's work on large-lattice-mismatch superlattices was terminated at IBM. Early failure to achieve device-quality strained-layer superlattices seemed to confirm the conventional wisdom that lattice mismatch was to be avoided in the optically and electrically active regions of any useful III-V heterostructures. QW and SL studies through the early 1980s focused on material systems with the lowest achievable lattice mismatch, i.e. AlGaAs/GaAs and InAs/GaSb.

C   STRAINED-LAYER HETEROSTRUCTURE RESEARCH: A PARADIGM SHIFT

The feasibility and potential advantages of employing large lattice-mismatched interfaces in the optically active and electrically active areas of high quality heterostructures were first proposed by the author [8,9]. This work originated as a desire to carry out a tight-binding band structure study of III-V lattice-mismatched heterostructures. A tight binding approach, developed by the author and D.L. Smith [10], for treating single interfaces required that the heterostructure layers have the same periodicities in the planes parallel to the interface. This is not true for mismatched heterostructures with dislocation-accommodation of the mismatch, i.e. for most of the mismatched structures studied up to that time. However, the author was aware of the literature cited above, and realised that thin-layer structures with strain-accommodated mismatch did satisfy this requirement. Strained-layer structures were immediately of special interest as these represented the only class of mismatched heterostructures that were amenable to the intended theoretical analysis. It was also clear that the large elastic strains associated with large mismatches would have significant effects on the band structures of these heterostructures which would be absent or negligible in closely lattice-matched systems. Such effects could be both interesting and useful, and the interplay of these effects and quantum confinement effects had not been considered in previous QW and SL work. Finally, it was obvious that intentionally considering mismatched materials for heterostructure work would greatly broaden the available combinations of bandgaps, conduction and valence band offsets, transport properties, etc. to be examined. The only constraint was that large lattice mismatches corresponded to very thin layers - from atomically thin to a few tens of nm. There seemed to be no reason to avoid such systems for QW and SL structures, as layers in this range were also needed to observe novel quantum effects. Yet, there were no published studies of the electrical and optical properties of large-mismatch QWs and SLSs. It became apparent that lattice mismatch was being avoided in QWs and SLs because of the conventional wisdom in the field. Poor quality material was expected for strained-layer structures, as no distinction was made between the effects of lattice mismatch and the effects of misfit dislocations.

The author recognised that the problems associated with mismatched heterostructures were in fact directly attributable to the misfit dislocations which usually resulted from the mismatch. Thin strained-layers can accommodate mismatch without producing the problematic misfit dislocations. Further, elastic distortion of the entire lattice does not introduce localised lattice defects or recombination centres. Thus, there were no existing data which suggested that strained-layers should have inferior material properties to closely lattice matched structures. Further, for strained-layers with thicknesses less than hc, there are no lower-energy dislocated structural states for the elastically strained layers to 'relax' to. In other words, such structures are thermodynamically stable against dislocation generation when designed and grown to meet the hc constraint.

D   EARLY EXPERIMENTAL SUPPORT

Experimental and computational studies at Sandia National Labs established or verified a number of key properties of III-V strained-layer systems (see the references contained in [9,11]). Early contributors to the work described here included R.M. Biefeld, P.L. Gourley, I.J. Fritz, L.R. Dawson, T.E. Zipperian, T.J. Drummond, S.T. Picraux, S.R. Kurtz, J.E. Schirber, E.D. Jones, B. Dodson, J. Tsao, and the author. The ability to grow SLSs by CVD and MBE with lattice mismatches in the percent range was clearly established in the GaAsP, InGaAs and InAsSb alloy systems. The accommodation of these large mismatches by biaxial elastic strain was also clearly established through X-ray diffraction and ion beam channelling studies. The critical thickness limits for elastic strain accommodation as a function of lattice mismatch were experimentally established for the InGaAs system, and were found to be consistent with earlier work in the GaAsP system by Matthew and Blakeslee. Material quality was established first by TEM studies which revealed dislocation-free SLS material. Further studies demonstrated that respectable luminescence efficiencies, carrier lifetimes and modulation-doping-enhanced mobilities were routinely achievable in active regions containing multiple interfaces with over 1% lattice mismatch. Novel effects on the SLS band structures associated with the large biaxial layer strains were also observed. The strain-enhanced splitting of the light and heavy hole valence bands in InGaAs/GaAs SLSs was shown to enable the preferential occupation of holes in highly nonparabolic, in-plane, light-mass valence bands.

Prototype device structures were also demonstrated. LEDs and modulation-doped FETs with large lattice mismatch were first demonstrated in the InGaAs system at Sandia. Electrically-injected InGaAs strained QW lasers were first demonstrated by Laidig and co-workers [12], and soon after at Sandia. None of these devices broke existing performance records, yet they clearly demonstrated that the conventional wisdom regarding large mismatch devices was wrong for properly-designed QW and SLS structures.

E   CONTROVERSY

The early work recounted above was at odds with the experience and expectations of the III-V community, and soon encountered widespread scepticism. The author received private expressions of disbelief of the Sandia results by many leading crystal growers and device physicists in the 1983-1985 time frame. Many felt that the physics of mismatch accommodation by elastic strain either did not extend to large mismatches or did not allow the possibility of achieving high quality material. The intuitive association of large mismatch with poor material quality was strong, so that many experts still expected that dislocations or some type of strain-related defects would in some way destroy the material quality. TEM demonstrations of dislocation-free interfaces and prototype device demonstrations to the contrary were initially not widely believed, and were occasionally attributed to misrepresentations of the data or the amount of mismatch in the samples, i.e. to fraud! Another common criticism was that lattice-mismatched systems would have no performance advantages to offer over lattice-matched systems. While a variety of advantages were suggested by theoretical work, clear superiority of strained-layer systems for some device applications was not experimentally established until some years later.

The expectation that large-mismatched heterostructures are inherently unstable at high optical and electrical power levels was the most difficult aspect of the flawed conventional wisdom to overturn. Holonyak and co-workers photopumped a series of SLS samples and reported rapid failure of their laser action [13]. They demonstrated that the failed laser material contained a large network of dislocations which was not present initially in the SLS part of the structure. Many researchers attributed these results to the 'intrinsic instability' of all strained layers, i.e. strained layers would 'relax' by dislocation formation at the high current and high optical powers in devices such as semiconductor lasers. The lasing experiments clearly confirmed that strained-layer structures could exhibit high quality at first, but also convinced many that strained-layer structures could not fulfil the useful technological roles suggested by the author.

Later studies directly demonstrated long-lived strained-layer injection lasers [14,15]. Why did the early structures degrade through the apparent introduction of dislocations? There are at least two possible explanations. First, the layer thicknesses of the samples may have exceeded the hc for the actual mismatches in the sample. Structures which violate the hc constraint are indeed metastable against relaxation in high power applications. Second, the pressure of a graded buffer layer can promote dislocations beneath the SLS. Other samples were grown without buffers, but the published layer compositions were improperly chosen for lattice matching of the thick SLS as a whole with the underlying substrate, so that dislocations would be expected in all of the SLS samples reported. Thick SLSs must also be carefully lattice matched to underlying buffer layers (or the substrate) to avoid dislocation-accommodation of the SLS itself. Existing dislocations in the passive regions of III-V (including SLS) laser structures provide degradation mechanisms for lasers via dark-line defect climb into the active region under lasing operation. These considerations suggest the advantages of a few strained QWs over a thick SLS for some device applications.

F   RESURGENCE OF STRAINED-LAYER RESEARCH

Widespread interest in III-V strained-layer research returned [16,17] when advantages for laser and modulation-doped FET (MODFET) applications became more clear [18]. Potential laser performance advantages were described [19-21], and were based on the advantageous modifications of the band structures by both compressive and tensile strain. Experimental studies established improved performance of strained-QW lasers compared to similar unstrained QW lasers [22]. Lifetest studies of populations of strained InGaAs QW laser devices did not produce rapid catastrophic failures, and in fact indicated that the strained lasers might be more robust than the lattice-matched AlGaAs/GaAs QW lasers [14,15]. The development of Er-doped optical fibre amplifiers motivated interest in strained InGaAs QW lasers, as these lasers operate at an optimal pump wavelength for the fibre amplifiers [23].

Performance advantages of AlGaAs/strained InGaAs QW MODFETs over similar AlGaAs/GaAs QW devices were proposed and demonstrated [24-26]. Higher frequency operation resulted from the possibility of obtaining deeper QWs and better carrier confinement while keeping AlGaAs alloy concentrations small enough to avoid DX centre problems. Careful studies also refined the hc constraint values for these device structures, and demonstrated that devices which obey the hc constraint can survive ion-implantation and annealing IC processing steps [27]. This approach has yielded world-record high-frequency, low-noise FET performance.

Other SLS material and device research directions, too numerous to describe exhaustively here, have also been pursued. The use of SLSs to achieve high material quality in low-bandgap mismatched III-V materials systems such as InAsSb [28,29] and InAs/InGaSb [30] has motivated the study of III-V SLS IR detector and emitter concepts. SLSs grown on substrate orientations other than (100) were shown to have large built-in piezoelectric fields which yield interesting nonlinear optical properties [31]. The early demonstrations of preferential light-hole conduction in InGaAs SLSs suggested the possibility of a high-speed, low-power AlGaAs/strained InGaAs complementary logic technology. Interest in non-III-V strained systems has also flourished, with much attention focused on SiGe and high bandgap II-VI alloys.

G   CONCLUSION

The novel strain-related properties, flexibility in the choice of materials combinations, and device performance advantages offered by strained-layer heterostructures have been firmly established.

REFERENCES

[1]   F.C. Frank, J.H. van der Merwe [*Proc. R. Soc. Lond. A (UK)* vol.198 (1949) p.216-65]
[2]   J.H. van der Merwe [*CRC Crit. Rev. Solid State Mater. Sci. (USA)* vol.7 (1978) p.209-31]

[3]   J.W. Matthews, A.E. Blakeslee [*J. Cryst. Growth (Netherlands)* vol.27 (1974) p.118-25]
[4]   J.W. Matthews, A.E. Blakeslee [*J. Cryst. Growth (Netherlands)* vol.29 (1975) p.273-80]
[5]   J.W. Matthews, A.E. Blakeslee [*J. Cryst. Growth (Netherlands)* vol.32 (1976) p.265-73]
[6]   L. Esaki, R. Tsu [IBM Res. Note, RC-2418 March (1969)]
[7]   A.E. Blakeslee, C.F. Aliotta [*IBM J. Res. Dev. (USA)* vol.14 (1970) p.686-8]
[8]   G.C. Osbourn [*J. Appl. Phys. (USA)* vol.53 (1982) p.1586-9]
[9]   G.C. Osbourn, P.L. Gourley, I.J. Fritz, R.M. Biefeld, L.R. Dawson, T.E. Zipperian [*Semicond. Semimet. (USA)* vol.24 (Academic, San Diego, 1987) ch.8 and references therein]
[10]   G.C. Osbourn, D.L. Smith [*Phys. Rev. B (USA)* vol.19 (1979) p.2124-33]
[11]   R.M. Biefeld (Ed.) [*Compound Semiconductor Strained-Layer Superlattices* (Trans Tech Pub., Zurich, Switzerland, 1989) and references contained therein]
[12]   W.D. Laidig, P.J. Caldwell, Y.F. Lin, C.K. Peng [*Appl. Phys. Lett. (USA)* vol.44 (1984) p.653-5]
[13]   M.D. Camras et al [*J. Appl. Phys. (USA)* vol.54 (1983) p.6183-9]
[14]   W. Stutius, P. Gavrilovic, J.E. Williams, K. Meehan, J.H. Zarrabi [*Electron. Lett. (UK)* vol.24 (1988) p.1493]
[15]   S.E. Fischer, R.G. Waters, D. Fekete, J.M. Ballantyne, Y.C. Chen, B.A. Soltz [*Appl. Phys. Lett. (USA)* vol.54 (1989) p.1861-2]
[16]   T.P. Pearsall (Ed.) [*Semicond. Semimet. (USA)* vol.32 (Academic, San Diego, 1991) and references therein]
[17]   T.P. Pearsall (Ed.) [*Semicond. Semimet. (USA)* vol.33 (Academic, San Diego, 1991) and references therein]
[18]   H. Morkoc, B. Sverdlov, G. Gao [*Proc. IEEE (USA)* vol.81 (1993) p.493-556 and references therein]
[19]   E. Yablonovitch, E.O. Kane [*J. Lightwave Technol. (USA)* vol.4 (1986) p.961]
[20]   A.R. Adams [*Electron. Lett. (UK)* vol.22 (1986) p.250]
[21]   E.P. O'Reilly, G. Jones, A. Ghiti, A.R. Adams [*Electron. Lett. (UK)* vol.27 (1991) p.1417-9]
[22]   P.J.A. Thijs, E.A. Monte, T. Van Dongen, C.W.T. Bulle-Lieuwma [*J. Cryst. Growth (Netherlands)* vol.105 (1990) p.339-47]
[23]   R.I. Laming et al [*Electron. Lett. (UK)* vol.25 (1989) p.12-4]
[24]   T.E. Zipperian, T.J. Drummond [*Electron. Lett. (UK)* vol.21 (1985) p.823-4]
[25]   J.J. Rosenberg, M. Benlamri, P.D. Kirchner, J.M. Woodall, G.D. Pettit [*IEEE Electron Device Lett. (USA)* vol.6 (1985) p.491-3]
[26]   A. Ketterson et al [*IEEE Electron Device Lett. (USA)* vol.6 (1985) p.628-30]
[27]   T.E. Zipperian, E.D. Jones, B.W. Dodson, J.F. Klem, P.L. Gourley, T.A. Plut [*Inst. Phys. Conf. Ser. (UK)* no.96 (1989) p.365-70]
[28]   G.C. Osbourn [*J. Vac. Sci. Technol. B (USA)* vol.2 (1984) p.176]
[29]   C. Mailhiot, D.L. Smith [*J. Vac. Sci. Technol. A (USA)* vol.7 (1989) p.445]
[30]   S.R. Kurtz, R.M. Biefeld [*Phys. Rev. B (USA)* vol.44 (1991) p.1143-9]
[31]   D.L. Smith, C. Mailhiot [*Phys. Rev. Lett. (USA)* vol.58 (1987) p.1264]

CHAPTER 2

THEORETICAL ASPECTS

2.1 Band structures of III-V quantum wells and superlattices

Yia-Chung Chang

July 1995

A   INTRODUCTION

We review various theoretical methods used in the literature for calculating the band structures of III-V quantum wells and superlattices, giving an assessment of the usefulness of each method. These include the effective-mass method [1,2], **k.p** method [3-9], bond-orbital method [10,11], empirical tight-binding method [12-15], empirical pseudopotential method [16-18], and first-principles methods [19-22]. Representative band structures for different kinds of quantum wells and superlattices are discussed.

B   EFFECTIVE-MASS METHOD

The effective-mass method is the simplest and physically most intuitive. It works well when the states of interest are derived from a single parabolic band. In this method the conduction or valence band edges of the two constituent semiconductors form piecewise constant potentials, when plotted in the growth direction (z), and the electron or hole in each region behaves like a free particle with an effective mass determined from the curvature of the bulk band structure at the band edge. The difference in energy of the band edges for the two constituent semiconductors is called the band offset. The band offset gives rise to a quantum well which confines the electron (hole) in the material with lower (higher) conduction (valence) band edge. For a quantum well, this is also called a 'particle-in-a-box model'. For a superlattice, the potential profile is given by the Kronig-Penney model. TABLE 1 lists the experimentally determined band offsets for the conduction and valence bands (ΔE_c and ΔE_v) for many III-V semiconductor heterostructures.

TABLE 1. Band offsets for conduction and valence bands ΔE_c and ΔE_v in units of eV.

	ΔE_c	ΔE_v	Ref
GaAs/AlAs	0.98	0.63	[23]
GaAs/Ga$_{1-x}$Al$_x$As ($x < 0.3$)	0.75x	0.50x	[24]
InAs/GaAs	1.03	0.07	[25]
InAs/GaSb	0.92	0.52	[26]
GaAs/InP	-0.43	0.34	[25]
GaSb/AlSb	1.26	0.3	[27]
HgTe/CdTe	1.56	0.35	[28]

For states confined in a deep quantum well, the 'particle-in-a-box' model gives energy levels according to the simple expression

$$E_n = \frac{(\hbar n\pi)^2}{2m^*w^2}$$

where m^* is the effective mass, w is the well width, and $n = 1,2,3,...$ is the principal quantum number. The expression is accurate as long as E_n is much less than the depth of the quantum well, V_0. For states with energy not too small compared with V_0, the energy should be determined by solving a transcendental equation [29]

$$k \tan(kw/2)/m^* = \kappa/m^{*\prime} \text{ for even parity states} \tag{1}$$

$$k \cot(kw/2)/m^* = -\kappa/m^{*\prime} \text{ for odd parity states} \tag{2}$$

where

$$k = \sqrt{2m^*E} / \hbar,$$

$$\kappa = \sqrt{2m^{*\prime}(V_0-E)}/\hbar$$

and $m^*(m^{*\prime})$ is the effective mass for the well (barrier) material. Here the boundary condition is that both the wave and particle current are continuous across the interface. For $\mathbf{k}_\parallel \neq 0$, the energy is given by the above transcendental equation with

$$k = \sqrt{2m^*E/\hbar^2 - k_\parallel^2}$$

and

$$b = \sqrt{2m^{*\prime}(V_0-E)/\hbar^2 + k_\parallel^2}$$

It is sometimes approximated by

$$E_n(\mathbf{k}_\parallel) = E_n^{(0)} + \frac{\hbar^2 k_\parallel^2}{2m^*} \tag{3}$$

where $E_n^{(0)}$ denote the energies at $\mathbf{k}_\parallel = 0$. This expression is adequate when the wave function is mostly confined in the well. For a superlattice, the 'Kronig-Penney' solution (with the same boundary condition as used above) is given by [3]

$$2\cos q(w+b) = 2\cos kw \cos k'b - (\xi+\xi^{-1})\sin kw \sin k'b \tag{4}$$

where q is the superlattice wave vector along z, $k^2 = 2m^*E/\hbar^2$, $k'^2 = 2m^{*\prime}(E-V_0)/\hbar^2$, $\xi = (k/k')(m^{*\prime}/m^*)$, and $w(b)$ is the width of the well (barrier) material in a superlattice unit cell.

C   k.p AND BOND-ORBITAL METHODS

The effective-mass method described above is not adequate for valence-band states with $k_\parallel \neq 0$ when the heterostructure states are derived from more than one bulk band (a phenomenon known as valence-band mixing) and for narrow-gap semiconductor heterostructures when the states of interest are derived from both conduction and valence bands. In this case, a more sophisticated method such as the k.p [3-9] or bond-orbital method [10,11] is needed. The k.p method describes bulk bands accurately near the zone centre [30,31]; thus, it is a good method when the heterostructure states of interest are derived mainly from bulk bands near the zone centre. The wave function in either the well or barrier region is expanded in terms of Bloch states ($\psi_{\alpha,kz}$) of constituent bulk materials (α labels different bulk bands involved),

$$\psi(\mathbf{r}) = \sum_{\alpha,k_z} F_\alpha(k_z)\psi_{\alpha,kz}(\mathbf{r}) \tag{5}$$

The Fourier transform of $F_\alpha(k_z)$ is called the envelope function $F_\alpha(z)$. The band structure is solved by matching the envelope functions in different regions of the heterostructures at the interfaces. Alternatively, one can expand the envelope functions in terms of a set of trial wave functions (either gaussians [6] or plane waves [9]) and solve them based on the variational principle. For an infinitely deep quantum well, the boundary condition is simply

$$F_\alpha(z) = 0 \text{ for all } \alpha \tag{6}$$

and the band energies can be obtained by simple transcendental equations [4,32]. For the general case, many bulk bands are involved and the boundary conditions can become quite complicated. One approximation often used is to ignore the difference in band parameters for different regions, but keep the quantum-confinement effects due to the band offsets [6]. This approximation works fairly well for states which are well confined in the quantum well region. FIGURE 1 shows a set of valence-band structures for GaAs/Al$_{0.25}$Ga$_{0.75}$As quantum wells obtained from the k.p method with this approximation. The complicated shape of the bands is due to the anti-crossing effects of states with the same symmetry which contain an admixture of bulk states derived from heavy and light hole bands. The band energies are anisotropic functions of $k_\parallel$ as one can see that the band dispersion is different for $k_\parallel$ along different directions. For many III-V heterostructures (including GaAs/AlGaAs systems) the in-plane anisotropy in band structures is small: an axial approximation [5] can often be used to simplify the description.

The bond-orbital method can be used in a general sense [10] or as a discretised version of the k.p method [11]. In the latter case, it is also referred to as the effective bond orbital method (EBOM) [33] and it can be used as effectively as the k.p method. Furthermore it uses the tight-binding formalism to determine the boundary conditions, which are much simpler to implement. The method can also be applied to heterostructures with composition variation or any slow-varying z-dependent potentials which may be caused by an external electric field or strain.

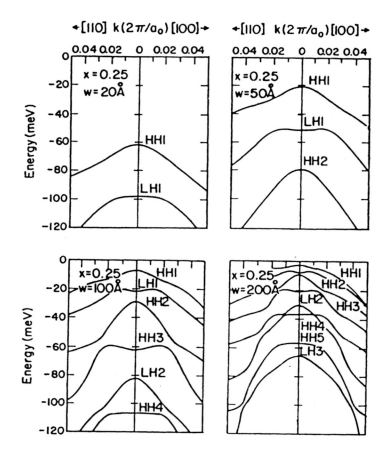

FIGURE 1. Valence band structures of GaAs/Ga$_{0.75}$Al$_{0.25}$As quantum wells obtained by the **k.p** method for well widths of 20, 50, 100, and 200 Å. From [6].

FIGURE 2 shows the band structures of an InAs/GaSb superlattice obtained by EBOM [11]. InAs/GaSb forms a type-II superlattice, in which the electron and hole are confined in different regions of the superlattice [34]. This property gives rise to long carrier lifetimes and weak inter-band optical absorption [34,35]. However, due to the band-mixing between conduction and light-hole bands, the superlattice displays a strong p-type doped inter-subband optical transition for in-plane polarisation, which may find application in normal-incidence far-infrared optical devices [36].

FIGURE 3 shows a set of band structures of In$_{1-x}$Ga$_x$As/InP quantum wells obtained by EBOM [37]. In$_{1-x}$Ga$_x$As/InP superlattices can be tensile-strained (for x > 0.47), lattice-matched (for x ≈ 0.47), or compressive-strained (for x < 0.47). The feasibility and usefulness of strained-layered superlattices was first studied by Osbourn [38]. FIGURE 3 displays a variety of valence-band structures available for In$_{1-x}$Ga$_x$As/InP quantum wells.

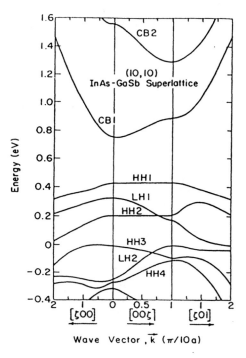

FIGURE 2. Band structure of (InAs)$_{10}$(GaSb)$_{10}$ superlattice obtained by the bond-orbital method. From [11].

In the lattice-matched case, the band structures have similar behaviour to the $GaAs/Ga_{1-x}Al_xAs$ quantum wells. For the compressive-strained case, the first heavy hole band becomes almost parabolic which displays very little band mixing. This is because the strain causes enlarged separation of heavy-hole and light-hole band energies at $k_\parallel = 0$, thus reducing the heavy- and light-hole mixing at finite $k_\parallel$. As a result, the top valence band has a smaller effective mass, which tends to reduce the Auger recombination rate in the laser structures made of these systems and lower the threshold current [39]. For the tensile-strained case, the opposite effect occurs. With the proper amount of strain (see the $x = 0.64$ case), the heavy-hole and light-hole band energies at $k_\parallel = 0$ almost coincide, which leads to a large level repulsion effect at finite $k_\parallel$ and gives rise to a negative effective mass for the hole near the zone centre. The result is an enlarged joint density of states for optical transitions, and it also has useful applications in laser structures [40].

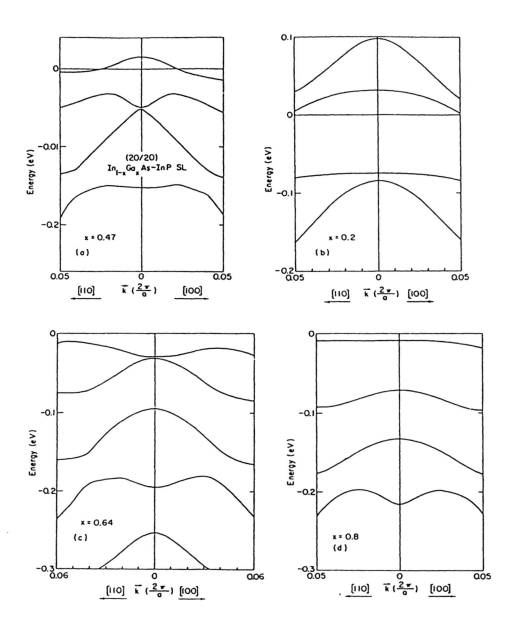

FIGURE 3. Valence band structures of $In_{1-x}Ga_xAs/InP$ quantum wells with
(a) $x = 0.47$, (b) $x = 0.2$, (c) $x = 0.64$, (d) $x = 0.8$. From [37].

D   EMPIRICAL FULL-ZONE METHODS

When the states derived from bulk bands away from the zone centre are of interest such as in superlattices made of indirect semiconductors, a full-zone method such as the tight-binding [12-15] or pseudopotential method [16-18] is needed. A full-zone **k.p** method has also been developed [41], which gives essentially the same results as the pseudopotential method. Phenomena like the Γ-X mixing and the inter-valley interference [42] can be studied by these methods. Simplified model calculations such as the Wannier-orbital method [43] and the anti-bonding orbital method [44] have also been introduced for such purposes.

E   FIRST-PRINCIPLES METHODS

To predict the band offsets or interface bond lengths for lattice-mismatched materials, the first-principles methods [19-22] must be called for. First-principles theoretical calculations of the band offset are usually reliable to about ± 0.1 eV. More accurate values can only be determined by fitting to optical and/or electronic measurements on heterostructure devices. These methods are also needed for calculating band structures of small-period superlattices (such as monolayer superlattices) when the effects of charge transfers at interfaces become significant. However, because the calculation is usually performed within the local density approximation [45], which is not capable of getting the correct bandgaps [46], an effective potential [47] which mimics the effects of self-energy correction should be added in order to produce reliable superlattice band structures for the conduction bands.

F   CONCLUSION

To summarise, the effective-mass method is applicable for calculating conduction bands in general and valence bands at $\mathbf{k}_\parallel = 0$ for most wide-gap ($E_g > 1$ eV) III-V superlattices. For narrow-gap superlattices and for valence bands in general, the **k.p** and bond-orbital methods are adequate unless the states of interest are derived from bulk bands far away from the zone centre (i.e. X or L valleys), for which the full-zone empirical methods can be used. For all calculations, the band offsets are treated as input parameters which are usually determined experimentally. In cases where the band offsets are unknown experimentally, the first-principles calculations can be used to predict them with an accuracy of ± 0.1 eV.

REFERENCES

[1]   R. Dingle [*Festkorperprobleme* Ed. J. Trensch (Pergamon, New York, 1975) vol.15 p.21]
[2]   G.A. Sai-Halasz, L. Esaki, W.A. Harrison [*Phys. Rev. B (USA)* vol.18 (1978) p.2812]
[3]   G. Bastard [*Phys. Rev. (USA)* vol.25 (1982) p.7584]
[4]   A. Fasolino, M. Altarelli [*Two-Dimensional Systems, Heterostructures, and Superlattices* Eds G. Bauer, F. Kucher, H. Heinrich (Springer-Verlag, New York, 1984)]
[5]   M. Altarelli, U. Ekenberg, A. Fasolino [*Phys. Rev. B (USA)* vol.32 (1985) p.5138]
[6]   G.D. Sanders, Y.C. Chang [*Phys. Rev. B (USA)* vol.31 (1985) p.6892; *Phys. Rev. B (USA)* vol.32 (1985) p.4282; *Phys. Rev. B (USA)* vol.35 (1987) p.1300]
[7]   R. Lassnig [*Phys. Rev. (USA)* vol.31 (1985) p.8076]
[8]   A. Broido, L.J. Sham [*Phys. Rev. B (USA)* vol.34 (1986) p.3917]

[9]   B. Zhu, K. Huang [*Phys. Rev. B (USA)* vol.36 (1987) p.8102]

[10]   See, for example, W.A. Harrison [*Solid State Theory* (McGraw-Hill, 1970) p.141]

[11]   Y.C. Chang [*Phys. Rev. B (USA)* vol.37 (1988) p.8215]

[12]   J.N. Schulman, T.C. McGill [*Phys. Rev. Lett. (USA)* vol.39 (1977) p.1680]

[13]   Y.C. Chang, J.N. Schulman [*Appl. Phys. Lett. (USA)* vol.43 (1983) p.536]

[14]   J.N. Schulman, Y.C. Chang [*Phys. Rev. B (USA)* vol.31 (1985) p.2056; *Phys. Rev. B (USA)* vol.33 (1986) p.2594]

[15]   A. Madhukar, S. Das Sarma [*J. Vac. Sci. Technol. (USA)* vol.17 (1980) p.1120]

[16]   Ed Caruthers, P.J. Lin-Chung [*Phys. Rev. B (USA)* vol.17 (1978) p.2705; *Phys. Rev. Lett. (USA)* vol.39 (1977) p.1543]

[17]   M.A. Gell, D. Ninno, M. Jaros, D.C. Herbert [*Phys. Rev. B (USA)* vol.34 (1986) p.2416]

[18]   M.A. Gell, D. Ninno, M. Jaros, D.J. Wolford, T.F. Keuch, J.A. Bradley [*Phys. Rev. B (USA)* vol.35 (1987) p.1196]

[19]   J. Ihm, P.K. Lam, M.L. Cohen [*Phys. Rev. B (USA)* vol.20 (1979) p.4120]

[20]   C.G. Van de Walle, R.M. Martin [*Phys. Rev. B (USA)* vol.35 (1987) p.8154]

[21]   A. Baldereschi, S. Baroni, R. Resta [*Phys. Rev. Lett. (USA)* vol.61 (1988) p.734]

[22]   W.R. Lamberecht, B. Segall, O.K. Andersen [*Phys. Rev. B (USA)* vol.41 (1990) p.2813]

[23]   D.J. Wolford et al [*Phys. Rev. (USA)* vol.31 (1985) p.4056]

[24]   R.C. Miller, A.C. Gossard, D.A. Kleinman, D. Muntaneu [*Phys. Rev. B (USA)* vol.29 (1984) p.3470]

[25]   R.S. Bauer, G. Margaritondo [*Phys. Today (USA)* vol.40 (1987) p.27]

[26]   G.A. Sai-Halasz, L.L. Chang, J.M. Welter, C.A. Chang, L. Esaki [*Solid State Commun. (USA)* vol.27 (1978) p.935]

[27]   C. Tejedor, J.M. Calleja, F. Meseguer, E.E. Mendez, C.-A. Chang, L. Esaki [*Phys. Rev. B (USA)* vol.32 (1985) p.5303]

[28]   S.P. Kowalczyk, J.T. Cheung, E.A. Kraut, R.W. Grant [*Phys. Rev. Lett. (USA)* vol.56 (1986) p.1605]

[29]   See, for example, E. Merzbacher [*Quantum Mechanics* (McGraw-Hill, New York, 1968) ch.6]

[30]   E.O. Kane [*J. Phys. Chem. Solids (UK)* vol.1 (1956) p.82]

[31]   J.M. Luttinger, W. Kohn [*Phys. Rev. (USA)* vol.97 (1956) p.869]

[32]   Y.C. Chang [*Appl. Phys. Lett. (USA)* vol.46 (1985) p.710]

[33]   G.T. Einevoll, Y.C. Chang [*Phys. Rev. B (USA)* vol.40 (1989) p.9683-97; *Phys. Rev. B (USA)* vol.41 (1990) p.1447-60]

[34]   P. Voisin et al [*Solid State Commun. (USA)* vol.39 (1981) p.79]

[35]   Y.C. Chang, J.N. Schulman [*Phys. Rev. B (USA)* vol.31 (1985) p.2069]

[36]   H.H. Chen, M.P. Houng, Y.H. Wang, Y.C. Chang [*Appl. Phys. Lett. (USA)* vol.61 (1992) p.509-11]

[37]   M.-P. Houng, Y.C. Chang [*J. Appl. Phys. (USA)* vol.65 (1989) p.3092]

[38]   G.C. Osbourn [*J. Vac. Sci. Technol. (USA)* vol.21 no.2 (1982) p.469; *J. Appl. Phys. (USA)* vol.53 (1982) p.1586; *Phys. Rev. B (USA)* vol.27 (1983) p.5126]; G.C. Osbourn, R.M. Riefield, P.L. Gourley [*Appl. Phys. Lett. (USA)* vol.41 (1982) p.172]

[39]   E. Yablonovitch, E.O. Kane [*J. Lightwave Technol. (USA)* vol.4 (1986) p.504]

[40]   D. Ahn [*Appl. Phys. Lett. (USA)* vol.66 (1995) p.30]

[41]   C. Mailhiot, D.L. Smith [*Phys. Rev. B (USA)* vol.33 (1986) p.8360]

[42]   Y.C. Chang, D.Z.-Y. Ting [*J. Vac. Sci. Technol. B (USA)* vol.1 no.2 (1983) p.435]

[43]   D.Z.-Y. Ting, Y.C. Chang [*Phys. Rev. B (USA)* vol.36 (1987) p.4357]

[44]   Y.C. Chang, A.E. Chiou, M. Khoshnevissan [*J. Appl. Phys. (USA)* vol.71 (1992) p.1349]

[45]   P. Hohenberg, W. Kohn [*Phys. Rev. B (USA)* vol.136 (1964) p.864]; W. Kohn, L. Sham [*Phys. Rev. A (USA)* vol.140 (1965) p.1133]

[46]   M.S. Hybertsen, S.G. Louie [*Phys. Rev. B (USA)* vol.34 (1986) p.5390]

[47]   N.E. Christensen [*Phys. Rev. B (USA)* vol.30 (1984) p.5753]

2.2 Transport properties of III-V semiconductor quantum wells and superlattices

Yia-Chung Chang

July 1995

A   INTRODUCTION

We review the basic theoretical aspects of transport properties of III-V semiconductor heterostructures. Electron transport in a heterostructure can be divided into two categories: (1) conduction in the plane (CIP) and (2) conduction perpendicular to the plane (CPP). For CIP, the electron transport is determined by scattering of electrons from phonons, impurities, and interface roughness. For CPP in double-barrier heterostructures, the transport characteristics are dominated by the resonant tunnelling effect in addition to the above scattering effects. For hot-carrier transport perpendicular to the interface, one also needs to consider the tunnelling-assisted impact ionisation effects [2]. Emphasis in the present Datareview will be put on the theories of electron-phonon scattering and resonant tunnelling. The theory of impurity scattering is essentially the same as in bulk materials. The theory for interface-roughness scattering is mostly phenomenological [1], as the roughness of interfaces varies from sample to sample.

B   BASIC PROPERTIES OF PHONONS IN SUPERLATTICES

To understand the electron-phonon interaction, it is necessary to know the basic properties of phonons. Phonons in III-V semiconductor superlattices have been studied with various models. The linear-chain model [3,4] was used to analyse phonon modes with the wave vectors perpendicular to the interfaces. The dielectric continuum model was used to explain the interface and slab modes and the anisotropy of the optical phonons [5-8] from a macroscopic point of view. Microscopic calculations based on the adiabatic bond-charge model [9], rigid-ion model [10-12], shell model [13], dipole superlattice model [14,15], valence-force model [16], and first-principles method [17] have been reported. The long-range Coulomb interaction was found to be essential for understanding the anisotropy of optical modes and the interface modes of superlattices [12,13]. In the long wavelength limit, simple envelope-function analysis [18] can be used to study the optical phonons with results in excellent agreement with the microscopic calculations [12].

For **k** in the growth direction, the Coulomb part of the dynamic matrix decays rapidly with interlayer distance [18] and it is adequate to describe the equation of motion for ions with the linear-chain model. The linear-chain model for superlattices can be solved easily by using a slab method or transfer-matrix method. The main results can be summarised as follows:

(a)   When the frequencies of phonon branches in both constituent materials coincide (e.g. most parts of acoustical branches in GaAs and AlAs), the corresponding superlattice phonon modes can be obtained approximately by a zone-folding procedure. The mini gaps at the mini-zone boundary are usually small.

(b)   When the phonon branches in the two constituent materials are misaligned (e.g. all optical branches in GaAs and AlAs), the corresponding superlattice modes are confined in either material, and the frequencies can be approximated by the corresponding bulk phonon frequencies at the quantised wave vector, $q_n = n\pi/d$, where n is the principal quantum number and d is the width of the material in which the mode is confined.

One can approximately write the 6-component (three each for cation and anion) displacement vector at a diatomic layer (or 'bilayer' for short) labelled by J as [18]

$$U_{vn}^{(\lambda)}(J) \approx f_n^{(\lambda)}(J)P_{vn}^{(\lambda)}$$

where $f_n^{(\lambda)}(J)$ is called the 'envelope function', $\lambda = 1,2$ refers to modes confined in well and barrier materials, respectively, n is the principal quantum number ($n = 1,2,...,N_\lambda$), $v = T,L$ (or x,z) denotes the transverse and longitudinal modes (or direction of vibration), and $P_{vn}^{(\lambda)}$ is the polarisation vector for bulk λ at $q_n = n\pi/(N+1)$. The envelope function $f_n^{(\lambda)}(J)$ for an optical mode confined in a particular slab is approximately given by the function

$$\sin \frac{n\pi J}{(N_\lambda+1)}$$

where N_λ is the number of bilayers inside material λ.

For **k** in the general direction (with polar angle $\theta \neq 0$), the Coulomb part of the dynamic matrix contains an irregular term $k_i k_j/k^2$ which takes on different values as $\mathbf{k} \to 0$ from different directions [19]. This term in conjunction with the lack of rotational invariance of the dynamic matrix for a superlattice lead to an interesting anisotropic behaviour for the optical modes in the long wavelength limit [12,18]. For optical phonons (which are confined in either the well or barrier region) we expand the displacement vectors for $\theta \neq 0$ in terms of the $\theta = 0$ solutions

$$U(J) = \sum_{n,v} C_{vn}f_n(J)P_{vn}$$

Substituting this expansion into the original equation of motion yields a simple matrix equation for C_{vn} [18]

$$(\omega^2 - \omega_{vn}^2) C_{vn} = \beta d_n \sum_{v'n'} d_{n'} C_{v'n'} S_{vv'}$$

where ω_{vn} denotes the $\theta = 0$ optical-phonon frequencies, β and d_n are simple constants defined in [18] and $S_{vv'} = \delta_{v,z}\delta_{v',z} - k_v k_{v'}/k^2$. This equation allows a simple solution to the angular dependence of the long-wavelength optical-phonon frequencies and polarisation vectors, which are needed for the evaluation of electron-phonon interactions. In the dispersionless limit (i.e. both ω_{vn} and P_{vn} independent of n), the results reduce to those of the dielectric continuum model [7]. In this limit the envelope function (defined as $f_v(J) = \Sigma_n C_{vn}f_n(J)$) for the principal (nodeless) mode changes continuously from a sinusoidal function to a constant as θ varies from 0 to $\pi/2$. At $\theta = \pi/2$, the phonon wave vector becomes parallel to the interface, and if we allow the magnitude of $\mathbf{k}_{\parallel}$ to deviate from zero, the envelope function becomes localised at the interface with a decay length inversely proportional to $\mathbf{k}_{\parallel}$. Thus the principal optical-phonon modes are also called 'interface' modes when the wave vector is finite and parallel to the interface.

In general, one can expand the superlattice phonon displacement vectors for mode j in terms of a complete set of bulk displacement vectors for the well material ($\mathbf{U}^0$),

$$\mathbf{U}_q^{(j)}(J) = \sum_n f_q^{(j)}(v,g_n)\mathbf{U}^0_{v,q+g_n \hat{z}}(J)$$

where the Fourier transform of f corresponds to the envelope function mentioned above. The dynamic matrix defined in the bulk basis can be diagonalised to yield all superlattice phonon solutions. This approach is particularly simple, if the Coulomb parts of the two constituent materials are treated the same, and then the perturbation to the bulk dynamic matrix contains only the short-range part [20]. If the perturbation in the dynamic matrix (after transformation with the inverse of the mass matrix) is replaced by the difference in squared optical phonon frequencies for two constituent materials ($\Delta\omega_0^2$), the above reduces to the dipole superlattice model of Huang and Zhu [14].

C   ELECTRON-PHONON SCATTERING

The electron-phonon scattering determines the intrinsic carrier lifetime. We shall formulate it for a superlattice. The quantum well case can be obtained by taking the limit of infinite barrier thickness. We write electronic states and phonon polarisation vectors of the superlattice in terms of the corresponding bulk states and polarisation vectors. The resulting electron-phonon coupling can also be written in terms of the corresponding bulk electron-phonon coupling constant. A detailed discussion of bulk electron-phonon coupling can be found in [21].

C1   Deformation-potential Mechanism

In the rigid-ion model, the electron-phonon interaction due to the deformation of electronic potential caused by the lattice vibration is given by [22]

$$H_{el-ph} = -\sum_{\alpha S} \frac{1}{\sqrt{N M_\alpha}} \sum_{qj} Q_{qj} e^{iq.S} \hat{\xi}^{(j)}(\mathbf{R}_\alpha) \bullet \nabla V_\alpha(\mathbf{r} - \mathbf{S} - \mathbf{R}_\alpha) \qquad (1)$$

where $\mathbf{S}$ labels the superlattice unit cells (SUCs), N is the total number of SUCs in the sample, α labels the different ions in an SUC, M_α and $\mathbf{R}_\alpha$ denote the mass and position of ion α, $\hat{\xi}^{(j)}(\mathbf{R}_\alpha)$ denotes the polarisation vector at position $\mathbf{R}_\alpha$ of phonon mode j, and V_α describes the potential for an electron interacting with ion α. Q_{qj} is the normal mode coordinate of mode j which takes the second quantisation form

$$Q_{qj} = \sqrt{\frac{\hbar}{2\omega_{qj}}} (a_{qj}^+ + a_{qj})$$

where ω_{qj} denotes the frequency of mode j.

We expand the superlattice electronic states (with wave vector $\mathbf{k}$) in terms of bulk Bloch states $\phi_{v,k_z}(\mathbf{r})$ of the well material, where v labels the bulk bands involved.

$$\psi_k(\mathbf{r}) = \sum_{\mu,s} F_k(\mu,g_s)\phi_{\mu,k_z+g_s}(\mathbf{r})$$

where the $\mathbf{k}_\parallel$ dependence in the bulk states has been kept implicit. Here g_s denotes the z-component of the superlattice reciprocal lattice vector. The matrix element of H_{el-ph} between two electronic states with wave vectors $\mathbf{k}$ and $\mathbf{k}'$ is given by

$$<\mathbf{k}'|H_{el-ph}|\mathbf{k}> =$$

$$-i \sum_{nj} \sqrt{\frac{\hbar}{2\omega_{qj}N\,M_{cell}}} \sum_{q,vs,\mu's'} F^*_{k'}(\mu',s') F_k(\mu,s) f_q^{(j)}(v,g_n)$$

$$\delta_{k'-k\pm q,g_n\hat{z}}\delta_{s'-s,n}D^{(v)}_{\mu,\mu'}(\mathbf{q} + g_n\,\hat{z}) \tag{2}$$

where

$$D^{(v)}_{\mu,\mu'}(\mathbf{q}) \equiv \int d^3r\phi_{\mu,k}(\mathbf{r})\nabla U_{v,q}(\mathbf{r})\phi_{\mu',k'}(\mathbf{r}) \tag{3}$$

and

$$U_{v,n}(\mathbf{r}) \equiv \sum_{\alpha,S} \sqrt{\frac{M_{cell}}{M_\alpha}} e^{i\mathbf{q}.\mathbf{R}_\alpha}\mathbf{P}_{v,q,\alpha} \bullet \nabla V^{(\alpha)}(\mathbf{r} - \mathbf{S} - \mathbf{R}_\alpha)$$

where $\mathbf{P}_{v,q,\alpha}$ denotes the α (cation or anion) component of the polarisation vector at wave vector $\mathbf{q}$ for bulk mode v and M_{cell} is total mass per bulk unit cell for the corresponding bulk material. The +(-) sign in the above equation denotes a phonon absorption (emission) process. Note that $D_{v,v'}(\mathbf{q})$ in EQN (3) is just the electron-phonon coupling constant for a bulk material. For optical phonons near the zone centre $D(\mathbf{q})$ is approximately independent of $\mathbf{q}$, while for acoustic modes $D(\mathbf{q})$ is proportional to $\mathbf{P}_{v,q,\alpha} \bullet \mathbf{q}$, with the proportionality constant being called the deformation potential.

C2   Polar-optical Scattering

For III-V semiconductors, the optical phonon induces a relative motion between ions within a bulk unit cell which leads to dipole oscillation. The interaction of the electron with the dipole field is called the polar-optical scattering (or Frölich scattering) [23]. The electron-phonon interaction in this case is again given by EQN (2) with the coupling constant $D(\mathbf{q})$ replaced by

$$D^{(v)}_{\mu,\mu'}(\mathbf{q}) = \delta_{\mu,\mu'} \sum_\alpha \sqrt{\frac{M_{cell}}{M_\alpha}} \frac{ee^*_\alpha}{\upsilon_c\varepsilon_\infty} \mathbf{P}_{v,q,\alpha} \bullet \hat{q}/q$$

where e is the free electron charge, e^*_α is the effective dynamic charge of ion α within a bulk unit cell, υ_c is the volume of the bulk unit cell, and ε_∞ is the high-frequency dielectric constant. Using the dipole superlattice model, Huang and Zhu [24] derived simple analytic expressions for the electron-phonon coupling potentials for polar-optical scattering involving confined optical phonons.

Once the electron-phonon coupling constants are calculated, the carrier lifetimes can be evaluated based on Fermi's golden rule and the mobility can be calculated based on procedures similar to those for bulk semiconductors as described in [21]. The Huang and Zhu model [24] has been used frequently to evaluate the LO phonon scattering rates in quantum wells [25-28]. Using the electron-phonon scattering rates, the Raman spectra for III-V semiconductor superlattices have also been calculated [27-30]. Calculations of scattering rate with interface phonons with and without a magnetic field have also been reported [25,31-33].

D   RESONANT TUNNELLING

For electron transport perpendicular to the superlattice, resonant tunnelling plays an important role. Most studies on resonant tunnelling behaviour are concentrated on double-barrier structures (DBS) due to their ability to produce a large negative differential resistance. Very large peak-to-valley current ratios have been reported for double-barrier tunnelling structures made of III-V semiconductors. Many theoretical calculations have been performed for III-V double-barrier structures [34-48]. The transfer-matrix method is used in most theoretical calculations within various models to obtain the transmission coefficient $T(k_\parallel, E)$ as a function of in-plane wave vector $\mathbf{k}_\parallel$ and carrier energy E. The tunnelling current is related to $T(\mathbf{k}_\parallel, E)$ by [49]

$$J = \frac{e}{4\pi^3 \hbar} \int T(k_\parallel, E) \, [f(E) - f(E + eV)] \, dEd^2 k_\parallel$$

where V is the voltage difference across the heterostructure and f(E) is the Fermi distribution function for the carrier. For the time-dependent resonant-tunnelling characteristics, a quantum transport theory is needed [36-38].

For electron tunnelling in wide-gap heterostructures, the effective-mass model is usually used [34,35]. However, this model sometimes gives a poor estimate of the transmission coefficient, since the decay length of an electron in the barrier material can be quite different from that given by EMA when the electron energy is far away from the band edge. A two-band $\mathbf{k.p}$ or tight-binding model [39-41] is needed to give the correct result. For hole tunnelling in general or electron tunnelling for narrow-gap heterostructures the discretised $\mathbf{k.p}$ model [42] or the effective bond-orbital model [43] is needed to take into account the band-mixing effects. Interesting interference effects exist when a heavy-mass band and light-mass band are mixed in the tunnelling processes [47]. For electron tunnelling in heterostructures made of indirect materials (e.g. GaP, Si, or Ge), either the full-band model [44,45], Wannier-orbital model [46] or the anti-bonding model [48] should be used to take into account the inter-valley mixing effect. The effects of phonons and interface-roughness on the tunnelling current or tunnelling time have been reported [50,51,57]. Theoretical studies of the lifetimes of quasi-bound states in double-barrier heterostructures can be found in [53-57].

E   CONCLUSION

We have reviewed the theoretical aspects of phonons, electron-phonon coupling, and resonant tunnelling in III-V semiconductor superlattices. They play important roles in determining the carrier dynamics and transport properties.

REFERENCES

[1]   H. Sakaki, T. Noda, K. Hirakawa, M. Tanaka, T. Matsusue [*Appl. Phys. Lett. (USA)* vol.51 (1987) p.1934]

[2]   S.L. Chuang, K. Hess [*J. Appl. Phys. (USA)* vol.61 (1987) p.1510]

[3]   A.S. Barker Jr., J.L. Merz, A.C. Gossard [*Phys. Rev. B (USA)* vol.17 (1978) p.3181]

[4]   C. Colvard et al [*Phys. Rev. B (USA)* vol.31 (1985) p.2080]

[5]   S.M. Rytov [*Zh. Eksp. Teor. Fiz. (Russia)* vol.29 (1955) p.605; *Sov. Phys. - TEPT (USA)* vol.2 (1956) p.466]

[6]   R. Fuchs, K.L. Kliewer [*Phys. Rev. A (USA)* vol.140 (1965) p.2076]

[7]   E.P. Pokatilov, S.I. Beril [*Phys. Status Solidi B (Germany)* vol.118 (1983) p.567]

[8]   M. Babiker [*J. Phys. C. Solid State Phys. (UK)* vol.19 (1986) p.683]

[9]   S.K. Yip, Y.C. Chang [*Phys. Rev. B (USA)* vol.30 no.12 (1984) p.7037]

[10]   A. Kobayashi [PhD Thesis, Dept. of Phys., Univ. of Illinois at Urbana-Champaign, 1977]

[11]   T. Toriyama, N. Kobayashi, Y. Horikoshi [*Jpn. J. Appl. Phys. (Japan)* vol.25 (1986) p.1895]

[12]   S.-F. Ren, H. Chu, Y.-C. Chang [*Phys. Rev. Lett. (USA)* vol.50 (1987) p.1841; *Phys. Rev. B (USA)* vol.37 (1988) p.8899]

[13]   E. Richter, D. Strauch [*Solid State Commun. (USA)* vol.64 (1987) p.867]

[14]   K. Huang, B. Zhu [*Phys. Rev. B (USA)* vol.38 (1988) p.2138]

[15]   B. Zhu [*Phys. Rev. B (USA)* vol.38 (1988) p.7694]

[16]   T. Tsuchiya, H. Akera, T. Ando [*Phys. Rev. B (USA)* vol.39 (1989) p.6025]

[17]   E. Molinari, S. Baroni, P. Giannozzi, S. de Gironcoli [*Phys. Rev. B (USA)* vol.45 (1992) p.4280]

[18]   H. Chu, S.-F. Ren, Y.-C. Chang [*Phys. Rev. B (USA)* vol.37 (1988) p.10746]

[19]   K. Kunc, M. Balkanski, M. Nusimovici [*Phys. Status Solidi B (Germany)* vol.71 (1975) p.341; *Phys. Rev. B (USA)* vol.12 (1975) p.4346]

[20]   Y.C. Chang, S. Ren, H. Chu [*Superlattices Microstruct. (UK)* vol.4 (1988) p.303]

[21]   B.K. Ridley [*Quantum Process in Semiconductors* (Oxford, New York, 1982)]

[22]   O.E. Madelung [*Introduction to Solid State Theory* (Springer-Verlag, New York, 1978) ch.4]

[23]   H. Frölich [*Proc. R. Soc. A (UK)* vol.160 (1937) p.230]

[24]   K. Huang, B. Zhu [*Phys. Rev. B (USA)* vol.38 (1988) p.13377]

[25]   S. Rudin, T.L. Reinecke [*Phys. Rev. B (USA)* vol.41 (1990) p.7713]

[26]   G. Weber, A.M. de Paula, J.F. Ryan [*Semicond. Sci. Technol. (UK)* vol.6 (1991) p.397]

[27]   B. Zhu, K. Huang, H. Tang [*Phys. Rev. B (USA)* vol.40 (1989) p.6299]

[28]   G. Wen, Y.C. Chang [*Phys. Rev. B (USA)* vol.45 (1992) p.13562]

[29]   K. Huang, B. Zhu, H. Tang [*Phys. Rev. B (USA)* vol.41 (1990) p.5825]

[30]   S. Ren, Y.C. Chang, H. Chu [*Phys. Rev. B (USA)* vol.47 (1993) p.1489-99]

[31]   R. Lassnig [*Phys. Rev. B (USA)* vol.30 (1984) p.7132]

[32]   J.K. Jain, S. Das Sarma [*Phys. Rev. Lett. (USA)* vol.62 (1989) p.2305]

[33]   J.S. Bhat, B.G. Mulimani, S.S. Kubakaddi [*J. Appl. Phys. (USA)* vol.74 (1993) p.4561]

[34]   R. Tsu, L. Esaki [*Appl. Phys. Lett. (USA)* vol.22 (1973) p.562]

[35]   M.O. Vassell, J. Lee, H.F. Lockwood [*J. Appl. Phys. (USA)* vol.54 (1983) p.5206]

[36]   D.D. Coon, H.C. Liu [*J. Appl. Phys. (USA)* vol.58 (1985) p.2230]

[37]   W.J. Frensley [*Appl. Phys. Lett. (USA)* vol.51 (1987) p.448]

[38]   H.C. Liu [*Phys. Rev. B (USA)* vol.43 (1993) p.12538; vol.48 (1993) p.4977]

[39] J.R. Söderström, E.T. Yu, M.K. Jackson, Y. Rajakarunanayake, T.C. McGill [*J. Appl. Phys. (USA)* vol.68 (1990) p.1372]

[40] D.Z.-Y. Ting, E.T. Yu, D.A. Collins, D.H. Chow, T.C. McGill [*J. Vac. Sci. Technol. B (USA)* vol.8 (1990) p.810]

[41] J.N. Schulman, C.L. Anderson [*Appl. Phys. Lett. (USA)* vol.48 (1986) p.1684]

[42] C.Y. Chao, S.L. Chuang [*Phys. Rev. B (USA)* vol.43 (1991) p.7027]

[43] D.Z.-Y. Ting, E.T. Yu, T.C. McGill [*Phys. Rev. B (USA)* vol.45 (1992) p.3583]

[44] G.C. Osbourn, D.L. Smith [*Phys. Rev. B (USA)* vol.19 (1979) p.2124]

[45] C. Mailhiot, T.C. McGill, J.N. Schulman [*J. Vac. Sci. Technol. B (USA)* vol.1 no.2 (1983) p.439]

[46] M.K. Jackson, D.Z.-Y. Ting, D.H. Chow, D.A. Collins, J.R. Söderström, T.C. McGill [*Phys. Rev. B (USA)* vol.43 (1991) p.4856]

[47] J.C. Chiang, Y.C. Chang [*J. Appl. Phys. (USA)* vol.73 (1993) p.2402]

[48] J.C. Chiang, Y.C. Chang [*Phys. Rev. B (USA)* vol.47 (1993) p.7140]

[49] C.B. Duke [*Solid State Physics. Suppl. 10. Tunneling in Solids* (Academic Press, New York, 1969)]

[50] G.Y. Wu, T.C. McGill [*Phys. Rev. B (USA)* vol.40 (1989) p.9969]

[51] H.C. Liu [*J. Appl. Phys. (USA)* vol.67 (1990) p.593]

[52] D.Z.-Y. Ting, T.C. McGill [*Appl. Phys. Lett. (USA)* vol.64 (1994) p.2004]

[53] M. Büttiker, R. Landauer [*Phys. Rev. Lett. (USA)* vol.49 (1982) p.1739]

[54] N. Harada, S. Kuroda [*Jpn. J. Appl. Phys. (Japan)* vol.25 (1986) p.L871]

[55] S. Collins, D. Lowe, J.R. Barker [*J. Phys. C (UK)* vol.20 (1987) p.6213]

[56] H. Guo, K. Diff, G. Neofotistos, J.D. Gunton [*Appl. Phys. Lett. (USA)* vol.53 (1988) p.131]

[57] D.Z.-Y. Ting, T.C. McGill [*J. Vac. Sci. Technol. B (USA)* vol.7 (1989) p.1031]

2.3   Miniband parameters in superlattices

B.R. Nag

May 1995

A   INTRODUCTION

Superlattices have been constructed with the GaAs/Al$_{0.3}$Ga$_{0.7}$As [1-3], GaAs/Al$_{0.26}$Ga$_{0.74}$As [4], GaAs/Al$_{0.28}$Ga$_{0.72}$As [5], GaAs/AlAs [5-10], Ga$_{0.47}$In$_{0.53}$As/InP [11], Ga$_{0.47}$In$_{0.53}$As/Al$_{0.48}$In$_{0.52}$As [12], InAs/GaSb [13], InAs/Ga$_{0.75}$In$_{0.25}$Sb [14,15], GaSb/Ga$_{0.85}$In$_{0.15}$Sb [16], GaSb/AlSb [17], GaP/AlP [18,19], GaAs/Ga$_{0.95}$In$_{0.05}$As [20], GaAs/Ga$_{0.88}$In$_{0.12}$As [21] and GaAs/Ga$_{0.84}$In$_{0.16}$As [22] heterostructures.   Miniband parameters have, however, been determined experimentally for only a few of these structures.   Even for such structures only the combined bandwidths of the electrons and holes have been obtained from the experiments The individual bandwidths of electron minibands have been determined from infrared measurements only for the GaAs/Al$_{0.3}$Ga$_{0.7}$As superlattice.   Calculated values of miniband parameters however have been reported for many of the structures.   Available experimental results are reviewed in this Datareview together with the calculated values and the methods used for the calculations.

B   CALCULATION OF MINIBAND PARAMETERS

The following dispersion relation, derived by using the so-called envelope-function Kronig-Penny model [23-25], has been used extensively for studying the miniband parameters.

$$\cos k_w L_w \cosh k_B L_B - (1/2)\,(r-1/r)\sin k_w L_w \sinh k_B L_B = \cos k_z\,(L_w+L_B);$$

$$k_w{}^2 = [2m_w{}^*(E)/\hbar^2]\,(E-E_w),\ k_B{}^2 = [2m_B{}^*(E)/\hbar^2]\,(E_B-E), \tag{1}$$

$$r = k_w m_B{}^*(E)/k_B m_w{}^*(E)$$

where E is the carrier energy, and k_i, E_i, $m_i{}^*(E)$ and L_i are respectively the wavevector, band edge energy, energy-dependent effective mass and the layer width.   The subscript 'i' is replaced by w for the well and by B for the barrier material.   The wavevector for the superlattice is k_z The energy-dependent effective mass is taken to include the band non-parabolicity according to the **k.p** theory.

Band edge energies and the miniband width are evaluated by solving EQN (1), putting $k_z = 0$ and $k_z = \pi/(L_w+L_B)$.

It may be seen that for large values of $k_B L_B$, EQN (1) reduces to

$$\cos k_w L_w - (1/2)\,(r-1/r)\sin k_w L_w = 0 \tag{2}$$

The structure then behaves as a multiple quantum well (MQW) system, for which the energy eigenvalues are discrete and the miniband width is zero. The position of the energy levels and the separation between the successive levels are, however, controlled by both L_w and (E_B-E_w). The miniband parameters, i.e. the position and the width, may, therefore, be studied by varying L_B, L_w and (E_B-E_w). Widths are controlled by the growth conditions, while (E_B-E_w) is varied by altering the composition. Experiments have therefore been carried out by using different materials and different structures to realise different values of miniband parameters. Values of these parameters have been compared mostly with those obtained from EQN (1).

Results are also reported from more detailed calculations in which the coupled equations for the envelope-functions corresponding to the conduction band, heavy-hole band, light-hole band and the split-off band are solved either by the transfer matrix method [26-28] or by the finite element method [29]. For type II superlattices, e.g. InAs/GaSb, results have been obtained by applying the method of linear combination of atomic orbitals [30] or the tight-binding method [31].

C   EXPERIMENTS

Signatures of minibands have been found in absorption spectra [32], photoconduction spectra [33,10], photoluminescence excitation spectra [34,10], electro-modulated reflection spectra [35], photomodulated reflection spectra [36] and in piezo-modulated reflectivity spectra [1,37]. In all these experiments, either the incident light signal causes excitation of electrons from the valence-band minibands to the conduction-band minibands or the excess electrons from the conduction-band minibands recombine with the holes in the valence-band minibands. Kinks are caused in the spectrum because of the van Hove singularities in the joint density-of-states at the Γ point and at the folded zone-edge saddle point [1]. The difference in energy between two such kinks gives the sum of the bandwidths of the electron miniband (Δe) and the heavy-hole miniband (ΔHH) or the light-hole miniband (ΔLH). Similar experiments on infrared absorption spectra [2] give the separation between the edges of the lower and the higher minibands in the same conduction or valence band. The combined bandwidth of successive minibands may be obtained from such experiments. In some cases even the bandwidths of individual minibands may be obtained by using the impurity level excitations [2].

Results of the experiments are collected in TABLE 1 and TABLE 2. In TABLE 1 are reviewed the reported values of widths of minibands, mostly the combined widths. For many experimental superlattices [39-41] the minibands have very small dispersion. Interband transition energies, as reported for these structures, are reviewed in TABLE 2.

It should be mentioned that in most of the studies it is reported that the experimental values of the parameters agree with those calculated by using the Kronig-Penny model and solving the coupled envelope functions. Such values, wherever reported, are also included in TABLES 1 and 2 for comparison. For the sake of completeness, in some cases values calculated by the present author are also included. Material parameters were taken from [38] for these calculations. It is seen that the calculated values are, in general, very close to the experiments.

TABLE 1. Minibandwidth of superlattices (L_w, L_B are the widths of the well and the barrier layers; Δe_i, ΔHH_i and ΔLH_i are the bandwidths of the ith miniband of electrons, heavy holes and light holes in meV; Ex - Experiment; Th - Theory; ML - Monolayer).

1. Material: GaAs/Al$_x$Ga$_{1-x}$As

Ref	L_w(A)	L_B(A)	x	Δe_1		Δe_2	
				Ex	Th	Ex	Th
[2]	75	25	0.35	17	18	54	74

Ref	L_w(ML)	L_B(ML)	x	$\Delta e_1 + \Delta HH_1$		$\Delta e_1 + \Delta LH_1$	
				Ex	Th	Ex	Th
[1]	18	10	0.31	33	33	56	65
	27	10	0.28	-	17	29	33
	9	7	0.34	158	159	219	243

Ref	L_w(A)	L_B(A)	x	$\Delta e_1 + \Delta HH_1$		$\Delta e_1 + \Delta HH_1$	
				Ex	Th	Ex	Th
[4]	50	50	0.26	6	7	12	14

Ref	L_w(A)	L_B(A)	x	$\Delta e_1 + \Delta HH_1$		$\Delta e_2 + \Delta HH_2$		$\Delta e_3 + \Delta HH_3$		$\Delta e_4 + \Delta HH_4$	
				Ex	Th	Ex	Th	Ex	Th	Ex	Th
[5]	275	17	0.28	0	1.4	8	5.4	22	12	17	21

$\Delta e_5 + \Delta HH_5$		$\Delta e_6 + \Delta HH_6$		$\Delta e_1 + \Delta LH_1$		$\Delta e_2 + \Delta LH_2$	
Ex	Th	Ex	Th	Ex	Th	Ex	Th
28	32.6	38	46	5	2.2	3	8.7

$\Delta e_3 + \Delta LH_3$		$\Delta e_4 + \Delta LH_4$		$\Delta e_5 + \Delta LH_5$		$\Delta e_6 + \Delta LH_6$	
Ex	Th	Ex	Th	Ex	Th	Ex	Th
12	19.7	43	35	59	54.2	102	77.2

2. Material: GaAs/AlAs

Ref	L_w(A)	L_B(A)	$\Delta e_1 + \Delta HH_1$	Ref	L_w(A)	L_B(A)	$\Delta e_1 + \Delta HH_1$	Δe_1
			Ex				Ex	Th
[5]	31.3	2.83	300	[9]	31.3	5.7	186	
					31.3	11.4	70	
[9]	62.6	8.6	30		31.3	17.3	26	
	39.2	8.6	70					
	31.3	8.6	136	[7]	62	23		4
	26.1	8.6	146		54	20		9
					62	14		16

Table 1 continued.

3. Material: $Ga_{0.47}In_{0.53}As/Al_{0.49}In_{0.51}As$

Ref	$L_w(A)$	$L_B(A)$	e_1	Δe_1
			Th	Th
[12]	45	40	140	17.7
	45	35	136	25.5
	45	30	131	37.1
	45	25	124	54.8
	45	20	114	80.6

4. Material: $GaSb/Ga_{1-x}In_xSb$

Ref	$L_w(A)$	$L_B(A)$	x	$\Delta e_1 + \Delta HH_1$	
				Ex	Th
[6]	92	60	0.118	27	29
	80	60	0.148	32	33
	80	58	0.165	33	35

5. Material: $GaAs/Ga_{1-x}In_xAs$

Ref	$L_w(A)$	$L_B(A)$	x	Δe_1	Ref	$L_w(A)$	$L_B(A)$	$\Delta e_1 + \Delta HH_1$	
				Th				Ex	Th
[20]	50	55	0.05	36.7	[22]	50	50	25	33
	50	104.1	0.05	12.7					
	50	146	0.05	5.9	[21]	27	110	16	16
	50	202.7	0.05	2.5					

TABLE 2. Energy bandgap in superlattices (L_w, L_B are the widths of the well layer and the barrier layer, e_i and HH_i are the energy levels of electrons and heavy holes in meV; Ex - Experiment; Th - Theory).

1. Material: GaAs/AlAs

Ref	$L_w(A)$	$L_B(A)$	$e_1 - HH_1$	
			Ex	Th
[39]	17.5	9.5	1.85	1.87
[40]	50	10	1.66	1.63
	53	25	1.65	1.64
	30	50	1.78	1.81
[41]	45	45	1.63	1.68

2. Material: $GaSb/Ga_{1-x}In_xSb$

Ref	$L_w(A)$	$L_B(A)$	x	$e_1 - LH_1$		$e_1 - HH_3$	
				Ex	Th	Ex	Th
[16]	92	60	0.118	794 ± 7	795	794 ± 7	794
	80	60	0.148	790 ± 7	791	790 ± 7	792
	80	58	0.165	787 ± 9	788	787 ± 9	787

Table 2 continued.

3. Material: $Ga_{0.47}In_{0.53}As/InP$

Ref	$L_w(A)$	$L_B(A)$	$e_2 - e_1$		$e_3 - e_1$		$e_4 - e_1$		$e_5 - e_1$	
			Ex	Th	Ex	Th	Ex	Th	Ex	Th
[11,42]	65	367	155	159	170	169	180	185	200	195

D   CONCLUSION

Available experimental and theoretical values of miniband parameters have been reviewed. Experimental results are explained for most of the superlattices by the envelope function approximation and the multi-band **k.p** perturbation theory.   It is also found that various combinations of III-V compounds and layer widths have been used to construct superlattices, but few experiments have been conducted to determine the miniband parameters of these structures.

REFERENCES

[1]   C. Parks, A.K. Ramdas, M.R. Melloch, L.R. Ram-Mohan [*Phys. Rev. B (USA)* vol.48 (1993) p.5413-21]

[2]   M. Helm, W. Hilber, T. Fromherz, F.M. Peeters, K. Alavi, R.N. Pathak [*Solid-State Electron. (UK)* vol.37 (1994) p.1277-80]

[3]   M. Yamaguchi et al [*Solid-State Electron. (UK)* vol.37 (1994) p.839-42]

[4]   H. Shen, S.H. Pan, F.H. Pollak, M. Dutta, T.R. Ancoin [*Phys. Rev. B (USA)* vol.36 (1987) p.9384-7]

[5]   M.W. Peterson et al [*Appl. Phys. Lett. (USA)* vol.53 (1988) p.2666-9]

[6]   K.H. Schmidt, N. Linder, G.H. Dohler, H.T. Grahn, K. Ploog, H. Schneider [*Phys. Rev. Lett. (USA)* vol.72 (1994) p.2769-72]

[7]   K.H. Schmidt et al [*Solid-State Electron. (UK)* vol.37 (1994) p.1337-40]

[8]   F. Aristone, J.F. Palmier, A. Sibille, D.K. Maude, J.C. Portul, F. Mollot [*Solid-State Electron. (UK)* vol.37 (1994) p.1015-9]

[9]   K. Fujiwara, K. Kawashima, T. Yamamoto, K. Ploog [*Solid-State Electron. (UK)* vol.37 (1994) p.889-92]

[10]   K. Fujiwara et al [*Phys. Rev. B (USA)* vol.49 (1994) p.1809-12]

[11]   J. Oiknine-Schlesinger, E. Ehrenfreund, D. Gershoni, D. Ritter, M.B. Panish, R.A. Hamm [*Appl. Phys. Lett. (USA)* vol.59 (1991) p.970-2]

[12]   J.F. Palmier et al [*Solid-State Electron. (UK)* vol.37 (1994) p.697-700]

[13]   B.R. Bennett, B.V. Shanabrook, R.J. Wagner, J.L. Davis, J.R. Waterman, M.E. Twigg [*Solid-State Electron. (UK)* vol.37 (1994) p.733-7]

[14]   C.A. Hoffman, J.R. Meyer, E.R. Youngdale, F.J. Bartoli, R.H. Miles, L.R. Ram-Mohan [*Solid-State Electron. (UK)* vol.37 (1994) p.1203-6]

[15]   M. Lakrimi, T.A. Vaughan, R.J. Nicholas, D.M. Symons, N.T. Mason, P.J. Walker [*Solid-State Electron. (UK)* vol.37 (1994) p.1227-30]

[16]   S.L. Wong, R.J. Warburton, R.J. Nicholas, N.J. Mason, P.J. Walker [*Phys. Rev. B (USA)* vol.49 (1994) p.11210-21]

[17]   G. Scamarcio et al [*Solid-State Electron. (UK)* vol.37 (1994) p.625-8]

[18]   H. Asahi et al [*Appl. Phys. Lett. (USA)* vol.58 (1991) p.1407-9]

[19]   A. Mori et al [*Solid-State Electron. (UK)* vol.37 (1994) p.649-52]

[20]   R.J. Warburton, J.G. Michels, P. Peyla, R.J. Nicholas, K. Woodbridge [*Solid-State Electron. (UK)* vol.37 (1994) p.831-4]

[21]   K.J. Moore, G. Duggan, K. Woodbridge, C. Roberts, N.J. Pulsford, R.J. Nicholas [*Phys. Rev. B (USA)* vol.42 (1990) p.3024-9]

[22]   C. Vasquesz-Lopez, E. Ribeiro, F. Cerdeira, P. Motisuke, M.A. Sacilotti [*J. Appl. Phys. (USA)* vol.69 (1991) p.7837-43]

[23]   D. Mukherji, B.R. Nag [*Phys. Rev. B (USA)* vol.12 (1975) p.4338-45]

[24]   G. Bastard [*Phys. Rev. B (USA)* vol.24 (1981) p.5693-7]

[25]   M. Altarelli [in *Interfaces, Quantum Wells and Superlattices* Eds C.R. Leavens, R. Taylor (Plenum Press, 1988) p.43]

[26]   L.R. Ram-Mohan, K.H. Yoo, R.L. Aggarwal [*Phys. Rev. B (USA)* vol.38 (1988) p.6151-9]

[27]   K.H. Yoo, L.R. Ram-Mohan, D.F. Nelson [*Phys. Rev. B (USA)* vol.39 (1989) p.12808-13]

[28]   D.C. Hutchings [*Appl. Phys. Lett. (USA)* vol.55 (1989) p.1082-4]

[29]   J. Shertzer, L.R. Ram-Mohan, D. Dossa [*Phys. Rev. A (USA)* vol.40 (1989) p.4777-80]

[30]   G.A.S. Halasz, L. Esaki, W.A. Harrison [*Phys. Rev. B (USA)* vol.18 (1978) p.2812-8]

[31]   J.N. Schulman, Y.-C. Chang [*Phys. Rev. B (USA)* vol.24 (1981) p.4445-7]

[32]   R. Dingle, W. Wiegmann, C.H. Henry [*Phys. Rev. Lett. (USA)* vol.33 (1974) p.827-30]

[33]   D.C. Rogers, R.J. Nicholas [*J. Phys. C (UK)* vol.18 (1985) p.L891-6]

[34]   R.C. Miller, D.A. Kleinman, W.A. Nordland Jr., A.C. Gossard [*Phys. Rev. B (USA)* vol.22 (1980) p.863-71]

[35]   M. Erman, J.B. Theeten, P. Frizlink, S. Gaillard, Fan Jia Hia, C. Alibert [*J. Appl. Phys. (USA)* vol.56 (1984) p.3241-9]

[36]   O.J. Glembocki, B.V. Shanabrook, N. Bottka, W.T. Beard, J. Comas [*Appl. Phys. Lett. (USA)* vol.46 (1985) p.970-2]

[37]   Y.R. Lee, A.K. Ramdas, F.A. Chambers, J.M. Messe, L.R. Ram-Mohan [*Appl. Phys. Lett. (USA)* vol.50 (1987) p.600-2]

[38]   O. Madelung (Ed.) [*Semiconductors* (Springer, 1991)]

[39]   A.C. Gossard, P.M. Petroff, W. Wiegmann, R. Dingle, A. Savage [*Appl. Phys. Lett. (USA)* vol.29 (1976) p.323-5]

[40]   B.A. Vojak, W.D. Laidig, N. Holonyak Jr., M.R. Camras, J.J. Coleman, P.D. Dapkus [*J. Appl. Phys. (USA)* vol.52 (1981) p.621-6]

[41]   R. Tsu, A. Koma, L. Esaki [*J. Appl. Phys. (USA)* vol.46 (1975) p.842-5]

[42]   B.R. Nag, S. Mukhopadhyay [*Appl. Phys. Lett. (USA)* vol.58 (1991) p.1056-8]

2.4 Excitons in quantum wells

K.K. Bajaj

June 1995

A INTRODUCTION

There has been a great deal of interest in both the theoretical and experimental investigations of the behaviour of Wannier excitons in semiconductor quantum well structures in recent years. This work has been motivated from the point of view of both basic understanding and applications. Excitonic transitions have been observed in a variety of III-V and II-VI semiconductor quantum well structures both in absorption and in emission and their behaviour has been studied in the presence of external perturbations. A great deal of useful information concerning their properties has been obtained from these studies.

In this Datareview we briefly review the properties of excitons in quantum well structures. Though we focus most of our attention on GaAs/AlGaAs systems, the discussion presented here is also applicable to other similar systems. We shall briefly review various calculations of the binding energies of excitons in a variety of quantum well structures such as rectangular, non-rectangular, type II and coupled systems and present a comparison of their properties in these systems.

B RECTANGULAR QUANTUM WELLS

During the past two decades it has become possible to grow systems consisting of alternate layers of two different semiconductors with controlled thicknesses and relatively sharp interfaces using well developed epitaxial crystal growth techniques such as molecular beam epitaxy (MBE) and metal organic chemical vapour deposition (MOCVD). Though a variety of systems have been grown, the most extensively studied structure is the one consisting of alternate layers of GaAs and $Al_xGa_{1-x}As$. Depending on the Al concentration in $Al_xGa_{1-x}As$ its bandgap can be made considerably larger than that of GaAs thus leading to discontinuities of the conduction and valence band edges at the interfaces. Thus, electrons and holes in the GaAs matrix find themselves in potential wells whose heights depend on the Al concentration in the surrounding $Al_xGa_{1-x}As$ layers. This results in discrete energy levels for electrons and holes and, in addition, the lifting of the degeneracy of the valence band leading to the formation of heavy- and light-hole subbands. The Coulomb interaction between the electrons and the holes thus leads to the formation of heavy- and light-hole excitons.

The Hamiltonian of an exciton in a structure consisting of a layer of GaAs sandwiched between two semi-infinite layers of $Al_xGa_{1-x}As$ grown along the (001) direction can be expressed using an effective mass approximation as

$$H = \frac{\hbar^2}{2m_e} + T_h\,(-i\nabla_h) - \frac{e^2}{\varepsilon_0|\underline{r}_e - \underline{r}_h|} + V_e(z_e) + V_h(z_h) \qquad (1)$$

Here the first term denotes the kinetic energy of the conduction electron with effective mass m_e and the second term is the kinetic energy of the hole as first described by Luttinger [1]. In the present case we can ignore the split-off valence band and express this as

$$T_h(\underline{k}) = \begin{bmatrix} L_+ & M & N & O \\ M^* & L_- & O & N \\ N^* & O & L_- & -M \\ O & N^* & -M^* & L_+ \end{bmatrix} \tag{2}$$

where

$$L_\pm = \frac{\hbar^2}{2m_0}\left[(\gamma_1 \pm \gamma_2)\left(k_x^2 + k_y^2\right) + (\gamma_1 \mp 2\gamma_2)k_z^2 \right] \tag{3}$$

$$M = \frac{\hbar^2}{2m_0}\sqrt{3}\,\gamma_3 k_z(k_x - ik_y) \tag{4}$$

and

$$N = \frac{\hbar^2}{2m_0}\sqrt{3}\,[\gamma_2(k_x^2 - k_y^2) - 2i\gamma_3 k_x k_y] \tag{5}$$

Here m_0 is the free electron mass and γ_1, γ_2 and γ_3 are three material parameters which describe the valence band structure. The positions of the electron and the hole are designated by $\underline{r}_e$ and $\underline{r}_h$ respectively and ε_0 is the static dielectric constant of the material assumed to be the same for GaAs and $Al_xGa_{1-x}As$. The potential wells for the conduction electrons $V_e(z_e)$ and for the holes $V_h(z_h)$ are assumed to be square wells of width L,

$$V_e(z_e) = \left\{ \begin{array}{ll} 0, & |z_e| < L/2 \\ V_e, & |z_e| > L/2 \end{array} \right\} \tag{6a}$$

$$V_h(z_h) = \left\{ \begin{array}{ll} 0, & |z_h| < L/2 \\ V_h, & |z_h| > L/2 \end{array} \right\} \tag{6b}$$

Here we have chosen, without any loss of generality, the origin of the coordinate system to be the centre of the GaAs well. The values of the potential heights V_e and V_h are determined from the Al concentration in $Al_xGa_{1-x}As$, using the following expression [2] for the total bandgap discontinuity:

$$\Delta E_g = 1.155x + 0.37x^2 \tag{7}$$

in units of eV. The values of V_e and V_h are assumed to be 60 and 40% of ΔE respectively.

The first attempt to calculate the binding energies of the ground state (1s-like, hereafter referred to as 1s) and of the first excited state (2s-like, referred to as 2s) of excitons associated with the lowest hole subbands was made by Miller et al [3]. They assumed that the heavy- and the light-hole subbands were completely decoupled, i.e. they ignored the contributions of the off-diagonal terms of the Luttinger Hamiltonian (EQN (2)). This leads to the formation of two types of excitons, one associated with the heavy-hole subband and the other with the light-hole subband. With this approximation, the Hamiltonian of a heavy (light) exciton in a quantum well reduces to

$$H = \frac{\hbar^2}{2\mu_\pm} \left[\frac{1}{\rho} \frac{\partial}{\partial \rho} \rho \frac{\partial}{\partial \rho} + \frac{1}{\rho^2} \frac{\partial^2}{\partial \phi^2} \right] - \frac{\hbar^2}{2m_e} \frac{\partial^2}{\partial z_e^2}$$

$$- \frac{\hbar^2}{2m_\pm} \frac{\partial^2}{\partial z_h^2} - \frac{e^2}{\varepsilon_0 | r_e - r_h |} + V_e(z_e) + V_h(z_h)$$

(8)

where we have used cylindrical coordinates and have used the following definitions:

$$\frac{1}{\mu_\pm} = \frac{1}{m_e} + \frac{1}{m_0} (\gamma_1 \pm \gamma_2)$$

(9)

and

$$\frac{1}{m_\pm} = \frac{1}{m_0} (\gamma_1 \mp 2\gamma_2)$$

(10)

In EQNS (9) and (10) the upper sign refers to the $J_z = \pm 3/2$ (heavy-hole) band and the lower sign to the $J_z = \pm 1/2$ (light-hole) band. Fairly accurate values of the Luttinger parameters γ_1, γ_2 and γ_3 for GaAs are now available [4]. An exact solution of the Schrodinger equation corresponding to the exciton Hamiltonian (EQN (8)) is clearly not possible. They therefore followed a variational approach and used the following form of the trial wave function:

$$\psi = f_e(z_e)f_h(z_h)g(\rho,z,\phi)$$

(11)

where $z = z_e - z_h$, $f_e(z_e)$ and $f_h(z_h)$ are the ground-state solutions of the electron and the hole respectively in the quantum well and $g(\rho,z,\phi)$ describes the internal motion of the exciton. In order further to simplify their calculations, they assumed infinite potential barriers and used for g wave functions which are appropriate for thin wells (<200 Å). Using the known values of the dielectric constant and the various mass parameters they minimised the expectation values (E) of the Hamiltonian (EQN (8)) with respect to the variational parameters of the trial wave function (EQN (11)). The binding energy of the exciton, say for the 1s state [E_{1s}], was then obtained by subtracting E from the total ground-state-energy of the electron and the hole in the wells, i.e. the sum of the electron and the hole subband energies. They found that the binding energies of the 1s and 2s states of both the heavy- and the light-hole excitons increased as the well size (L) was reduced and reached their respective two-dimensional values as the well size went to zero. Bastard et al [5] and Shinozuka and Matsuura [6] also calculated the binding energies of the ground state of these excitons using infinite potential barriers and obtained results similar to those derived by Miller et al [3].

Greene and Bajaj [7] were the first to calculate the binding energies of the 1s state of both the heavy- and the light-hole excitons as a function of the well size using a more realistic case of finite potential barriers for electrons and holes. They solved the Hamiltonian (EQN (8)) following a variational approach using a trial wave function of the form given by EQN (11) where $f_e(z_e)$ and $f_h(z_h)$ were now the well-known solutions of the particle in a box problem with finite barriers. They used the following expression for the function g

$$g(\rho,z,\phi) = (1 + \alpha z^2)e^{-\sigma(\rho^2+z^2)^{\frac{1}{2}}} \tag{12}$$

where α and σ were the variational parameters.

In order to improve the accuracy of their results Greene et al [8] later used the following expression for g

$$g(\rho,z,\phi) = \rho^{|m|} e^{im\phi} \sum_j a_j g_j(\rho,z) \tag{13}$$

where m is the projection of the angular momentum along the z-axis and the basis functions g_j are taken to be

$$g_1(\rho,z) = e^{-\alpha(\rho^2+z^2)^{\frac{1}{2}}} \tag{14a}$$

$$g_2(\rho,z) = z^2 e^{-\alpha(\rho^2+z^2)^{\frac{1}{2}}} \tag{14b}$$

$$g_3(\rho,z) = \rho e^{-\beta(\rho^2+z^2)^{\frac{1}{2}}} \tag{14c}$$

The quantities α and β are nonlinear variational parameters and are adjusted to minimise the total energy. The coefficients a_j are determined in the usual fashion by solving the matrix eigenvalue equation.

The variational binding energies of the 1s, 2s and $2p_+$ states were obtained by subtracting from the lowest electron and hole subband energies (E_e and E_h) the eigenvalues of EQN (8).

Greene et al [8] calculated the values of the binding energies of 1s, 2s and $2p_+$ states of the heavy-hole exciton and the light-hole exciton as a function of L for values of Al concentration $x = 0.15$ and 0.30. In order to compare their results with those of Miller et al [3] and Bastard et al [5] they also calculated the binding energies of these levels for the case of an infinite potential barrier. The values of the various physical parameters used in this calculation are given in [8]. The variation of the binding energies of the 1s state of a heavy-hole exciton $E_{1s}(h)$ (solid lines) and light-hole exciton $E_{1s}(l)$ (dashed lines) as a function of the well-size L for three different values of the potential-barrier heights are displayed in FIGURE 1. One finds that for a given value of x, the value of $E_{1s}(h)$ increases as L is reduced until it reaches a maximum and then drops rapidly. The value of L at which $E_{1s}(h)$ reaches a maximum is smaller for larger x. Essentially the same behaviour is exhibited by $E_{1s}(l)$. The reason for this is quite simple. As L is reduced the exciton wave function is compressed in the quantum well, leading to increased binding. However, below a certain value of L, which depends on the Al concentration, the spread of the exciton wave function into the surrounding $Al_xGa_{1-x}As$ layers becomes more important. This forces the binding energy of the exciton to go over to the value typical of bulk $Al_xGa_{1-x}As$ as L is reduced further. In the case of the infinite potential barriers the values of $E_{1s}(h)$ and $E_{1s}(l)$ increase montonically as L is reduced and go over to their respective two-dimensional values, i.e. four times their bulk values, as L goes to zero.

Similar behaviour is found for the 2s and $2p_{\pm}$ states and is displayed in FIGURES 2 and 3 respectively.

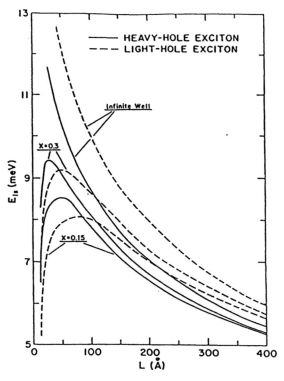

FIGURE 1. Variation of the binding energy of the ground state, E_{1s}, of a heavy-hole exciton (solid lines) and a light-hole exciton (dashed lines) as a function of the GaAs quantum-well size (L) for Al concentration x = 0.15 and 0.3, and for an infinite potential well.

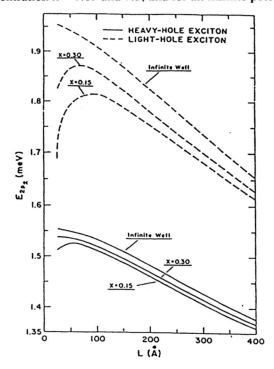

FIGURE 2. Variation of the binding energy of the 2p± state, $E_{2p\pm}$, of a heavy-hole exciton (solid lines) and a light-hole exciton (dashed lines) as a function of the GaAs quantum-well size (L) for Al concentrations x = 0.15 and 0.30, and for an infinite potential well.

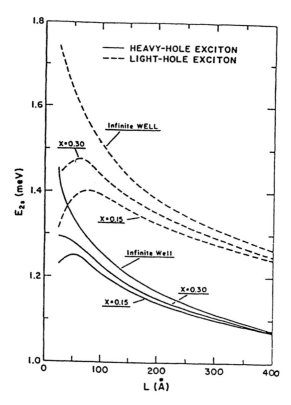

FIGURE 3. Variation of the binding energy of the 2s state, E_{2s}, of a heavy-hole exciton (solid lines) and of a light-hole exciton (dashed lines) as a function of the GaAs quantum-well size (L) for Al concentrations $x = 0.15$ and 0.30, and for an infinite potential well.

In the foregoing calculations the effect of the off-diagonal terms in the exciton Hamiltonian (EQN (1)) on its binding energy has been ignored. In the case of an exciton in bulk GaAs, the contribution of the off-diagonal terms to its binding energy is known to be small (<5%) [9]. This may, however, not be the case in quantum wells. It is known that in quantum wells, even in the absence of the Coulomb term, the inclusion of the off-diagonal terms results in strong mixing between the heavy- and the light-hole states. In addition, this hybridisation leads to hole subband structure which is highly anisotropic and non-parabolic in the transverse direction. Several of the hole subbands can have even negative zone-centre masses. These features, when included in the calculations of the exciton binding energies, can lead to results different from those obtained by using decoupled parabolic valence subbands. In this picture. the heavy-hole subband states, for instance, have a mixture of light-hole states and vice versa. Even though the heavy-hole and the light-hole subbands are no longer purely heavy-hole or light-hole in character, we still refer to them as heavy-hole and light-hole subbands for the sake of convenience. In addition, we also have not included the Coulomb coupling between excitons belonging to different subbands, the non-parabolicity of the bulk conduction band, and differences between the mass parameters and the dielectric constants between the well and the barrier materials. Each of these effects has been studied, mostly separately, and has been found to enhance the values of the exciton binding energies by almost comparable amounts (typically $\leq$ 1 meV for thin wells). Recently Andreani and Pasquarello [10] have calculated the values of the binding energies of the 1s and the 2s states of both the heavy- and the light-hole excitons in GaAs/Al$_x$Ga$_{1-x}$As quantum wells including the above mentioned effects. They follow a variational approach and expand their trial wave function g in terms of two-dimensional hydrogenic and Gaussian wave functions. Their results for the 1s state for the heavy-hole exciton are shown in FIGURE 4 and those for the light-hole exciton are displayed in FIGURE 5 for GaAs/Al$_{0.4}$Ga$_{0.6}$As quantum well structures. It is clear that the inclusion of the effects originally ignored by Greene et al [8] leads to considerably larger values of the exciton binding energies especially for narrow wells.

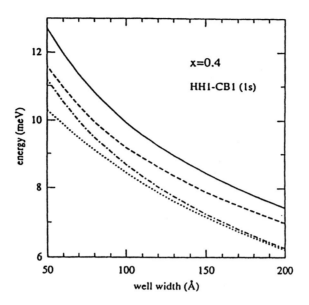

FIGURE 4. Binding energy of the ground-state HH1-CB1 exciton in GaAs/Ga$_{0.6}$Al$_{0.4}$As quantum wells. Dotted line: two-band approximation, parabolic CB, equal dielectric constants. Dashed-dotted line: including CB nonparabolicity. Dashed line: including nonparabolicity and Coulomb coupling. Solid line: full calculation, including also the dielectric mismatch.

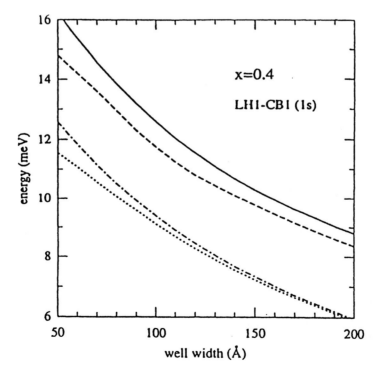

FIGURE 5. Binding energy of the ground-state LH1-CB1 exciton in GaAs/Ga$_{0.6}$Al$_{0.4}$As quantum wells. The meaning of the different curves is the same as in FIGURE 4.

In FIGURE 6, we display the variation of the binding energy of the 1s state of the heavy-hole exciton as a function of the well width L for several values of the Al concentration in the barrier layers. As expected, for a given value of the well width the value of the binding energy increases with Al concentration. The value of the binding energy exceeds the two-dimensional limit, namely 4 times the bulk Rydberg ($\approx$ 4 meV) for narrow wells with AlAs barriers. This enhancement is largely due to the contribution of the electrostatic potential due to the image charges to the Coulomb potential. Similar results also are obtained for the light-hole exciton. The behaviour of the binding energy of the 2s excited states of excitons is similar to that of the binding energy of the 1s state.

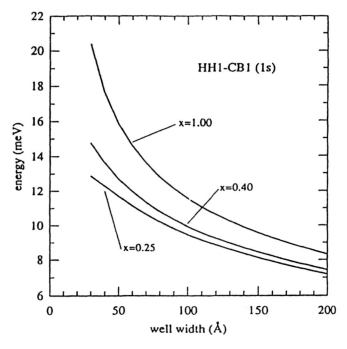

FIGURE 6. Binding energy of the ground-state HH1-CB1 exciton in GaAs/Ga$_{1-x}$Al$_x$As quantum wells for different values of x.

C   NON-RECTANGULAR QUANTUM WELLS

Though an extensive effort has been devoted to calculating the binding energies of excitons in rectangular quantum wells, relatively little work has been reported in the non-rectangular quantum wells. Sanders and Bajaj [11,12] have calculated the binding energies of the 1s state of excitons in parabolic and asymmetric triangular quantum wells using a variational approach and have compared their results with those in rectangular quantum wells. They have included the effects of the valence band mixing but have neglected the Coulomb coupling between excitons associated with different subbands, non-parabolicity of the conduction band and the differences in dielectric constants and mass parameters between the well and the barrier materials. For further information, the reader is referred to [11,12].

D   TYPE II QUANTUM WELLS

In the quantum well structures described in Section C, the electrons and the holes reside in the same layer. Such structures are often referred to as type I. However, there are quantum well structures where the electrons and the holes are confined in different layers due to a particular type of conduction and valence band line up. One such structure which has been the subject of considerable study since the early 1970s consists of alternate layers of InAs and GaSb where the electrons reside in InAs layers and holes in GaSb layers. Another structure which has received considerable attention in recent years consists of alternate layers of GaAs and AlAs where the GaAs layers are so narrow (<30 Å) that the energy of the first Γ electron subband level is higher than that of the X subband in AlAs thus leading to electron confinement in AlAs and hole confinement in GaAs. Such structures where the electrons and the holes are confined in different layers are often referred to as staggered or type II.

Several groups have reported calculations of the exciton binding energies in these structures assuming infinite [13,14,16] and finite [15] potential barriers between GaAs and AlAs layers. In all these calculations, the effects of the valence band mixing and different dielectric constants have been ignored. In addition, the effects of electron subband and hole subband mixing, which are important in these structures which lack inversion symmetry, also are not included in several of these calculations. Recently Cen and Bajaj [17] have calculated the exciton binding energies in these structures taking into account such subband mixing effects. In the following we briefly review the results of their calculations.

The Hamiltonian of an exciton in a type II structure is described essentially by EQN (8) with the difference that the electron mass in AlAs is anisotropic and the dielectric constant ε_0 now denotes an average value. To solve for the binding energy of the 1s state of, say, a heavy-hole exciton, Cen and Bajaj [17] assume the following form of the trial wave function:

$$\psi = (z_e, z_h; \rho, \phi) = \sum_{k,l} A_{kl} F_e^k(z_e) F_h^l(z_h) g(\rho, \phi, z_e - z_h) \tag{15}$$

where $F_e^k(z_e)$ and $F_h^l(z_h)$ are the eigenfunctions of the kth electron and the lth hole band respectively, $g(\rho, \phi, z)$ describes the internal motion of the exciton and A_{kl} determines the strength of coupling between the electron in the kth subband level to a hole in the lth subband level through the Coulomb potential between them.

Cen and Bajaj [17] have calculated the binding energies of both the heavy-hole (HH) and light-hole (LH) excitons in GaAs/AlAs type II QW structures. They express the variational function $g(\rho, \phi, z)$ in terms of a Gaussian basis set and consider the coupling between the two lowest conduction subbands and the two lowest valence subbands. The values of the various physical parameters used in their calculation are given in [17].

In FIGURE 7(a), we show the calculated binding energies of the HH exciton as a function of the AlAs layer thickness L_e, with GaAs layer thickness L_h as a parameter. For comparison, results calculated with only one electron and one hole subband are also plotted. For $L_h = 20$ Å, there are two confined HH subbands which couple with the confined electron states and an increase of ~1.5 meV is obtained at $L_e \approx 20$ Å, and ~2.5 meV at $L_e \approx 100$ Å. A similar increase is also obtained for $L_h \approx 30$ Å. In FIGURE 7(b), we also compare results of the binding energy of the LH exciton, as a function of L_e with L_h as a parameter. For $L_h = 20$ Å, the second LH subband is above the confining potential; coupling between the confined electron states to the first hole subband makes the dominant contribution to the exciton binding energies which are slightly higher (~1 - 2 meV) than the previous results due to coupling of the second confined electron state to the more diffuse hole state of the first hole subband. At $L_h = 30$ Å, both confined LH levels couple to the electron states, and the increase in the exciton binding energy is about 4 meV.

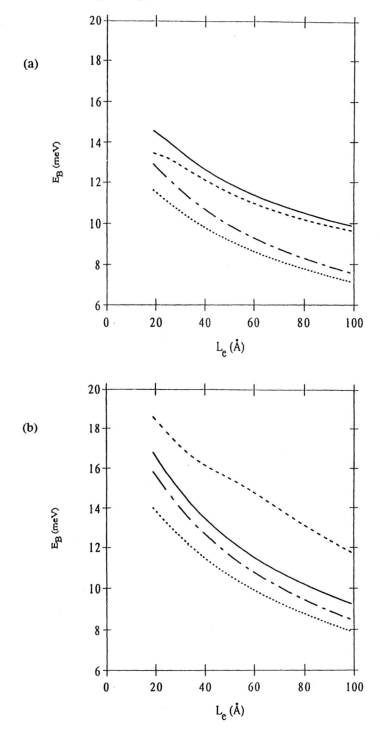

FIGURE 7. (a) Variation of the binding energy E_B of the HH exciton as a function of the AlAs layer thickness L_e. For $L_h = 20$ Å, results obtained with effects of subband mixing are indicated by the solid lines, results obtain without subband mixing by long-dash-dotted lines; for $L_h = 30$ Å, results obtained with effects of subband mixing are indicated by the dashed lines, results obtained without subband mixing by dotted lines. (b) Variation of the LH exciton binding energy E_B as a function of L_e. The meaning of the different curves is the same as in (a).

We have also calculated the exciton binding energies as a function of the GaAs layer thickness L_h, with L_e as a parameter. The results are displayed in FIGURE 8. A minimum in the binding energy exists for the HH exciton at $L_h \approx 14$ Å (not shown here), and for the LH exciton at $L_h \approx 25$ Å, when the second hole subband level changes in character from an unconfined continuum state to a confined state. Since a hole in a spatially extended continuum state would spend very little time in the vicinity of the confined electron, its coupling with the electron states is minimal, and only the first subband makes a contribution to the exciton binding energy. As the GaAs layer becomes wider, the second hole subband becomes confined and its coupling with the electron states becomes non-negligible. The average electron-hole distance becomes smaller as a result of the larger overlap between the electron and hole wave functions, and the binding energy increases as a function of L_h. However, as L_h further increases, the overlap of the wave functions would decrease since electron and hole wave functions would be more concentrated in separate wells. The binding energies would then again decrease as a function of L_h. So in a type II quantum well structure, a minimum or a maximum in the exciton binding energy as a function of the GaAs layer thickness could exist as a result of the varying degree of coupling between the confined electron states and the second hole subband state as the latter changes in character from an unconfined state to a confined state. Such an effect does not exist in the absence of the subband coupling.

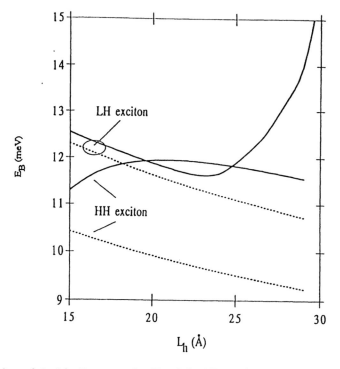

FIGURE 8. Variation of the binding energies E_B of the HH exciton and the LH exciton as functions of the GaAs layer thickness L_h, with $L_e = 50$ Å. Solid lines indicate results obtained with subband mixing. dotted lines indicate results obtained without subband mixing.

E   DOUBLE QUANTUM WELLS

A double quantum well (DQW) is a semiconductor structure in which two single quantum wells are separated by only a thin potential barrier across which electrons and holes in one well can tunnel into the other. As in single quantum wells, the electrons and holes confined in a DQW can form excitons due to their mutual Coulomb attraction. One of the advantages that

DQW structures offer over the single quantum wells is the enhanced exciton electro-optical response.

With the effects of the confinement and inter-well coupling (through tunnelling across the potential barrier) provided by a DQW, we have an interesting physical system in which these competing factors influence those exciton characteristics determining the exciton electro-optical properties of the DQW. Although a number of authors have done a considerable amount of work concerning excitons in DQWs [18-22], there still remain some conflicting results obtained by these various groups as to how inter-well coupling would qualitatively affect exciton binding energies in a DQW in the weak and strong inter-well coupling limits. A schematic band diagram of a symmetric GaAs/AlGaAs double quantum well is given in FIGURE 9. The Hamiltonian of an exciton associated with a heavy (light) hole subband is described essentially by EQN (8) except that there are now four interfaces where we have conduction and valence band discontinuities and the electron and the hole wave functions reside in five different regions. Cen and Bajaj [23] have recently calculated the binding energies of heavy and light hole excitons in DQW structures following a variational approach similar to that used in the case of type II quantum wells. They include the subband mixing between the lowest even and odd parity wave functions both for the electrons and for the holes and express the wave function describing the internal motion of the exciton g in terms of a Gaussian basis set.

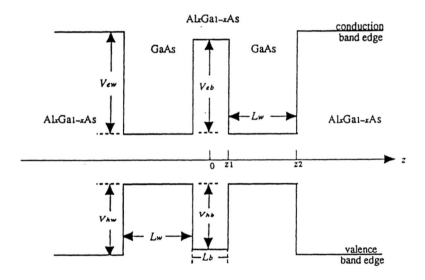

FIGURE 9. Schematic band diagram of a symmetric GaAs/Al$_x$Ga$_{1-x}$As double quantum well.

In FIGURE 10, we compare binding energies of the heavy-hole exciton in a DQW calculated by Cen and Bajaj with those obtained by Kamizato and Matsuura (KM) [19] with and without subband mixing, and those by Dignam and Sipe (DS) [21] with subband mixing, as a function of the barrier thickness L_b. It is evident that the two-subband treatment by DS seriously underestimates the binding energy in the strong inter-well coupling limit ($L_b \leq 0.2\ a_B$) and can not recover the fact that at $L_b = 0$, the DQW is simply a single well of width $2L_w$. On the other hand, the two-subband DS result in the weak inter-well coupling limit approaches that obtained by KM without subband mixing, which can not recover the single-well result at large barrier thickness either. It appears that although the DS two-subband treatment works in the intermediate-inter-well coupling strengths, it overestimates the strength of the inter-well coupling in both the strong and the weak inter-well coupling limits. Our result agrees with that of KM including subband mixing at both strong and weak inter-well coupling limits. It is also evident that our formalism gives higher binding energies for all inter-well coupling strengths. It is clear that the proper inclusion of the subband mixing is absolutely essential to obtain the correct behaviour of excitons in DQW structures.

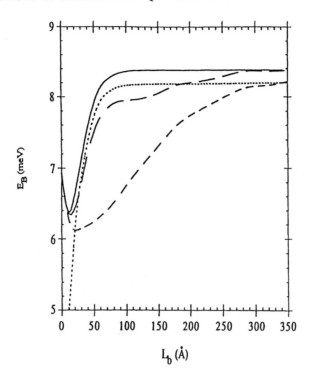

FIGURE 10. Comparison of binding energies of the heavy-hole exciton calculated, with and without subband mixing, as function of barrier thickness L_b. The well width is fixed at $L_w = 0.6\ a_B$. Solid line is the result with subband mixing included (Cen and Bajaj); dotted line by DS with subband mixing; short-dashed line is by KM without subband mixing; long-dashed line by KM with subband mixing. Material parameters are as in KM.

F   IONIC QUANTUM WELLS

There has been an enormous resurgence of interest in the study of the properties of ionic, wide-bandgap III-V semiconductors such as InN, GaN, AlN and their alloys and heterostructures and of the II-VI semiconductors such as ZnSe, ZnS, CdSe, CdS and their alloys and heterostructures during the past few years. This interest has been motivated largely by the fact that these wide bandgap semiconductors offer a promising class of materials for a variety of opto-electronic device applications in the highly desirable visible and ultraviolet (UV) wavelength regions, and in high temperature electronics. These semiconductors are highly ionic in character and therefore have a strong interaction between electrons/holes and the longitudinal optical (LO) phonon field of the crystal lattice. We expect these interactions to play an important role in determining the electronic and optical properties of quantum structures based on wide bandgap semiconductors.

During the past few years several groups [24-26] have attempted to calculate the binding energies of excitons in quantum well structures taking into account the effects of electron and hole-LO phonon interactions. They assume that the interactions between the electron and the hole and the LO phonon field are described by the well-known Fröhlich Hamiltonian [27] which is valid for bulk systems (or wide wells). They use a variational approach and find that the inclusion of the electron and hole-optical phonon interactions leads to enhanced values of the exciton binding energies. The values calculated by Ercelbi and Ozdinger [24] and by Degani and Hipolito [25] are, however, considerably larger than those reported by Matsuura [26]. This is due to the fact that the first two groups do not take account of the electron- and the hole-self energy terms correctly. The use of the usual bulk Fröhlich Hamiltonian, though it helps to make the problem somewhat easier to solve, may not, at first sight, appear quite appropriate for quantum well structures, especially narrow ones. This is because the presence of the hetero-interfaces modifies the phonon spectra and gives rise to confinement of optical phonons in each layer, half space LO phonons in the barrier layers, as well as interface modes which are localised in the vicinity of the interfaces. Calculations of the binding energies of excitons should therefore include the interactions of electrons and holes with all these phonon modes in an appropriate manner. However, it has been shown by Mori and Ando [28] that the form factors of electrons (holes) interacting with interface, half space, and confined phonons add up to the form factor of electrons (holes) interacting with bulk LO phonons. As the form factors determine the effective interaction between the electron and the hole, the contribution of the electron (hole) interactions with interface, half space, and confined phonons to the exciton binding energies may not be too different from that of electron (hole) interactions with the bulk optical phonons due to the above mentioned sum rule. However, Xie and Chen [29] have calculated the change in the binding energy of excitons in GaAs/$Al_xGa_{1-x}As$ quantum well structures due to electron-hole interactions with confined, half space and interface phonons following a variational approach similar to that used in [24-26]. Their calculations yield values of the changes in the exciton binding energies which are much too large to be physically meaningful.

Antonelli et al [30] have recently calculated the binding energies of excitons in ionic quantum well structures using Haken [31], Aldrich-Bajaj (AB) [32] and Pollmann-Buttner (PB) [33] effective potentials following a variational approach. These effective potentials have been derived by taking into full account the interactions of electrons and holes with the LO phonon field. Even though these effective interaction potentials have been derived for the bulk case, the results obtained using AB and PB potentials, for instance, should be quite reliable. The Haken potential is known to give values of the exciton binding energies which are considerably larger than their experimental values.

Antonelli et al have calculated the values of the binding energies of the heavy-hole excitons as functions of the well width in GaAs/$Al_{0.3}Ga_{0.7}As$, GaN/$Al_{0.3}Ga_{0.7}N$ and ZnSe/ZnS quantum well structures. In FIGURE 11 is displayed the variation of the exciton binding energies as a function of well width in a GaN/$Al_{0.3}Ga_{0.7}N$ quantum well structure. They have used the following values of the various physical parameters [34]: $m_e = 0.2m_0$, $m_h = 0.8m_0$, $\varepsilon_0 = 9.5$, $\varepsilon_\infty = 5.35$ and $\hbar\omega = 93$ meV. An isotropic value of the heavy hole mass was used since the values of the Kohn-Luttinger parameters in GaN are not known. The energy gap of $Al_{0.3}Ga_{0.7}N$ was determined by a linear interpolation between the bandgaps of GaN and AlN. The band offsets between GaN and $Al_{0.3}Ga_{0.7}N$ are not known. For the purpose of this calculation the electron and the hole potential barriers were chosen to be, in analogy with the GaAs/AlGaAs structure, 60 and 40% of the total bandgap discontinuity respectively. As shown in FIGURE 11, the values of the exciton binding energies obtained using Haken potential are the largest and those obtained using screened Coulomb potential are the lowest.

The values calculated using AB and PB potentials are the same and are about 12 to 15% larger than those obtained using screened Coulomb potential except for narrow wells where the enhancement is as much as 20%. Similar results are also obtained in the case of ZnSe/ZnS quantum well structures. The enhancement of the exciton binding energies in GaAs/ $Al_{0.3}Ga_{0.7}As$ quantum wells is only a few %.

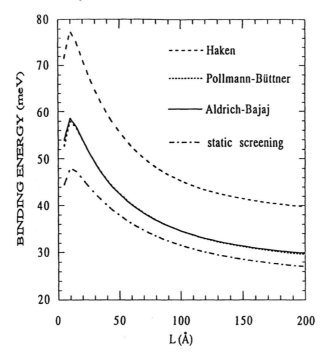

FIGURE 11. Variations of the heavy-hole exciton binding energies as a function of the well width in a GaN/$Ga_{0.7}Al_{0.3}$N quantum well structure for the various effective potentials.

G   GENERAL REMARKS

As mentioned earlier, a great deal of effort has been devoted to the study of the properties of excitons in quantum wells during the past two decades. Much less work has been done on the properties of excitons in superlattices. This is primarily due to the fact that the confinement effects provided by the quantum wells dramatically modify the properties of excitons. In superlattices however, there are no confinement effects: there is a formation of electron and hole minibands from which the excitons are formed. The behaviour of excitons in superlattices, therefore, is not significantly different from that in the bulk, except that their electronic structure becomes highly anisotropic.

The confinement effects not only lead to larger values of the exciton binding energies but also enhance the value of the excitonic oscillator strength which is related to the overlap of the electron and the hole wave functions. Oscillator strength is a measure of the excitonic absorption. Several calculations of the excitonic oscillator strength have been reported in the literature. These are not described here due to limitation of space.

H   CONCLUSION

We have provided a brief review of the results of various calculations of the binding energies of excitons in a variety of quantum well structures such as rectangular, non-rectangular, type II and coupled systems. We have discussed the effect of various approximations on the values of the binding energies of excitons.

REFERENCES

[1]   J.M. Luttinger [*Phys. Rev. (USA)* vol.102 (1956) p.1030-5]

[2]   H.J. Lee, L.Y. Joravel, J.C. Wolley, A.J. Springthorpe [*Phys. Rev. B (USA)* vol.21 (1980) p.659-69]

[3]   R.C. Miller, D.A. Kleinman, W.T. Tsang, A.C. Gossard [*Phys. Rev. B (USA)* vol.24 (1981) p.1134-6]

[4]   K.K. Bajaj, C.H. Aldrich [*Solid State Commun. (USA)* vol.35 (1980) p.163-7]

[5]   G. Bastard, E.E. Mendez, L.L. Chang, L. Esaki [*Phys. Rev. B (USA)* vol.26 (1982) p.1974-9]

[6]   Y. Shinozuka, M. Matsuura [*Phys. Rev. B (USA)* vol.28 (1983) p.4878-81]

[7]   R.L. Greene, K.K. Bajaj [*Solid State Commun. (USA)* vol.45 (1983) p.831-5]

[8]   R.L. Greene, K.K. Bajaj, D.E. Phelps [*Phys. Rev. B (USA)* vol.29 (1984) p.1807-12]

[9]   A. Baldereschi, N.O. Lipari [*Phys. Rev. B (USA)* vol.3 (1971) p.439-43]

[10]   L.C. Andreani, A. Pasquarello [*Phys. Rev. B (USA)* vol.42 (1990) p.8928-38 and references cited therein]

[11]   G.D. Sanders, K.K. Bajaj [*Phys. Rev. B (USA)* vol.36 (1987) p.4849-57]

[12]   G.D. Sanders, K.K. Bajaj [*J. Appl. Phys. (USA)* vol.68 (1990) p.5348-56]

[13]   G. Duggan, H.I. Ralph [*Phys. Rev. B (USA)* vol.35 (1987) p.4152-4]

[14]   M. Matsuura, Y. Shinozuka [*Phys. Rev. B (USA)* vol.38 (1988) p.9830-7]

[15]   B.R. Salmassi, G.E. Bauer [*Phys. Rev. B (USA)* vol.39 (1989) p.1970-2]

[16]   M.H. Degani, G.A. Farias [*Phys. Rev. B (USA)* vol.42 (1990) p.11701-7]

[17]   J. Cen, K.K. Bajaj [*Phys. Rev. B (USA)* vol.45 (1992) p.14380-3]

[18]   I. Galbraith, G. Duggan [*Phys. Rev. B (USA)* vol.40 (1989) p.5515-21]

[19]   T. Kamizato, M. Matsuura [*Phys. Rev. B (USA)* vol.40 (1989) p.5378-84]

[20]   J. Lee, M.O. Vassell, E.S. Koteles, B. Elman [*Phys. Rev. B (USA)* vol.39 (1989) p.10133-43]

[21]   M.M. Dignam, J.E. Sipe [*Phys. Rev. B (USA)* vol.43 (1991) p.4084-96]

[22]   F.M. Peeters, J.E. Golub [*Phys. Rev. B (USA)* vol.43 (1991) p.5159-62]

[23]   J. Cen, K.K. Bajaj [*Phys. Rev. B (USA)* vol.46 (1992) p.15280-9]

[24]   A. Ercelbi, U. Ozdinger [*Solid State Commun. (USA)* vol.57 (1986) p.441-4]

[25]   M.H. Degani, H. Hipolito [*Phys. Rev. B (USA)* vol.35 (1987) p.4507-10]

[26]   M. Matsuura [*Phys. Rev. B (USA)* vol.37 (1988) p.6977-82]

[27]   H. Frölich [*Adv. Phys. (UK)* vol.3 (1954) p.325-54]

[28]   N. Mori, T. Ando [*Phys. Rev. B (USA)* vol.40 (1989) p.6175-88]

[29]   X.J. Xie, C.Y. Chen [*J. Phys.. Condens. Matter (UK)* vol.6 (1994) p.1007-18]

[30]   A. Antonelli, J. Cen, K.K. Bajaj [to be published]

[31]   H. Haken [*Nuovo Cimento (Italy)* vol.10 (1956) p.1230-49]

[32]   C. Aldrich, K.K. Bajaj [*Solid State Commun. (USA)* vol.22 (1977) p.157-60]

[33]   J. Pollman, H. Buttner [*Solid State Commun. (USA)* vol.17 (1975) p.1171-4]

[34]   S. Strite, H. Morkoc [*J. Vac. Sci. Technol. B (USA)* vol.10 (1992) p.1237-66]

2.5 Effects of electric fields in quantum wells and superlattices

J.P. Loehr

January 1996

A INTRODUCTION

In layered semiconductor structures, such as quantum wells and superlattices, external electric fields applied normal to the layer boundaries can alter the behaviour of charged carriers (Section B). Piezoelectric effects can also produce internal electric fields in strained structures grown on (111) substrates (Section C).

B THE QUANTUM CONFINED STARK EFFECT IN QUANTUM WELLS

When an electric field is applied across a quantum well the confinement profile is changed from square to triangular. Electrons and holes are driven to opposite sides of the well, resulting in the so-called quantum confined Stark effect (QCSE) [1-4]. The effect is actually comprised of three primary changes to the optical absorption spectrum:

(1)   The electron subband levels drop and the hole subband levels rise, decreasing the interband transition energy.

(2)   Interband transition selection rules are relaxed, allowing 'forbidden' transitions to usurp oscillator strength from the canonical 'allowed' ($\Delta n = 0$) transitions. The cumulative effect is to smooth out the absorption spectrum.

(3)   The $\Delta n = 0$ exciton binding energies and oscillator strengths are diminished, because of reduced electron-hole overlap, and the exciton absorption line is broadened because of field-induced carrier tunnelling.

The net result is that the band edge absorption is reduced, broadened, and shifted to lower energy; the exciton peak positions shift quadratically with the applied electric field (FIGURE 1). These effects significantly modify the optical properties of quantum wells and superlattices [5], and have been studied at length both experimentally and theoretically [6]. Particular attention has been given to variational [7,8] and direct [9] calculations of the excitonic states in an applied electric field.

C STRAIN-INDUCED PIEZOELECTRIC FIELDS

Shear strains in polar crystals separate charge and through the piezoelectric effect [10] induce an electrical polarisation $P_i = 2e_{14}\varepsilon_{jk}$ in the material, i.e.

$$P_x = 2e_{14}\varepsilon_{yz}, \quad P_y = 2e_{14}\varepsilon_{zx}, \quad P_z = 2e_{14}\varepsilon_{xy} \tag{1}$$

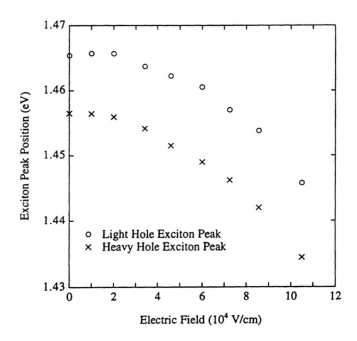

FIGURE 1. Field-dependence of the exciton absorption peak energies in a 95 Å GaAs quantum well confined by $Al_{0.32}Ga_{0.68}As$; the data (from [2]) were taken at room temperature.

Here e_{14} is the piezoelectric tensor element (TABLE 1), and the off-diagonal strain tensor elements ε_{jk} must be referred to the crystal axes $\hat{x}$, $\hat{y}$, and $\hat{z}$. Typically these strains are built into a superlattice or quantum well structure by lattice-mismatched growth on (111) substrates [11-17]. In this case all off-diagonal strain tensor elements are equal, and given by

$$\varepsilon_{xy} = \varepsilon_{yz} = \varepsilon_{zx} = -\left(\frac{a_{substrate} - a_{layer}}{a_{layer}}\right)\left[\frac{c_{11} + 2c_{12}}{c_{11} + 2c_{12} + 4c_{44}}\right] \qquad (2)$$

The polarisation direction is opposite for compressive and tensile strain and produces an electric field across the strained layers. For the simplest case of a quantum well grown on a (111) substrate this electric field is given (in MKS units) by

$$\mathbf{E} = -2\sqrt{3}\, e_{14}\varepsilon_{xy}/(\varepsilon_0\varepsilon_r)\mathbf{u} \qquad (3)$$

where $\mathbf{u}$ is the unit normal to the (111)A surface pointing out of the crystal, ε_0 is the permittivity of free space, and ε_r is the dielectric constant of the material (TABLE 1).

TABLE 1. Piezoelectric [18] and dielectric [19] constants for bulk materials.

	e_{14} (C m^{-2})	ε_r
GaAs	-0.16	12.91
InAs	-0.045	14.54

It is possible to induce quite large (~100 kV/cm) electric fields by growing thin layers of $In_xGa_{1-x}As$ on (111)B GaAs [13-16] or $In_{0.53+x}Ga_{0.47-x}As$ on (111)B InP [17]. These fields modify the optical absorption and refraction of superlattices and quantum wells through the QCSE. Often these strained layers are embedded in the intrinsic region of a p-i-n diode and the piezoelectric field is designed to oppose the built-in field (FIGURE 2) [13-16]. In this case the total electric field across the quantum wells depends on the built-in field of the diode, the thickness of the intrinsic region, the interface charge present, and the applied bias. A reverse bias initially blue-shifts the absorption spectrum (as opposed to the red shift in the conventional QCSE), and by charting the absorption peak versus applied bias the piezoelectric field across the well can be determined. For unknown reasons, however, the observed piezoelectric fields in these p-i-n structures are ~30% lower than those predicted by the bulk values of the piezoelectric constants in TABLE 1 [14-16].

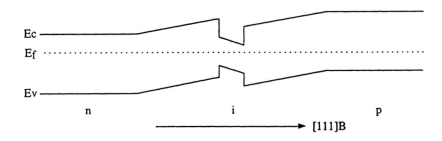

FIGURE 2. Schematic band diagram of a p-i-n strained quantum well diode grown on a (111)B substrate (not to scale). The [111]B direction is normal to the (111)B surface and points out of the crystal.

D   CONCLUSION

Externally-applied or piezoelectrically-induced electric fields can polarise the charge carriers in quantum wells and superlattices to change the band edge optical properties. Because of the transverse heterostructure confinement, these effects persist at much higher fields in layered structures than in bulk material.

REFERENCES

[1]   D.A.B. Miller et al [*Phys. Rev. Lett. (USA)* vol.53 (1984) p.2173-5]
[2]   D.A.B. Miller et al [*Phys. Rev. B (USA)* vol.32 (1985) p.1043-60]
[3]   P.W. Yu et al [*Phys. Rev. B (USA)* vol.40 (1989) p.3151-5]
[4]   C. Weisbuch, B. Vinter [*Quantum semiconductor structures: fundamentals and applications* (Academic Press, 1991)]
[5]   P.K. Bhattacharya [Datareview in this book: 6.5 Electro-optic effects in quantum wells and superlattices]
[6]   E.E. Mendez, F. Agullo-Rueda [*J. Lumin. (Netherlands)* vol.44 (1989) p.223-31]
[7]   G. Bastard, E.E. Mendez, L.L. Chang, L. Esaki [*Phys. Rev. B (USA)* vol.28 (1983) p.3241-5]
[8]   G.D. Sanders, K.K. Bajaj [*Phys. Rev. B (USA)* vol.35 (1987) p.2308-20]
[9]   J.P. Loehr, J. Singh [*Phys. Rev. B (USA)* vol.42 (1990) p.7154-62]
[10]   W.F. Cady [*Piezolectricity* (McGraw-Hill, 1946)]
[11]   D.L. Smith [*Solid State Commun. (UK)* vol.57 (1986) p.919-21]
[12]   C. Mailhiot, D.L. Smith [*Phys. Rev. B (USA)* vol.35 (1987) p.1242-59]
[13]   E.A. Caridi, T.Y. Chang, K.W. Goossen, L.F. Eastman [*Appl. Phys. Lett. (USA)* vol.56 (1990) p.659-61]
[14]   A.S. Pabla et al [*Appl. Phys. Lett. (USA)* vol.63 (1993) p.752-4]
[15]   R.L. Tober, T.B. Bahder [*Appl. Phys. Lett. (USA)* vol.63 (1993) p.2369-71]

[16]   J.L. Sanchez-Rojas, A. Sacedon, F. Gonzalez-Sanz, E. Calleja, E. Munoz [*Appl. Phys. Lett. (USA)* vol.65 (1994) p.2042-4]

[17]   H.C. Sun, L. Davis, Y. Lam, P.K. Bhattacharya, J.P. Loehr [*Inst. Phys. Conf. Ser. (UK)* no.129 (Institute of Physics, Bristol and Philadelphia, 1993) p.229-34]

[18]   R.M. Martin [*Phys. Rev. B (USA)* vol.5 (1972) p.1607-13]

[19]   B.R. Nag [*Electron Transport in Compound Semiconductors* (Springer-Verlag, 1980) p.372]

2.6   Effects of magnetic fields on quantum wells and superlattices

R.J. Nicholas

October 1995

A   INTRODUCTION

Broadly speaking two types of measurement use magnetic fields to study quantum wells and heterojunctions. First, there are the resonance (optical and infrared) experiments such as cyclotron and interband resonance which measure general material properties such as effective masses, exciton binding energies and band non-parabolicity. Secondly, there are more sample specific properties such as carrier density, mobility, occupied subband spacings and other scattering times which are usually measured by transport experiments. In some cases the transport measurements can also be used to study basic material properties, e.g. effective masses can be measured from the temperature dependence of the Shubnikov-de Haas oscillations.

A1   Landau Quantisation

With the exception of the low field Hall coefficient, most measurements are based on the quantisation conditions which lead to the formation of Landau levels for a free electron in a magnetic field [1]. The magnetic field is usually applied normal to the superlattice or quantum well (B∥z), so that the confinement potential is unaffected by the cyclotron motion in the x,y plane, which leads to a quantisation of the orbit motion in k-space so that

$$S_n = 2\pi(p+1/2) \, eB/\hbar \qquad (1)$$

and the electron energies are quantised with

$$E_n = (p+1/2)\hbar\omega_c = (p+1/2)\hbar \, eB/m^* \qquad (2)$$

Similar relations exist for the valence band, so the resulting energy levels in the magnetic field can be drawn schematically as in FIGURE 1, which shows the different optical transitions induced in cyclotron resonance (CR) and interband (IB) resonance.

B   CYCLOTRON RESONANCE

EQN (2) gives the most direct measurement of effective mass through experiments such as cyclotron resonance, which corresponds to a direct absorption transition between adjacent Landau levels with infrared or microwave radiation, and is subject to the selection rule $\Delta n = 1$, as shown in FIGURE 1. This is usually done at fairly low temperatures, since the accuracy of the quantisation is determined by the proportion of a cyclotron orbit which can be performed during the scattering time, given by the product $\omega_c\tau$.

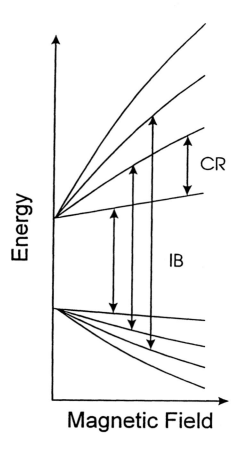

FIGURE 1. A schematic view of the electron and hole Landau levels showing the
transitions which give cyclotron resonance (CR) and interband (IB) transitions.

It is important to remember that the effective mass values can be a function of both the energy
of measurement, due to effects such as non-parabolicity of the bands, and of the crystal
orientation, where the effective mass is dependent on the direction of motion of the carriers.
In the majority of III-V materials the conduction band is located at the Γ-point ($k = 0$) and is
isotropic, and so the main effect which needs to be considered is non-parabolicity, which
causes the effective mass to increase weakly with energy, both within the plane perpendicular
to the magnetic field, and due to quantisation effects caused by confinement potentials such as
superlattice wells and heterojunction potentials. By contrast, the valence band can be much
more anisotropic, there are both heavy- and light-hole bands, and the hole masses can be a
strong function of the confinement potentials and strain splittings. Extensive reviews of
cyclotron resonance in low dimensional systems have been given by Nicholas [2] and by
McCombe and Petrou [3].

B1   Conduction Band

The electron effective mass m* has been measured by CR in a variety of heterojunctions and
quantum wells, and has been found to be given by the relation [4]

$$\frac{1}{m^*} = \frac{1}{m_0}\left(1 + 2\frac{K_2}{E_g}(\hbar\omega_c + <T_z>)\right) \tag{3}$$

for low density structures, where the cyclotron energy $\hbar\omega_c$ is higher than the Fermi energy E_F defined relative to the lowest confined electron level. m_0 is the band edge mass of the bulk quantum well material and $<T_z>$ is the kinetic energy due to the confinement potential, which for simple quantum wells is equal to the confinement energy, but can be substantially lower for more complex confinement (e.g. for a triangular potential well $<T_z> = E_c/3$). When considering relatively large energies a better relation is:

$$m^* = m_0 \left(1 + 2 \frac{|K_2|}{E_g} (\hbar\omega_c + <T_z>) \right) \tag{4}$$

For higher densities with $E_F > \hbar\omega_c$ we can write

$$m^* = m_0 \left(1 + 2 \frac{|K_2|}{E_g} (E_F + <T_z>) \right) \tag{5}$$

Theoretically the value of K_2 should be close to -1, using conventional two or three band **k.p** perturbation theory [5], but higher order corrections increase this value [6,7]. Experimentally the value of K_2 has been shown to be -1.4 for GaAs/GaAlAs in both doped heterojunctions [4] and undoped quantum wells, when the resonance is detected using optically detected cyclotron resonance [8]. For GaInAs heterostructures $K_2 = -1.3$ [9], and for InAs/GaSb $K_2 = -1.3$ [10]. Cyclotron resonance has been measured in a variety of other materials and structures, as reviewed by Nicholas [2].

Several additional effects can complicate the effective mass as measured by CR, in particular when measurements are made at either low energies, where disorder and localisation become very significant [11], or at energies close to the longitudinal optic (LO) phonon, when polaron effects occur [12]. For the latter case polaron coupling has been found to be strongly suppressed by high carrier densities [13].

B2   Valence Band

For p-type systems the behaviour is more complex, since the valence band in zinc blende and diamond-like materials is doubly degenerate in bulk materials, and this degeneracy is lifted by quantum confinement effects which make the effective mass values much more dependent on the structure studied. A typical example is shown in FIGURE 2 which shows the results of a calculation by Ekenberg and Altarelli [14] for a GaAs/GaAlAs heterojunction, compared with experimental data from Stormer et al

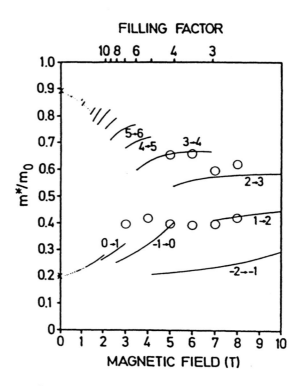

FIGURE 2. Effective masses as a function of magnetic field for the hole gas Landau levels of a GaAs/AlGaAs heterojunction. The open circles are the experimental results of [15]. (After [14].)

- 77 -

[15]. The data show two masses, with values of 0.4 and 0.6 m_e, together with the different masses which occur due to transitions between the different Landau levels. The two families of mass values correspond to the spin up and spin down levels associated with the heavy hole confined hole state which is the only occupied level. Extensive further studies have been reported for p-type GaAs systems, e.g. [16]. When p-type strained layers are studied, the in-plane hole mass can be strongly reduced; for example in the case of InGaAs/GaAs quantum wells the 'heavy' hole mass has been shown to fall to values as low as 0.15 m_e [17], and for GaSb/GaInSb wells masses as low as 0.075 m_e have been reported [18], but with relatively high non-parabolicities (FIGURE 3). For Si/SiGe, where for p-type systems the carriers are confined in the strained SiGe layers, reductions in the effective mass to values of 0.22 have been seen [19], but only for relatively low hole densities. For higher densities of the order 1×10^{16} m^{-2} the mass rises to ~0.4 m_e [20].

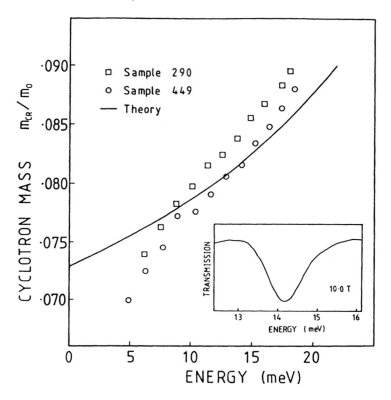

FIGURE 3. Cyclotron effective mass as a function of energy in a strained GaSb/Ga$_{0.85}$In$_{0.15}$Sb quantum well. The solid line shows the results of a calculation using eight band **k.p** theory. The inset shows a typical experimental recording. (After [18].)

B3   Tilted Fields

Tilting the magnetic field away from the normal to the confining potential introduces a magnetic force which mixes the cyclotron and quantum confinement motion. An anticrossing occurs as the cyclotron frequency equals the subband separation [2,3,21], which can be used as a spectroscopic tool to study the confinement potential [22].

C   INTERBAND MAGNETO-OPTICS

C1   Introduction

Further measurements of the electron and hole masses can be inferred from interband transitions where excitations occur from the valence band Landau levels to conduction band levels, also shown in FIGURE 1. The resulting features are a convolution of the mass values for both electrons and holes, and can be complicated by the presence of excitonic effects. As a result more care is required in the analysis, as discussed below and in considerably more detail in the review by McCombe and Petrou [3].

C1.1   Undoped structures

The simplest interband measurements are those on undoped structures, in which transitions are made directly from the Landau levels in the valence band to those in the conduction band. The selection rule for interband transitions is $\Delta n = 0, \pm 2, \pm 4, ...$ before the inclusion of spin effects, but transitions with $\Delta n \neq 0$ are generally quite weak. This leads to a fan of absorption transitions which extrapolate back close to the bandgap at $B = 0$. A typical example is shown in FIGURE 4 for a GaInAs/InP quantum well [9]. The energies of the strongest transitions are given by the relation

$$E = E_g^* + (n + 1/2)(\hbar\omega_c^e + \hbar\omega_c^h) - E_{n,X} \tag{6}$$

where $\hbar\omega_c^e$ and $\hbar\omega_c^h$ are the cyclotron energies of the electrons and holes respectively, which also need to include effects of non-parabolicity, and $E_{n,X}$ are the binding energies of the excitons. $E_g^* = E_g + E_{conf}(e) + E_{conf}(h)$ is the effective bandgap of the structure, where $E_{conf}(e)$ and $E_{conf}(h)$ are the confinement energies for the electron and hole. At zero field the excitonic binding energies are given by

$$E_{n,X} = \frac{R^*}{(n + 1/2)^2} \tag{7}$$

for a perfect 2-D system, where R^* is the effective Rydberg. For low fields the $n = 0$ (1s) transition is quadratic in B, but the excitonic corrections are small for the higher lying exciton states (2s, 3s ...) and their extrapolation back to $B = 0$ is a good approximation for the effective bandgap E_g^*. Numerical calculations of $E_{n,X}(B)$ are usually used for detailed fitting at high fields. The energy difference $E_g^* - E(1s)$ is a good experimental measure of the exciton binding energy [23,24]. Theoretically the exciton binding energy in a 2-D system is four times the 3-D value, but in a quantum well of finite depth the binding energy shows a maximum value, with the narrow well limit corresponding to the bulk value for the barrier material [25].

In a more detailed interpretation we also need to include the effects of electron and hole spin, which causes each of these transitions to split into two components, and the effects of non-parabolicity, which for electrons can be treated in the same way as described in the section above. This is particularly important in narrower gap systems, where interband measurements can be made at energies well above the bandgap. A typical example of this can be seen in FIGURE 4 in which the effects of non-parabolicity lead to a large amount of curvature in the fan diagram. In practice the effective mass of the electrons in III-V materials is considerably lower than that of the holes, making it necessary to use high quality structures with narrow linewidths in order to deduce accurate values of the valence band parameters. For a proper fitting of these it is usually necessary to do at least a five band **k.p** theory calculation of the Landau levels [26,27].

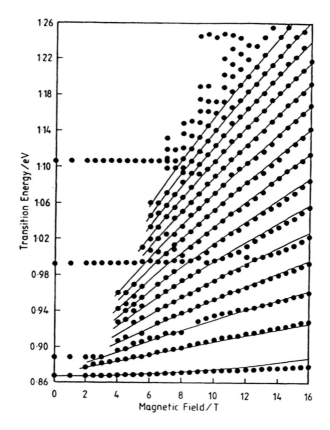

FIGURE 4. Transition energies as a function of magnetic field for a 10 nm InGaAs/InP multiquantum well. The solid lines are a theoretical fit to the data. (After [9].)

C1.2   Doped structures

For doped structures the behaviour is rather different, since it is now possible to observe transitions from filled Landau levels to the valence band using photoluminescence (PL). A typical example is shown in FIGURE 5, which shows the Landau levels for an InGaAs/InP quantum well [28]. In this system there is strong hole localisation, so that **k**-vector selection rules are relaxed and the full spectrum of electron Landau levels is seen. Similar measurements have been made for GaAs/GaAlAs quantum wells and heterojunctions; however, the restrictions imposed by **k**-vector conservation in this system limit the number of Landau levels which can be seen [29]. Nevertheless, some very detailed measurements have been made at low temperatures and high magnetic fields, which are used to investigate the formation of quantum liquids and solids [30-32]. A complicating factor in this type of measurement is the excitonic effects, which are quite small, but are dependent on screening and the filling factor of the Landau levels [33,34], leading to oscillations in the transition energy [28]. An alternative technique is to study photoconductivity, in which transitions to empty states can be observed, allowing the study of higher quantised subbands [35].

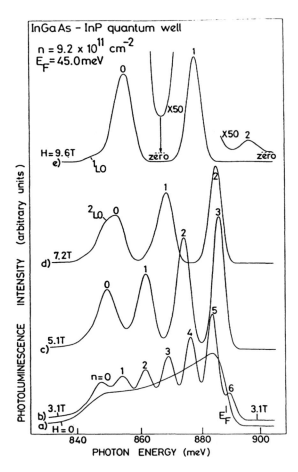

FIGURE 5. PL spectra at 2 K for an InGaAs/InP modulation doped quantum well with $n_e = 9.2 \times 10^{11}$ cm^{-2}, for several magnetic fields. (After [28].)

D   MAGNETOTRANSPORT

D1   Hall Effect

Hall effect measurements at low field are one of the most useful and straightforward ways of characterising semiconductor layers, since they measure both carrier density and mobility, and cover a very wide temperature range. More sophisticated analysis is possible when multiple carrier types or layers are present with differing mobilities, based on fits of the field dependence of the Hall voltage and magnetoresistance. However, the fitting procedures rapidly become unstable once many layers are used.

D2   Shubnikov-de Haas and Quantum Hall Effect

The majority of more sophisticated magnetotransport investigations study the oscillatory contributions to the transport coefficients which occur at high fields and low temperatures when the quantised Landau levels move through the Fermi energy. The best known of these are the Shubnikov-de Haas effect, which leads to large oscillations in the diagonal resistivity ρ_{xx}, and the quantum Hall effect. These occur when an integer multiple of the Landau level degeneracy, eB/h, is equal to the carrier density within a given layer. In the simplest case of a single layer with a single occupied level this leads to a single series of oscillations in ρ_{xx} with a periodicity in 1/B given by

$$(\Delta\,(1/B))^{-1} = n_e\,\frac{h}{2e} \tag{8}$$

where a factor of two has been included to take account of the spin degeneracy of the levels, which is only resolved at high fields. The low field oscillations are well described by the Ando formula [36]:

$$\Delta\rho/\rho_0 \sim \exp(-2\pi^2 k_B T_D/\hbar\omega_c)\,\frac{X}{\sinh(X)}\,\cos\!\left(\pi n_e\,\frac{h}{eB}\right) \tag{9}$$

where the exponential factor describes the decay of the amplitude of the oscillations with field. This can be used to deduce a value of the quantum lifetime τ_q ($= \hbar/2\pi^2 k T_D$) which measures the amount of short range scattering. The factor $X/\sinh(X)$, where $X = 2\pi^2 kT/\hbar\omega_c$, describes the temperature dependence of the oscillation amplitudes, and can be used to measure the carrier effective mass. At high fields and low temperatures the resistivity approaches zero at fields given by $peB/h = n_e$. A series of plateaux appear in the Hall resistivity at the points where the zeros occur in ρ_{xx}. This is the quantum Hall effect [37], and the plateaux values are given to a very high precision by h/pe^2 and are centred on an extrapolation of the low field Hall slope. When more than one level is occupied, or multiple doped layers are present, multiple series occur [38], as shown in FIGURE 6, and Fourier transforming the oscillations allows the occupancy of each of the levels to be measured separately [38-40]. Some care needs to be used in the analysis, since harmonics can appear due to peak sharpening and spin splitting. In some heavily doped structures as many as five separate series of oscillations have been detected due to multiple level occupancy [39,40].

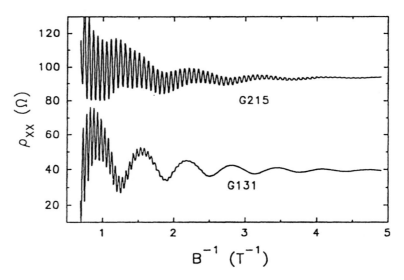

FIGURE 6. The resistivity as a function of inverse magnetic field at 0.55 K for two GaAs/AlGaAs heterojunctions. The two series of oscillations arise from electrons in the first two electric subbands in the heterojunctions. (After [38].)

By contrast, a classical measurement of the Hall coefficient will measure the total density of carriers within a given structure. A comparison of the carrier densities measured using the two techniques can give important information on the number of layers contributing to conduction in multiple layer structures.

Shubnikov-de Haas like oscillations occur in many different transport parameters, and have been observed for example in thermopower [41], magnetisation (the de Haas-van Alphen effect) [42], cyclotron resonance linewidth [43] and several others.

D3   Magnetophonon Effect

Resonant scattering has also been observed by LO phonons at higher temperatures. This gives rise to a series of peaks in the magnetoresistance which occur at the magnetophonon resonance condition

$$N\hbar\omega_c = \hbar\omega_{LO} \qquad\qquad (10)$$

where $\hbar\omega_{LO}$ is the energy of the LO phonon. The oscillations typically have a peak amplitude at $\sim kT \approx \hbar\omega_{LO}/3$. These oscillations can also be used as a measure of the effective mass, and have been observed in several different transport coefficients. The magnetophonon effect in 2-D structures has been reviewed by Nicholas [44].

D4   Tunnelling

Magnetic fields have also been used as a way of performing spectroscopic studies of both valence bands and quantum levels in double barrier resonant tunnelling heterojunctions [45,46]. The field is applied perpendicular to the current direction, causing the **k**-vector of the carrier which is tunnelling to be shifted by an amount $k_0 = eB\Delta x/\hbar$, where Δx is the effective tunnelling length. This allows a direct measurement of the carrier dispersion or wavefunction, with the energy scale being set by the tunnelling bias voltage at which a resonant peak occurs.

E   CONCLUSION

Magnetic fields can be used for a variety of assessment and characterisation purposes, as evidenced by the range of phenomena described above. In general, many of the measurements are quite sophisticated, but can be used for the establishment of accurate values for important material parameters such as effective masses. Magneto-electrical measurements are more useful as routine characterisation tools than magneto-optic ones.

REFERENCES

[1]   C. Kittel [*Quantum Theory of Solids* (John Wiley, 1963) p.217]
[2]   R.J. Nicholas [in *Handbook on Semiconductors vol.2: Optical Properties of Semiconductors, 2nd Edition* Ed. M. Balkanski (Elsevier-North Holland, 1994) p.385]
[3]   B.D. McCombe, A. Petrou [in *Handbook on Semiconductors vol.2: Optical Properties of Semiconductors, 2nd Edition* Ed. M. Balkanski (Elsevier-North Holland, 1994) p.285]
[4]   M.A. Hopkins, R.J. Nicholas, M.A. Brummell, J.J. Harris, C.T. Foxon [*Phys. Rev. B (USA)* vol.36 (1987) p.4789]
[5]   E.D. Palik, G.S. Picus, S. Teitleer, R.F. Wallis [*Phys. Rev. (USA)* vol.122 (1961) p.475]

[6]   C. Herman, C. Weisbuch [*Phys. Rev. B (USA)* vol.15 (1977) p.823]

[7]   U. Rossler [*Solid State Commun. (USA)* vol.49 (1984) p.943]

[8]   R.J. Warburton, J.G. Michels, R.J. Nicholas, J.J. Harris C.T. Foxon [*Phys. Rev. B (USA)* vol.46 (1992) p.13394]

[9]   J. Singleton et al [*J. de Physique C (France)* vol.48 no.C-5 (Nov. 1987) p.147-50]

[10]   D.J. Barnes, R.J. Nicholas, R.J. Warburton, N.J. Mason, P.J. Walker, N. Miura [*Phys. Rev. B (USA)* vol.49 (1994) p.10474]

[11]   J. Richter, H. Sigg, K. von Klitzing, K. Ploog [*Phys. Rev. B (USA)* vol.39 (1989) p.6268]

[12]   F.M. Peeters, J.T. Devreese [*Phys. Rev. B (USA)* vol.31 (1985) p.3689]

[13]   C.J.G.M. Langerak et al [*Phys. Rev. B (USA)* vol.38 (1988) p.13133]

[14]   U. Ekenberg, M. Altarelli [*Phys. Rev. B (USA)* vol.32 (1985) p.3712]

[15]   H.L. Stormer, Z. Schlesinger, A. Chang, D.C. Tsui, A.C. Gossard, W. Weigmann [*Phys. Rev. Lett. (USA)* vol.51 (1983) p.126]

[16]   S.J. Hawksworth et al [*Semicond. Sci. Technol. (UK)* vol.8 (1993) p.1465]

[17]   R.J. Warburton, R.J. Nicholas, L.K. Howard, M.T. Emeny [*Phys. Rev. B (USA)* vol.43 (1991) p.14124]

[18]   R.J. Warburton et al [*Surf. Sci. (Netherlands)* vol.228 (1990) p.270]

[19]   S.L. Wong, D. Kinder, R.J. Nicholas, T.E. Whall, R. Kubiak [*Phys. Rev. B (USA)* vol.51 (1995) p.13499]

[20]   J.P. Cheng, V.P. Kesan, D.A. Grutzmacher, T.O. Sedgwick [*Appl. Phys. Lett. (USA)* vol.64 (1994) p.1681]

[21]   Z. Schlesinger, J.C.M. Hwang, S.J. Allen [*Phys. Rev. Lett. (USA)* vol.50 (1983) p.2098]

[22]   J.G. Michels, R.J. Nicholas, G.M. Summers, D.M. Symons, J.J. Harris, C.T. Foxon [*Phys. Rev. B (USA)* vol.52 (1995) p.2688]

[23]   J.C. Maan [in *Two Dimensional Systems, Heterostructures and Superlattices* Eds G. Bauer, F. Kuchar, H. Heinrich (Springer-Verlag, Berlin, 1984) p.183]

[24]   D.C. Rogers, J. Singleton, R.J. Nicholas, C.T. Foxon, K. Woodbridge [*Phys. Rev. B (USA)* vol.34 (1986) p.4002]

[25]   R.L. Greene, K.K. Bajaj, D.E. Phelps [*Phys. Rev. B (USA)* vol.29 (1984) p.1807]

[26]   M.H. Weiler [in *Semiconductors and Semimetals* vol.16 Eds R.K. Willardson, A.C. Beer (New York: Academic Press, 1981) p.119]

[27]   L.R. Ram-Mohan, K.H. Yoo, R.L. Aggarwal [*Phys. Rev. B (USA)* vol.38 (1988) p.6151]

[28]   M.S. Skolnick, K.J. Nash, S.J. Bass, P.E. Simmonds, M.J. Kane [*Solid State Commun. (USA)* vol.67 (1988) p.637]

[29]   T.T.J.M. Berendschot, H.A.J.M. Reinen, H.J.A. Bluyssen [*Solid State Commun. (USA)* vol.63 (1987) p.873]

[30]   B.B. Goldberg, D. Heiman, A. Pinczuk, L. Pfeiffer, K. West [*Surf. Sci. (Netherlands)* vol.263 (1992) p.9]

[31]   A.J. Turberfield et al [*Surf. Sci. (Netherlands)* vol.263 (1992) p.1]

[32]   I.V. Kukushkin et al [*Surf. Sci. (Netherlands)* vol.263 (1992) p.30]

[33]   C. Delalande, G. Bastard, J. Orgonasi, J. Brum, H. Liu, M. Voos [*Phys. Rev. Lett. (USA)* vol.59 (1987) p.2690]

[34]   R. Stepniewski et al [in *The Physics of Semiconductors* Eds E.M. Anastassakis, J.D. Joannopoulos (World Scientific, Singapore, 1990) p.1282]

[35]   A.B. Henriques, E.T.R. Chidley, R.J. Nicholas, P. Dawson, C.T. Foxon [*Phys. Rev. B (USA)* vol.46 (1992) p.4047]

[36]   T. Ando, A.B. Fowler, F. Stern [*Rev. Mod. Phys. (USA)* vol.54 (1982) p.540]

[37]   K. von Klitzing, G. Dorda, M. Pepper [*Phys. Rev. Lett. (USA)* vol.45 (1980) p.494]

[38]   D.R. Leadley, R. Fletcher, R.J. Nicholas, F. Tao, J.J. Harris, C.T. Foxon [*Phys. Rev. B (USA)* vol.46 (1992) p.12439]

[39]   J. Singleton, F. Nasir, R.J. Nicholas, C.K. Sarkar [*J. Phys. C. Solid State Phys. (UK)* vol.19 (1986) p.35]

[40]   W. Zhao, F. Koch, J. Ziegler, H. Maier [*Phys. Rev. B (USA)* vol.31 (1985) p.2416]

[41]   R. Fletcher, J.C. Maan, K. Ploog, G. Weimann [*Phys. Rev. B (USA)* vol.33 (1986) p.7122]

[42]   J.P. Eisenstein, H.L. Stormer, V. Narayanamurti, A.J. Cho, A.C. Gossard, C.W. Tu [*Phys. Rev. Lett. (USA)* vol.55 (1985) p.875]

[43]   D. Heitmann [*Surf. Sci. (Netherlands)* vol.170 (1986) p.332]

[44]   R.J. Nicholas [in *Landau Level Spectroscopy* Eds G. Landwhr, E.I. Rashba in *Modern Problems in Condensed Matter Sciences* vol.27.2 (North-Holland, 1991) p.777]

[45]   R.K. Hayden et al [*Semicond. Sci. Technol. (UK)* vol.9 (1994) p.298]

[46]   J. Wang, P.H. Beton, N. Mori, L. Eaves, P.C. Main, M. Henini [in *The Physics of Semiconductors* Ed. D.J. Lockwood (World Scientific, 1995) p.1755]

2.7   Effects of biaxial strain in quantum wells and superlattices

J. Singh

January 1996

A   INTRODUCTION

Since the late 1970s there has been an increasing interest in the use of strain to modify electronic properties of semiconductor structures and as a result improve device performance. The driving force behind this development has been the gradual mastery of the art of strained hetero-epitaxy. By growing a semiconductor overlayer on top of a thick substrate, a large strain can be built into the overlayer, while still maintaining crystallinity and long-range order. The basic understanding behind this pseudomorphic growth was provided by Frank and van der Merwe in their classic papers in the late 1940s. These theories were extended by later workers (for a review see [1]). In essence, these theories show that as long as the lattice misfit is below approximately 10%, it is possible to grow an epitaxial film which is in complete registry with the substrate (i.e. it is pseudomorphic up to a thickness called the critical thickness). Beyond the critical thickness, dislocations are produced and the strain relaxes. It is important to note that the pseudomorphic films below critical thickness are thermodynamically stable, i.e. even though they have strain energy, the relaxed state with dislocations has a higher free energy. This allows one to have large built-in strain without serious concern that the film will 'disintegrate' after some time.

In this Datareview we will examine the effect the built-in strain has on the electronic properties of semiconductors. Before starting on some of the quantitative formalism let us point out some important points relevant to the study of strained effects:

- We will be interested in strained effects in thin structures (i.e. quantum wells) because of critical thickness limitations. Typical thicknesses of interest are 100 Å. Of course it is possible to have a number of such quantum wells by balancing the strain between the well region and the barrier region.

- The semiconductors of interest have an underlying cubic symmetry that leads to certain degeneracies in the band structure. For example the top of the valence band has a light hole-heavy hole degeneracy that arises from the x-, y- and z-symmetry. Pseudomorphic strain can remove this symmetry by physically distorting the crystal. Thus strain can lift the degeneracies at high symmetry points in the band structure and thus reduce the density of states. This is one of the key motivations for strained layer studies since essentially all physical phenomena in semiconductor devices are related to the density of states.

- Indirect gap materials like Si and Ge have a very high conduction band density of states due to the symmetry mentioned above. Strain produced by epitaxy can remove this symmetry and lower the conduction band density of states. Direct gap materials have a non-degenerate conduction band edge and therefore strain does not play much of a role in the conduction band.

B   STRAIN TENSOR IN LATTICE MISMATCHED EPITAXY

In order to study the effect of strain on electronic properties of semiconductors it is first essential to establish the strain tensor produced by epitaxy. As noted above, providing we are within the critical thickness, growth of an epitaxial layer whose lattice constant is close, but not equal, to the lattice constant of the substrate can result in a coherent strain, as opposed to polycrystalline or amorphous incoherent growth. If the strain is incorporated into the epitaxial crystal coherently, the lattice constant of the epitaxial layer in the direction parallel to the interface is forced to be equal to the lattice constant of the substrate. The lattice constant of the epitaxial layer perpendicular to the substrate will be changed by the Poisson effect. If the parallel lattice constant is forced to shrink, or a compressive strain is applied, the perpendicular lattice constant will grow. Conversely, if the parallel lattice constant of the epitaxial layer is forced to expand under tensile strain, the perpendicular lattice constant will shrink. These two cases are depicted in FIGURE 1. This type of coherently strained crystal is called pseudomorphic.

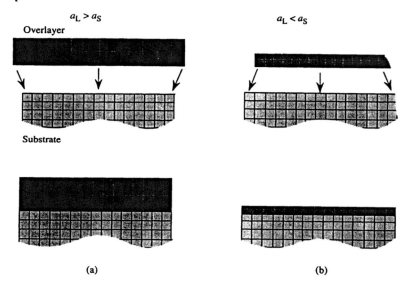

FIGURE 1. Pseudomorphic strain produced by epitaxy of an overlayer with a bulk lattice constant larger (a), or smaller (b) than the substrate. The overlayer must match the in-plane lattice constant of the substrate.

For systems of interest in the present work, the epitaxial semiconductor layer is biaxially strained in the plane of the substrate, by an amount $\varepsilon_{\parallel}$, and uniaxially strained in the perpendicular direction, by an amount $\varepsilon_{\perp}$. For a thick substrate, the in-plane strain of the layer is determined from the bulk lattice constants of the substrate material, a_S, and the layer material, a_L:

$$\varepsilon_{\parallel} = \frac{a_S}{a_L} - 1$$

$$= \varepsilon \tag{1}$$

Since the layer is subjected to no stress in the perpendicular direction, the perpendicular strain, $\varepsilon_{\perp}$, is simply proportional to $\varepsilon_{\parallel}$:

$$\varepsilon_{\perp} = \frac{-\varepsilon_{\parallel}}{\sigma} \tag{2}$$

where the constant σ is Poisson's ratio.

Noting that there is no stress in the direction of growth it can be simply shown that for the strained layer grown on a (001) substrate [2,3]

$$\sigma = \frac{c_{11}}{2c_{12}}$$

$$\varepsilon_{xx} = \varepsilon_{\parallel}$$

$$\varepsilon_{yy} = \varepsilon_{xx}$$

$$\varepsilon_{zz} = \frac{-2c_{12}}{c_{11}}\varepsilon_{\parallel} \tag{3}$$

$$\varepsilon_{xy} = 0$$

$$\varepsilon_{yz} = 0$$

$$\varepsilon_{xz} = 0$$

while in the case of a strained layer grown on a (111) substrate

$$\sigma = \frac{c_{11} + 2c_{12} + 4c_{44}}{2c_{11} + 4c_{12} + 4c_{44}}$$

$$\varepsilon_{xx} = \left[\frac{2}{3} - \frac{1}{3}\left(\frac{2c_{11} + 4c_{12} - 4c_{44}}{c_{11} + 2c_{12} + 4c_{44}}\right)\right]\varepsilon_{\parallel}$$

$$\varepsilon_{yy} = \varepsilon_{xx}$$

$$\varepsilon_{zz} = \varepsilon_{xx} \tag{4}$$

$$\varepsilon_{xy} = \left[-\frac{1}{3} - \frac{1}{3}\left(\frac{2c_{11} + 4c_{12} - 4c_{44}}{c_{11} + 2c_{12} + 4c_{44}}\right)\right]\varepsilon_{\parallel}$$

$$\varepsilon_{yz} = \varepsilon_{xy}$$

$$\varepsilon_{zx} = \varepsilon_{yz}$$

Here c_{11}, c_{12}, c_{44} are the force constants for the overlayer. The general strain tensor for arbitrary orientation is shown in FIGURE 2. In general, strained epitaxy causes a distortion of the cubic lattice and, depending upon the growth orientation, the distortions produce a new reduced crystal symmetry. It is important to note that for (001) growth, the strain tensor is diagonal while for (111), and several other directions, the strain tensor has non-diagonal terms. The non-diagonal terms can be exploited to produce built-in electric fields in certain heterostructures.

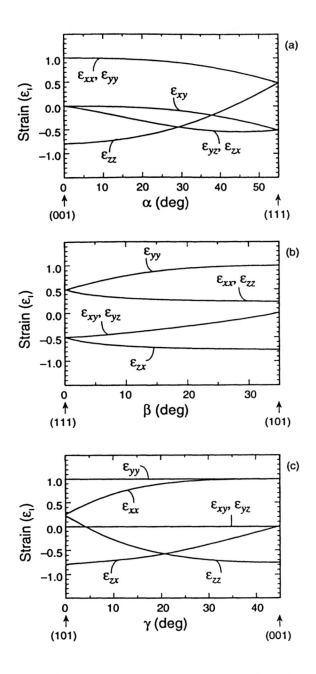

FIGURE 2. The general strain tensor produced when an overlayer is grown on different substrate orientations. The strain between the two materials is $\varepsilon_{\parallel}$.

C   EFFECT OF STRAIN ON BAND-EDGE STATES

Once the strain tensor is known, the deformation potential theory can be used to calculate the effects of strain on various eigenstates in the Brillouin zone [4-7]. The strain perturbation Hamiltonian is defined and its effects are calculated in the simple first order perturbation theory. We will summarise the results of the formalism. The deformation potential theory describes the effect of strain in terms of deformation potentials which are obtained for each band by pressure experiments.

Case 1: Let us first examine the nondegenerate Γ_2' state which represents the conduction band edge of direct bandgap semiconductors. This state is an s-type state and has the full cubic symmetry associated with it. There is only one deformation potential for this state and the effect of strain is simply to move the energy level. The shift in the energy is given by

$$\delta E^{(000)} = H_\varepsilon = D_{xx} (\varepsilon_{xx} + \varepsilon_{yy} + \varepsilon_{zz}) \tag{5}$$

Conventionally we write

$$D_{xx} = \Xi_d^{(000)} \tag{6}$$

where $\Xi_d^{(000)}$ represents the dilation deformation potential for the conduction band (000) valley.

Case 2: States along the [100] direction in k-space or Δ_1 symmetry states. These states are relevant for materials like Si which have the conduction band edge along this direction. In the unstrained system there is a degeneracy due to the symmetry between x, y, and z. This symmetry is broken and the shift in the various states is given by

$$\delta E^{(100)} = \Xi_d^{(100)} (\varepsilon_{xx} + \varepsilon_{yy} + \varepsilon_{zz}) + \Xi_u^{(100)} \varepsilon_{xx} \tag{7}$$

$$\delta E^{(010)} = \Xi_d^{(100)} (\varepsilon_{xx} + \varepsilon_{yy} + \varepsilon_{zz}) + \Xi_u^{(100)} \varepsilon_{yy} \tag{8}$$

$$\delta E^{(001)} = \Xi_d^{(100)} (\varepsilon_{xx} + \varepsilon_{yy} + \varepsilon_{zz}) + \Xi_u^{(100)} \varepsilon_{zz} \tag{9}$$

We note that if the strain tensor is such that the diagonal elements are unequal (as is the case in strained epitaxy), the strain will split the degeneracy of the Δ_1 branches.

Case 3: The L_1 symmetry or states along the [111] direction in k-space. This is the case for the conduction band edges for materials such as Ge and AlAs. In this case the effect of strain is given by

$$\delta E^{(111)} = D_{xx} (\varepsilon_{xx} + \varepsilon_{yy} + \varepsilon_{zz}) + 2D_{xy} (\varepsilon_{xy} + \varepsilon_{yz} + \varepsilon_{zx})$$

$$\delta E^{(11\bar{1})} = D_{xx} (\varepsilon_{xx} + \varepsilon_{yy} + \varepsilon_{zz}) + 2D_{xy} (\varepsilon_{xy} - \varepsilon_{yz} - \varepsilon_{zx})$$

$$\delta E^{(1\bar{1}1)} = D_{xx} (\varepsilon_{xx} + \varepsilon_{yy} + \varepsilon_{zz}) + 2D_{xy} (- \varepsilon_{xy} - \varepsilon_{yz} + \varepsilon_{zx}) \tag{10}$$

$$\delta E^{(\bar{1}11)} = D_{xx} (\varepsilon_{xx} + \varepsilon_{yy} + \varepsilon_{zz}) + 2D_{xy} (- \varepsilon_{xy} + \varepsilon_{yz} - \varepsilon_{zx})$$

where

$$D_{xx} = \Xi_d^{(111)} + \frac{1}{3} \Xi_u^{(11\bar{1})} \tag{11}$$

$$D_{xy} = \frac{1}{3} \Xi_u^{(111)} \tag{12}$$

<u>Case 4</u>: The triple degenerate states describing the valence band edge.

The valence band states are defined (near the band edge) by primarily p_x, p_y, p_z (denoted by x, y, z) basis states. In the angular momentum basis the top of the valence band has the $|3/2, \pm3/2\rangle$ (the heavy-hole state) and $|3/2, \pm1/2\rangle$ (the light-hole state) degeneracy. The general Hamiltonian describing the strained material is given by (the state ordering is $|3/2, 3/2\rangle$, $|3/2, -1/2\rangle$, $|3/2, 1/2\rangle$, and $|3/2, -3/2\rangle$)

$$H_\varepsilon = \begin{bmatrix} H_{hh}^\varepsilon & H_{12}^\varepsilon & H_{13}^\varepsilon & 0 \\ H_{12}^{\varepsilon*} & H_{lh}^\varepsilon & 0 & H_{13}^\varepsilon \\ H_{13}^{\varepsilon*} & 0 & H_{lh}^\varepsilon & -H_{12}^\varepsilon \\ 0 & H_{13}^{\varepsilon*} & -H_{12}^{\varepsilon*} & H_{hh}^\varepsilon \end{bmatrix} \tag{13}$$

where the matrix elements are given by

$$H_{hh}^\varepsilon = a(\varepsilon_{xx} + \varepsilon_{yy} + \varepsilon_{zz}) - b\left[\varepsilon_{zz} - \frac{1}{2}(\varepsilon_{xx} + \varepsilon_{yy})\right]$$

$$H_{lh}^\varepsilon = a(\varepsilon_{xx} + \varepsilon_{yy} + \varepsilon_{zz}) + b\left[\varepsilon_{zz} - \frac{1}{2}(\varepsilon_{xx} + \varepsilon_{yy})\right]$$

$$\tag{14}$$

$$H_{12}^\varepsilon = -d(\varepsilon_{xz} - i\varepsilon_{yz})$$

$$H_{13}^\varepsilon = \frac{\sqrt{3}}{2} b(\varepsilon_{yy} - \varepsilon_{xx}) + id\varepsilon_{xy}$$

Here, the quantities a, b, and d are valence band deformation potentials. As discussed earlier, strains achieved by lattice-mismatched epitaxial growth along the [001] direction can be characterised by $\varepsilon_{xx} = \varepsilon_{yy} = \varepsilon$, and $\varepsilon_{zz} = -(2c_{12}/c_{11})\varepsilon$. All of the diagonal strain terms are zero. Using this information, we get

$$H_{hh}^\varepsilon = \left[2a\left(\frac{c_{11} - c_{12}}{c_{11}}\right) + b\left(\frac{c_{11} + 2c_{12}}{c_{11}}\right)\right]\varepsilon \tag{15}$$

$$H_{lh}^\varepsilon = \left[2a\left(\frac{c_{11} - c_{12}}{c_{11}}\right) - b\left(\frac{c_{11} + 2c_{12}}{c_{11}}\right)\right]\varepsilon \tag{16}$$

with all other terms zero. For GaAs, InAs type systems, the values of a and b are such that the net effect is that the $(3/2, \pm3/2)$ state moves a total of 5.96ε eV and the $(3/2, \pm1/2)$ state moves by energy equal to 12.0ε eV [8].

The relationships between the deformation potentials used in the discussion above and what is measured from a hydrostatic pressure measurement are described as follows:

$$\Xi_d^{(000)} = \Xi_{hyd}^{(000)} + a$$

$$\Xi_d^{(100)} = \Xi_{hyd}^{(100)} - \frac{1}{3}\Xi_u^{(100)} + a \tag{17}$$

$$\Xi_d^{(111)} = \Xi_{hyd}^{(111)} - \frac{1}{3}\, \Xi_u^{(111)} + a$$

From hydrostatic compression

$$\delta E_g(\bar{k}) = \Xi_{hyd}^{(\bar{k})}\, (\varepsilon_{xx} + \varepsilon_{yy} + \varepsilon_{zz})$$

$$\Xi_{hyd}^{(\bar{k})} = -\frac{1}{3}\, (c_{11} + 2c_{12})\, \frac{dE_g}{dP}$$

(18)

The elastic stiffness constants, c_{11}, c_{12}, and the hydrostatic pressure coefficient of the bandgap, dE_g/dP, are fairly accurately measured by various experiments and we thus obtain the values of the individual deformation potentials for different bands.

D   SOME RESULTS ON STRAIN EFFECTS

The effect of strain on band structure for both conduction band and valence band states is illustrated by examining the direct bandgap material $In_xGa_{1-x}As$ grown on GaAs and the indirect bandgap material Ge_xSi_{1-x} alloy grown on Si.

For the conduction bands of direct bandgap materials, the strain tensor only moves the position of the band edge and has a rather small effect on the carrier mass. In FIGURE 3 we show the effect of alloying and strain on the electron effective masses for InGaAs grown on GaAs and InP. Note that for InP the composition $In_{0.53}Ga_{0.47}As$ is lattice-matched. We see from this figure that the decrease in carrier mass is primarily due to the addition of indium. The effect of strain is not very significant.

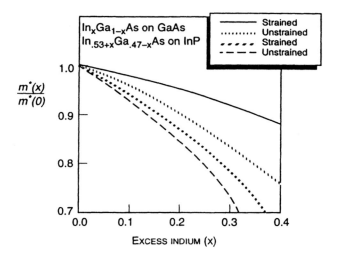

FIGURE 3.   Results on the effect of excess In on carrier masses in InGaAs grown on GaAs and InP substrates. The electron masses are shown for both unstrained and strained alloys. (After [9].)

In the case of the indirect bandgap $Si_{1-x}Ge_x$ alloy grown on Si, the conduction band is significantly affected according to EQNS (7-9). For (001) growth there is splitting in the 6 equivalent valleys. The results on the band edge states are shown in FIGURE 4. We see that a significant splitting occurs in both the conduction and valence band states as a result of strain. Since the splitting can approach ~100 meV, the results can be exploited for a number of electronic and optoelectronic devices.

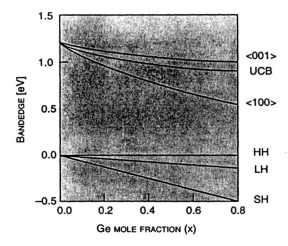

FIGURE 4. Splittings of the conduction band and valence band are shown as a function of alloy composition. UCB: Unstrained conduction band, HH: heavy hole, LH: light hole, SH: split-off hole.

Note that the biaxial compressive strain causes a lowering of the four-fold in-plane valleys below the 2 two-fold out-of-plane valleys [7]. The carrier masses are not expected to change significantly.

The effect of strain on the valence band structure is much more dramatic. In FIGURE 5 we show how the valence band structure of an InGaAs/AlGaAs quantum well is modified as it is subjected to compressive strain. We see that the valence band mass near the band edge is considerably reduced as a result of strain. This effect is widely used in semiconductor lasers to produce low threshold devices.

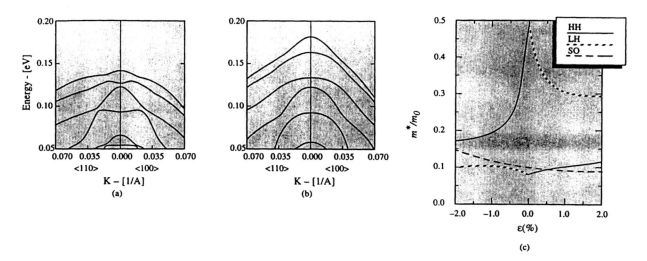

FIGURE 5. Hole band dispersions for a 100 Å quantum well made from (a) GaAs/Al$_{0.3}$Ga$_{0.7}$As well and (b) In$_{0.1}$Ga$_{0.9}$As/Al$_{0.3}$Ga$_{0.7}$As. (c) Change in the density of states mass as a function of strain at the band edge.

The effect of strain on the valence band states is also shown for the SiGe system in FIGURE 6. Here we show the density of states at the valence band edge for Si and Ge and the strained and unstrained alloy Si$_{0.75}$Ge$_{0.25}$.

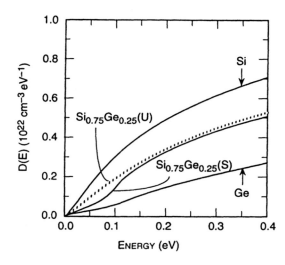

FIGURE 6. Effect of alloying and strain on valence band density of states. Note the decrease of the density of states in the strained structure ~100 meV from the band edge.

Notice that the density of states decreases as the Ge content increases since Ge has very low hole masses. For the strained alloy there is a large reduction in the density of states, especially up to 100 meV near the valence band edge.

E   CONCLUSION

It is quite evident from the discussions in this Datareview that strain can significantly affect the band structure of semiconductors. The level of strain we have considered can be incorporated in the semiconductor reasonably easily through strained epitaxy. Degeneracy splittings in the valence band can be of the order of 100 meV, accompanied by large changes in band curvatures. It is important to appreciate that such large uniaxial strains are extremely difficult to obtain by external apparatus. It is important to identify whether or not the changes produced by the strain have any impact on the physical properties of the semiconductors. The effects are, indeed, found to be significant.

REFERENCES

[1]   C.A.B. Ball, J.W. van der Merwe [*Dislocations in Solids* Ed. F.R.N. Nabarro (North-Holland, New York, 1983) vol.5]

[2]   G.C. Osbourn [*J. Appl. Phys. (USA)* vol.53 (1982) p.1586]

[3]   J.M. Hinckley, J. Singh [*Phys. Rev. B (USA)* vol.42 (1990) p.3546]

[4]   G.L. Picus, G.E. Bir [*Symmetry and Strain Induced Effects in Semiconductors* (John Wiley and Sons, New York, 1974)]

[5]   E.P. O'Reilly [*Semicond. Sci. Technol. (UK)* vol.4 (1989) p.121]

[6]   J. Singh [*Physics of Semiconductors and Their Heterostructures* (McGraw-Hill, New York, 1993)]

[7]   R. People [*IEEE J. Quantum Electron. (USA)* vol.22 (1986) p.1696]

[8]   H. Kato, N. Iguchi, S. Chika, M. Nakayama, N. Sano [*Jpn. J. Appl. Phys. (Japan)* vol.25 (1986) p.1327]

[9]   M.D. Jaffe [*Studies of Electronic Properties and Applications of Coherently Strained Semiconductors* PhD Thesis, University of Michigan, Ann Arbor, 1989]

CHAPTER 3

EPITAXIAL GROWTH AND DOPING

3.1 Molecular beam epitaxial growth and impurity energy levels of GaAs-based quantum wells and superlattices

A. Salvador and H. Morkoc

June 1995

A INTRODUCTION

Molecular beam epitaxial (MBE) growth of quantum wells (QWs) and superlattices (SLs) based on the GaAs/AlAs, InGaAs/GaAs and GaAs/InGaP material systems are discussed. The critical issues that are addressed are layer composition control, well width uniformity and the attainment of smooth heterojunction interfaces. Each of the material systems mentioned above is suited for different applications and in an effort to obtain improved layer quality, specialised MBE growth techniques have been developed. Doping of QWs and SLs and the corresponding impurity energy levels are also discussed. Unlike in bulk material systems, the binding energy of impurities in quantum wells is dependent on the spatial location of the dopant in relation to the barriers as well as the well width.

B MBE GROWTH OF GaAs BASED QWs AND SLs

B1 Classification of MBE Growth

MBE growth of III-V materials can be classified according to the type of sources used for the group III and group V elements [1,2]. In conventional MBE the elemental sources are used for group III cations. Arsenic is supplied as the tetramer (As_4) by sublimation from the solid or as the dimer (As_2), which can be obtained either by dissociation of As_4 using a two-zone cracker or by evaporating from GaAs. The group III elements are derived from evaporation of the respective liquid element. Over a wide range of substrate temperatures from 500 - 600°C Ga, Al and In sticking coefficients are near unity. Provided that an overpressure of As exists, the growth rate and the ternary alloy composition are then determined by the relative flux rate arrival of the group III elements. At higher substrate temperatures, Ga (>640°C) and In (>550°C) desorption occurs and the desorption rate has to be taken into account in the control of the ternary alloy composition.

In search of more stable sources of arsenic and for reproducible control of the composition of III-V-V ternary and quaternary compounds based on phosphorus and arsenic, variations of conventional MBE growth, classified under the generic name of gas source MBE, have been developed. While most of the work in GaAs/$Al_xGa_{(1-x)}$As heterostructures used conventional MBE growth, gas source MBE techniques have, in recent years, found increasing use particularly in the growth of high quality InP, InGaAs, GaAs [3] and InGaP bulk films and quantum well structures. By convention these types of MBE can be classified further as

(i)   hydride gas source MBE where only the elemental group V sources are replaced by the hydrides AsH_3 and PH_3 [5],

(ii) metalorganic MBE (MOMBE) where the metalorganics such as triethylgallium (TEG, $Ga(C_2H_5)_3$), triethyl aluminium (TEA, $Al(C_2H_5)_3$) and triethylindium (TEI, $In(CH_3)_3$) replace the elemental sources for the group III materials [6,7], and

(iii) chemical beam epitaxy (CBE), which is a combination of (i) and (ii) [8].

In hydride gas source MBE, thermal cracking of the hydrides produces the dimers As_2 and P_2, and the growth mechanism is similar to conventional MBE. In both MOMBE and CBE the substrate serves as the site both for pyrolysis of the metalorganics and for the epitaxial growth. The growth rate is thus highly dependent on the substrate temperature and the growth kinetics are far more complex than those of conventional MBE. An important consideration in the use of metalorganics in the MBE growth of Ga(Al)As is the high carbon incorporation in the resulting films. To circumvent this problem, alternative sources such as trimethylamine alane $((CH_3)_3NAlH_3)$ have been explored [9].

B2 Growth Conditions Used for Quantum Wells and Superlattices

B2.1 Substrate temperatures

In MBE growth of multilayers, particular attention should be given to the possibilities and limitations brought about by having different optimum growth conditions for the barrier and the well regions. To obtain improved surface morphology and reduce unwanted carbon and oxygen incorporation in $Al_xGa_{(1-x)}As$ films, with $x > 0.3$, growth temperatures of ~700°C are used [10,11]. However, at this temperature Ga desorption starts to occur and proper consideration should be given to the relative fluxes to ensure uniformity in the composition of the $Al_xGa_{(1-x)}As$ barriers as well as the GaAs layer thickness. While ramping of the substrate temperature as the well or barrier region grows is an alternative, it becomes cumbersome when the well and barrier widths are a few monolayers thick and the quantum well structure is to be repeated several times. Nevertheless, GaAs/$Al_xGa_{(1-x)}As$ MQWs with extremely sharp photoluminescence (PL) half widths grown at substrate temperatures of 700°C have been reported [12]. For lower Al mole fractions, $x < 0.3$, GaAs/$Al_xGa_{(1-x)}As$ as well as GaAs/AlAs SLs can be grown at lower temperatures of 550 - 630°C [13]. The low temperature PL measurements of the resulting GaAs/Al(Ga)As QW reveal that the full widths at half maximum (FWHM) of the exciton peaks are reasonable and indicate that the layer quality, as far as optical properties are concerned, is not compromised as a result of the use of lower growth temperatures.

The problem is more severe in the growth of strained $In_yGa_{(1-y)}As$/$Al_xGa_{(1-x)}As$ MQWs since In desorbs from the surface at a much lower temperature of 550°C, and thus both well and barrier composition uniformity can be compromised if higher growth temperatures are used. On the other hand, growth at lower temperatures in the range 500 - 550°C will result in increased interface roughness and intralayer well width non-uniformity. In $In_yGa_{(1-y)}As$ based layers used for 0.98 μm emission, only a few $In_yGa_{(1-y)}As$ quantum wells are needed and the disparity in the optimum substrate temperature between InGaAs and $Al_xGa_{(1-x)}As$ is remedied by use of an intermediate GaAs layer. The $In_yGa_{(1-y)}As$/GaAs quantum well can then be grown at 500 - 560°C and the $Al_xGa_{(1-x)}As$ layer, which now also serves as the cladding region, is grown at 700°C. Using this growth procedure low threshold current densities for $In_yGa_{(1-y)}As$/$Al_xGa_{(1-x)}As$ QW lasers have been reported [14].

An alternative to $Al_xGa_{(1-x)}As$ as a barrier material in $Al_xGa_{(1-x)}As$/GaAs and $In_yGa_{(1-y)}As$/$Al_xGa_{(1-x)}As$ MQWs is InGaP lattice-matched to GaAs. InGaP provides nearly the same confinement potential for a GaAs quantum well as does AlGaAs but can be grown at much lower temperatures [15]. Furthermore, unlike $Al_xGa_{(1-x)}As$, InGaP does not suffer from DX centres.

B2.2   Control of well width uniformity

In SLs and MQW structures it is important to keep well and barrier widths constant. Any deviation can result in the disruption of the intended superlattice periodicity, resulting in the loss or shortening of coherence length of the electrons or holes and in a decrease of the SL conduction band width. This has important implications in III-V SLs used for optical modulation where the degree of blue shift is dependent on the SL band width [16,17]. In GaAs/AlAs SL structures, one can use the oscillations of the RHEED intensity to monitor the layer-by-layer growth and thereby ensure control and reproducibility of the well/barrier width. It has been observed that the RHEED intensity oscillates with the periodicity corresponding to the time taken to deposit a lattice plane of GaAs (or AlAs) on the (001) surface, provided that the substrate temperature is below that where Ga (or Al) desorption from the surface begins [18]. At full coverage the RHEED intensity is at a maximum and, as 2D growth proceeds, islands, half a monolayer in height, begin to cover and roughen the surface thereby reducing the RHEED intensity. Beyond 50% coverage the RHEED intensity begins to increase as the islands begin to coalesce and finally reaches a maximum when the surface is fully covered. The RHEED intensity though is also damped indicating that a second island growth proceeds before full coverage is achieved.

B2.3   Smoothing of heterojunction interfaces

Abrupt, smooth heterojunction interfaces are essential for improved performance in SL and QW structures used for resonant tunnelling devices as well as other device structures. The observed large discrepancy in performance between GaAs/AlGaAs MODFETs grown with the AlGaAs on top of GaAs and those with GaAs on top of AlGaAs is partly due to the difficulty in obtaining a smooth AlGaAs surface over the GaAs film [19]. The use of growth interrupts has been shown to be an effective means of achieving smooth interfaces. In this method the group III flux is temporarily turned off allowing the adatoms on the surface to migrate and coalesce into large islands or incorporate at ledge edges. Using this approach in GaAs film growth, recovery of the RHEED intensity signal was obtained [20]. In a series of experiments by Tanaka and co-workers [21,22], a dramatic decrease in the photoluminescence linewidth of GaAs/AlAs QWs was observed when growth interruption was used and this was attributed to the smoothing effect of growth interrupt on interfaces.

A drawback of the growth interruption technique is that it may allow impurities to reach the surface during the interrupt and also lead to increased carbon incorporation in the film. Furthermore, it is believed that even with the use of growth interruption the interface between GaAs and $Al_xGa_{(1-x)}As$ is at best pseudosmooth with the lateral extent of the monolayer fluctuation smaller than that of the exciton Bohr radius. As an alternative to the growth interruption method, a thin (2 - 3 monolayers) AlAs layer can be inserted between the $Al_xGa_{(1-x)}As$ and GaAs to obtain a smooth interface in GaAs/$Al_xGa_{(1-x)}As$ QWs [23].

More recently, migration enhanced epitaxy (MEE) [24] and atomic layer epitaxy (ALE) [25] have been employed to obtain ultrathin-layer superlattices with very abrupt interfaces. In both methods the growing surface alternates between a group III completed and a group V saturated surface. In MEE both the group III and group V molecular beams are alternating with a periodicity corresponding to the atomic layer-by-layer growth. This, however, puts a high mechanical load on the shutters and has the further disadvantage of low growth rates. ALE on the other hand is the reverse of the growth interruption method, with the group III molecular beam continuously impinging on the surface and the group V molecular beam modulated with a periodicity coinciding with the monolayer growth rate. The other requirement in ALE is that the pulses of group V species supplied to the substrate surface should be short but intense enough to saturate the surface at each cycle. There is less mechanical load on the shutters in ALE and growth rates are comparable to those in conventional MBE.

C   DOPING OF QWs AND SLs

Depending on the desired application, the barrier or well region can be doped accordingly. Furthermore, the doping profile can be controlled to allow for either doping of a specified region within the well/barrier, so-called δ doping [26], or uniform doping of the entire well/barrier region. In most device applications it is the barriers that are doped while in certain laser applications it is desirable that the well region is also doped.

The most commonly used dopants for GaAs are:

(a)   Si for n-type doping, with a donor energy level of 5 meV,

(b)   Be for p-type doping, with an acceptor energy level of 19 meV.

The attractive features of Si as an n-type dopant in GaAs are its near unity incorporation coefficient, low diffusivity, and that it retains non-amphoteric behaviour on the (100) surface over a wide range of doping concentration. The controllable doping range for Si in GaAs has been reported as 10^{14} - 6×10^{18} cm^{-3} [27]. Beryllium has near unity incorporation and a controllable doping range from 1×10^{14} - 6×10^{19} cm^{-3}. However, at doping levels higher than 1×10^{19} cm^{-3} Be begins to diffuse interstitially [27].

Si is also the preferred dopant for $Al_xGa_{1-x}As$ although for increasing Al concentration the use of Si as an effective dopant diminishes due to a corresponding increase in the donor activation energy with the transition from a shallow to deep donor occurring at x = 0.235 [28]. In FIGURE 1 donor binding energy for Si impurity in $Al_xGa_{1-x}As$ is plotted as a function of Al mole fraction.

Recently, C has gained popularity as an alternative p-type dopant for GaAs and AlGaAs [29]. It has a lower diffusion coefficient than Be or Zn and high doping levels of 1×10^{19} cm^{-3} can be attained without degradation of the surface morphology of layers.

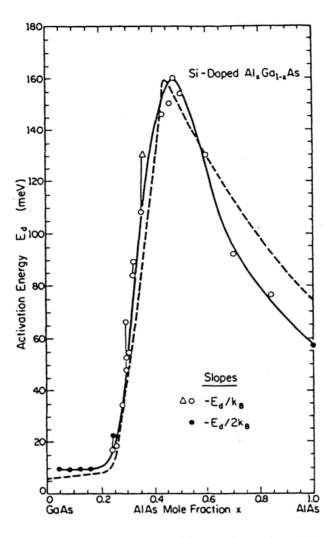

FIGURE 1. Plot of the activation energy of Si in AlGaAs (from [27]) showing the transition from a shallow to deep donor as the Al mole fraction is increased.

D   IMPURITY ENERGY LEVELS IN QWs AND SLs

The binding energy of the donor or acceptor state of an impurity in a QW or SL is dependent on the location of the impurity. This arises from the lack of translational invariance along the growth axis of the SL or QW structure. The two main parameters that affect the donor binding energy, defined as the energy difference between the ground state of the donor and the energy of the lowest conduction subband, for impurities located in the well region are:

(a)   the relative location of the impurity and the barrier region, and
(b)   the size of the well in relation to the Bohr radius of the impurity.

In calculating the binding energies of shallow donor impurities located at the centre of the well, Bastard [30] assumed an infinite barrier height potential and found a monotonic increase in the donor binding energy, as the well width is decreased. Mailhot and Chang [31] used instead a finite barrier height and found that the binding energy reaches a maximum at some non-zero well width. This behaviour can be explained as follows. In wide wells the binding energy of centre doped donor impurities should approach that corresponding to bulk GaAs. As the well width decreases, the energy of the first conduction subband increases accordingly due to the increased spatial confinement of the electron. The ground state energy of the donor also increases but not as fast as the conduction subband since the bound electron wavefunction is already confined by the Coulomb potential and is not affected by the barriers as much. Hence the binding energy increases. However, as the well width becomes increasingly narrower, the electron wavefunction of the donor ground state also begins to have appreciable penetration into the barriers and the binding energy reaches a maximum at some well width. Further reduction of the well width causes the binding energy to decrease, reaching the corresponding value for bulk AlGaAs at zero well width. A similar trend in the binding energy of C acceptor levels in GaAs/AlGaAs QWs was seen by Masselink et al [32] and the results of these calculations are shown in FIGURE 2.

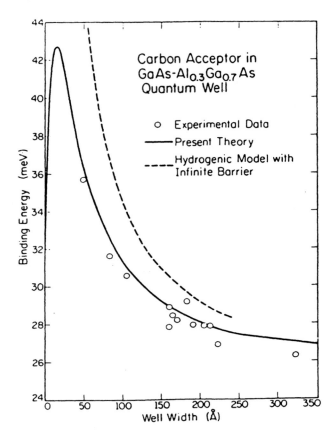

FIGURE 2. Calculated C acceptor binding energy of centre doped GaAs QW as a function of well width (from [32]). For comparison, the calculated values obtained using a hydrogenic model with infinite barriers are also plotted.

E   CONCLUSION

The advancements and developments in the molecular beam epitaxy of III-V semiconductors have extended the range of GaAs based material systems from GaAs/AlGaAs heterostructures to include GaAs/InGaAs and GaAs/InGaP as well as provided alternative sources for dopants. The present state in MBE growth allows one to grow high quality GaAs based QWs and SLs with sharp heterojunction interfaces and the desired periodicity. This is made possible through the use of appropriate substrate temperatures, RHEED oscillation patterns and growth interrupt related techniques. An important feature in doping of GaAs QWs is that unlike in bulk GaAs, the energy levels of impurities are position dependent.

ACKNOWLEDGEMENT

This work was supported by the Air Force Office of Scientific Research. The authors would like to thank Dr G.L. Witt for his support and encouragement.

REFERENCES

[1]   K. Ploog [in *Low-Dimensional Structures, From Basic Physics to Applications* Eds A.R. Peaker, H.G. Grimmeiss (Plenum Press, New York, 1990) p.47-670]

[2]   H. Sakaki [*Materials Processing Theory and Practice vol.7* (Elsevier Press, 1989) p.217-330]

[3]   J.E. Cunningham [*Mater. Res. Soc. Symp. Proc. (USA)* vol.281 (1992) p.3-17]

[4]   H.Y. Lee, M.J. Hafich, G.Y. Robinson [*J. Cryst. Growth (Netherlands)* vol.105 (1990) p.244]

[5]   M.B. Panish [*J. Electrochem. Soc. (USA)* vol.127 (1980) p.2729]

[6]   E. Veuhoff, W. Pletschen, P. Balk, H. Luth [*J. Cryst. Growth (Netherlands)* vol.55 (1981) p.30-3]

[7]   W.T. Tsang [*J. Cryst. Growth (Netherlands)* vol.95 (1989) p.121]

[8]   W.T. Tsang [in *VLSI Electronics: Microstructure Science* Ed. E.G. Einspruch (Academic Press, New York, 1989) vol.21 p.255]

[9]   C. Abernathy, A.S. Jordan, S.J. Pearton, F. Ren, F. Baiocchi [*J. Cryst. Growth (Netherlands)* vol.109 (1991) p.31]

[10]   N. Chang [*Thin Solid Films (Switzerland)* vol.231 (1993) p.143-57]

[11]   J.D. Grange [in *The Technology and Physics of Molecular Beam Epitaxy* Ed. E.H.C. Parker (Plenum Press, New York, 1985) p.55]

[12]   Y.L. Sun, W.T. Masselink, R. Fischer, M.V. Klein, H. Morkoc, K.K. Bajaj [*J. Appl. Phys. (USA)* vol.55 (1984) p.3554]

[13]   G. Danan et al [*Phys. Rev. B (USA)* vol.12 (1987) p.6207]

[14]   N. Chand, E.E. Becker, J.P. van der Ziel, S.N.G. Chu, N.K. Dutta [*Appl. Phys. Lett. (USA)* vol.58 (1991) p.1704]

[15]   F.G. Johnson, G.W. Wicks, R.E. Viturro, R. Laforce [*Mater. Res. Soc. Symp. Proc. (USA)* vol.281 (1992) p.49-54]

[16]   E.E. Mendez, F. Agullo-Rueda, J.M. Hong [*Phys. Rev. Lett. (USA)* vol.60 (1988) p.2426]

[17]   J. Bleuse, G. Bastard, P. Voisin [*Phys. Rev. Lett. (USA)* vol.60 (1988) p.220]

[18]   J.H. Neave, B.A. Joyce, P.J. Dobson, N. Norton [*Appl. Phys. A (Germany)* vol.31 (1983) p.1]

[19]   H. Morkoc, T.J. Drummond, R. Fischer [*J. Appl. Phys. (USA)* vol.53 (1982) p.1030-3]

[20]   M. Tanaka, H. Sakaki, J. Yoshino [*Jpn. J. Appl. Phys. (Japan)* vol.24 (1985) p.L417]

[21]   M. Tanaka, H. Sakaki [*J. Cryst. Growth (Netherlands)* vol.81 (1987) p.153]

[22]   M. Tanaka, H. Sakaki, J. Yoshino [*Jpn. J. Appl. Phys. (Japan)* vol.25 (1986) p.L155-8]

[23]   K. Ploog [in *Resonant Tunnelling in Semiconductors: Physics and Applications* Eds L.L. Chang, E.E. Mendez, C. Tejedor (Plenum Press, New York, 1991) p.26]

[24]   Y. Horikoshi, M. Kawashima [*J. Cryst. Growth (Netherlands)* vol.95 (1989) p.17]

[25]   F. Briones, L. Gonzales, A. Ruiz [*Appl. Phys. A (Germany)* vol.49 (1989) p.729]

[26]   C.E.C. Wood, G. Metze, J. Berry, L.F. Eastman [*J. Appl. Phys. (USA)* vol.51 (1980) p.383-7]

[27]   C.E. Wood [in *The Technology and Physics of Molecular Beam Epitaxy* Ed. E.H.C. Parker (Plenum Press, New York, 1985) p.66]

[28]   N. Chand et al [*Phys. Rev. B (USA)* vol.30 (1984) p.4481]

[29]   F. Ren, C.R. Abernathy, S.J. Pearton [*J. Appl. Phys. (USA)* vol.70 (1991) p.2885]

[30]   G. Bastard [*Phys. Rev. B (USA)* vol.24 (1981) p.4714-22]

[31]   C. Mailhot, Y.C. Chang, T.C. McGill [*Phys. Rev. B (USA)* vol.26 (1982) p.4449-57]

[32]   W.T. Masselink, Y.C. Chiang, H. Morkoc [*Phys. Rev. B (USA)* vol.28 (1983) p.7373]

3.2 Molecular beam epitaxy of InP-based quantum wells and superlattices

A.S. Brown

May 1995

A   INTRODUCTION

This Datareview summarises the properties of InP-based quantum wells and superlattices grown by molecular beam epitaxy (MBE). The degree to which process-controlled extrinsic effects, such as interface roughness and alloy homogeneity, determine the properties of heterostructures is emphasised. These effects, plus a brief review of the historical development of the growth of these structures, are first outlined (Section B). We then discuss the properties of lattice-matched AlInAs/GaInAs structures (Section C), strained AlInAs/GaInAs structures (Section D), and epitaxial InP-containing structures (Section E).

B   SUMMARY OF EXTRINSIC EFFECTS

Until the early 1980s, InP-based materials were grown primarily by liquid phase epitaxy (LPE). The use of InP substrates during growth by MBE was hindered initially due to the fact that the InP surface oxide desorption temperature is greater than the substrate congruent sublimation temperature. The development of the As-stabilisation technique for the removal of the InP oxide prior to growth was a key step in enabling the growth of arsenic-containing materials, i.e. GaInAs and AlInAs, by MBE [1]. The growth of the arsenides preceded that of the phosphides. Epitaxial solid source phosphide growth is more difficult than solid source arsenide growth due to the difficulty in controlling the evaporation from, and formation of, co-existing phases of phosphorus which exhibit vastly different vapour pressures. Consequently, gas sources in MBE [2] are extensively used for phosphorus growth, with phosphine being the most commonly used source gas. The first MBE growth with solid phosphorus occurred in the late 1970s [3]. Recent improvements in cracking source designs, which offer much more control of the source, have enabled a resurgence in the growth of phosphides by solid source MBE [4-7].

The properties of quantum wells and superlattices are determined by the intrinsic properties of the materials in which structures are implemented, as well as the extrinsic properties of the materials. These extrinsic effects can be changed, to a greater or lesser extent, by the process conditions. This Datareview focuses on the latter effects. Extrinsic properties of quantum wells and superlattices include the interface smoothness of the heterojunctions, and the alloy homogeneity of the barrier and well materials. In addition, the residual background carrier concentration can be great enough to introduce band-filling effects, which degrade the optical and electrical properties of heterostructures. The full width at half maximum (FWHM) of the photoluminescence peak from a quantum well or superlattice structure is a useful figure of merit for assessing material quality. The extent to which this property is perturbed by these individual effects is succinctly covered in an analysis by Singh [8], in which the linewidth broadening resulting from each of these individual effects is quantified. The interface roughness is modelled by island formation with various monolayer integral heights. An underlying assumption of this model is that the exciton radius (~10.0 nm) is similar to the size of the islands comprising the heterointerface. An additional criterion is added to the above requirements for the growth of high quality superlattices. Uniformity of the structure throughout the entire growth process is required. This places stringent constraints on the growth conditions in order to minimise any fluctuation in the well width and composition of the structure.

C   GaInAs/AlInAs LATTICE-MATCHED TO InP

The first GaInAs/AlInAs heterostructures were grown in 1981 by MBE [9]. Since the initial demonstration of this heterostructure, significant progress has been made in the refinement of the growth technique. A number of general observations related to the perfection of these materials can be made from the published reports on the production of these structures.

C1   GaInAs and AlInAs Alloy Quality

The alloy homogeneity of GaInAs is far superior to that of AlInAs due to the relatively small energy difference between the Ga-As and In-As bond strengths. In contrast to this, there is a large energy difference between the Al-As and In-As bond strengths. Previous work has indicated the existence of a miscibility gap in the AlInAs phase diagram, and a consequent degradation of electrical and optical properties of AlInAs, when the alloy is grown under conditions which do not provide kinetic limitations to clustering [10]. This phenomenon impacts on the expected properties of heterostructures containing AlInAs barriers. In particular, narrow quantum well structures, with significant penetration of the exciton into the barrier material, will have poor optical properties, dominated by the relatively poor quality AlInAs. The quality of the individual alloys can be assessed by the FWHM of the low temperature photoluminescence. Best values for FWHM for GaInAs and AlInAs vary from 1.5 - 3.0 meV and 8.0 - 10.0 meV, respectively [11]. Growth conditions affect the properties of the alloys. The maximum substrate temperature of the alloys is limited by the desorption of In which begins at a temperature of approximately 530°C for GaInAs and 560°C for AlInAs. Discrepancies exist in reports on the effects of V/III pressure ratio on the properties of the alloys, particularly for AlInAs. It appears that the properties of GaInAs are improved by the use of high V/III ratio [12,13]. For AlInAs, improving growth using higher As overpressure [14], as well as reduced As overpressure, has been reported [15,16]. Growth at high substrate temperature has been found [17] to improve material quality. In addition, AlInAs exhibits a forbidden growth temperature regime, around 400°C [18-20]. Enhanced clustering and In surface segregation are observed for films grown at this temperature. A report on AlInAs spontaneous compositional disordering induced by surface steps on vicinal substrates has also been made [21].

C2   Heterointerface Formation

For MBE growth of common anion heterointerfaces, the perfection of the interface is controlled predominantly by the cation migration kinetics. The disparity in the bond strengths of the binary endpoints for AlInAs results in a large difference in the migration lengths of the In and Al cations on the surface of the growing AlInAs film [10]. A higher substrate temperature can be used to increase the migration lengths of the Al cations, thereby improving the smoothness of the growth front. This solution, however, promotes the formation of clusters due to the lifting of the kinetic inhibitions of alloy clustering. Thus, the optimal growth conditions for AlInAs alloy quality and AlInAs/GaInAs interface quality appear mutually exclusive. This effect was experimentally demonstrated [22] by showing that an inverse relationship exists between the FWHM of photoluminescence from AlInAs buffers and FWHM of quantum wells grown directly on the buffer. Improvements in the quality of the barrier can be obtained by adding Ga to the AlInAs [23]. Quantum wells grown with AlInGaAs barriers and wells are comparable to the best GaInAs/InP quantum wells (as discussed below) [23,24].

Early reports of the optical properties of AlInAs/GaInAs quantum wells showed that poor interface quality, well thickness variations and a high electron density in the buffer layer resulted in large FWHMs of 20 - 80 meV for low temperature photoluminescence [25,26]. Interruption of the growth at the heterojunction interface is a known technique for increasing the smoothness of the surface via cation rearrangement before the initiation of the growth of the dissimilar material [27]. Experimental reports on the effect of growth interruption on AlInAs/GaInAs interface quality are not consistent. Juang et al [28] observe a significant improvement in the optical properties of single quantum well structures as a function of the growth interrupt time from 0 to 3 minutes. However, Reithmaier et al [29] observe no improvement for interruption after the growth of AlInAs, and the degradation of optical properties for interruption after the GaInAs growth. This discrepancy may result from variations in defect incorporation during the interruption step.

High quality GaInAs MQW structures have been prepared by MBE. The depth uniformity in the composition and thickness of the alloys is important in minimising broadening due to fluctuations in the vertical structure. Gupta et al [11] have produced structures with 10 nm wells and 10 nm barriers with an observed HWHM (half width at half maximum) of 2.8 meV. FIGURE 1 summarises data for FWHM versus well width for various single- and multi-quantum well structures.

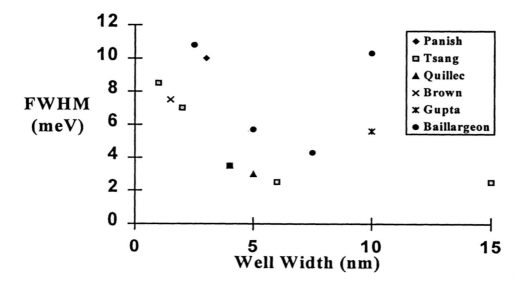

FIGURE 1. FWHM (meV) vs. well width (nm) for InP-based quantum wells: Panish GaInAs/InP [39]; Tsang GaInAs/InP [38]; Quillec GaInAs/AlGaInAs [23]; Brown GaInAs/AlInAs [22]; Gupta GaInAs/AlInAs MQW [11]; Baillargeon GaInAsP/GaInAs [40].

D   GaInAs/AlInAs STRAINED HETEROSTRUCTURES ON InP

The growth of strained structures is limited by defect generation arising from the strain energy accumulated during the growth of mismatched structures [30]. There are a number of reports on the growth of moderately strained GaInAs/AlInAs quantum wells by MBE. Asai and Kawamura [31] report limits of +0.9% strain for the InGaAs wells and -1.5% strain for the AlInAs ($In_{0.66}Ga_{0.34}As/Al_{0.7}In_{0.3}As$) for observing room temperature excitonic peaks in absorption spectra. A linewidth of 16 meV, as compared to 14 meV for lattice-matched material, is observed [31] for these strained structures. The specific growth conditions for strained structures affect the observed properties. For example, 5.0 nm-thick lattice-matched GaInAs quantum wells, with strained Al-rich, AlInAs barriers, exhibit improved properties when the barriers are grown at 400°C, as opposed to 500°C, due to the kinetic limitation of strain-induced defect nucleation with lowered substrate temperature [32]. This behaviour is in stark contrast to observations for lattice-matched materials due to the forbidden growth temperature regime for AlInAs.

In addition, highly strained InAs/AlInAs quantum wells on InP have been grown and characterised. Quantum wells with 1.0 - 3.0 nm thickness exhibit high intensity photoluminescence in the 1.2 - 1.6 µm range [33]. These structures are grown under conditions optimised for AlInAs. Photoluminescence linewidths of 50 meV were observed at 4 K [34]. Dramatic improvement in the observed photoluminescence FWHM from 48 to 27 meV was obtained by Tournie et al [35] for InAs/AlInAs quantum wells of 12 monolayers thickness grown under In stable conditions. They also report an FWHM of 8 meV for In-stable growth of GaInAs/InAs quantum wells, and 30 meV for As-stable growth. This technique is referred to as virtual surfactant epitaxy [36].

E   GaIn(As,P)/InP STRUCTURES

Mixed group V materials are relatively difficult to grow by MBE. For the group V elements, growth is typically performed in a regime where 100% of the cations are incorporated. This is in contrast to the incorporation kinetics of the anions in which only a small fraction of the molecules are incorporated. In addition, this process is extremely sensitive to growth conditions (such as the substrate temperature) [37], and the composition. Interface effects, such as strain due to bond length differences, and mixing, or exchange, of atoms driven by energy reduction during growth, can degrade the quality of these materials. Improvements, however, in the properties of GaInAs quantum wells are observed by replacing the relatively poor quality AlInAs (ternary) barriers with InP (binary barriers). Tsang and Schubert [38] have demonstrated by metalorganic MBE, GaInAs/InP quantum wells which still exhibit state-of-the-art linewidths for InP-based quantum wells (see FIGURE 1). High quality emission from GaInAs/InP (or GaInAsP barrier) quantum wells grown by gas source MBE [39] and solid source MBE [40] is also shown.

Reports on intrinsic interface strain, resulting from the different atomic bond lengths of As and P terminated materials, have been made for lattice matched GaInAs/InP heterointerfaces prepared by gas source MBE [41]. A high resolution X-ray diffraction study shows that by changing the beam, or switching sequences, at the heterointerface formation step, the atomic ordering and resulting interfacial strain can be modified. In addition, interdiffusion of As/P at the interface can degrade interface quality and the thermal stability of the materials after growth. Reports of improved interface quality via reduced As/P substitution have been made by assuring a group III stabilised surface during growth [42]. Improvements in the thermal stability of InGaAs/InP quantum wells have been made by using thicker buffer layers and a lower growth temperature [43]. An observed relaxation and lattice tilting occur for As/P exchange [44].

Linewidths (FWHM) of 8.9 meV were obtained for a 1.8 nm InGaAsP well (1.3 μm) with InGaAsP barriers (1.15 μm) [45]. Highly compressively strained MQW structures on InGaAs/InP have also been grown by MBE. The introduction of tensile strain in the barrier layers of the MQW enables dislocation-free material with excellent thermal stability to be produced. Without strain compensation, many misfit dislocations are created during thermal annealing [46].

Finally, reports on the properties of InAsP/InP have also been made [47]. These strained materials are useful for long-wavelength optical communications applications, and show promise for high frequency electronics [48].

F   CONCLUSION

Many extrinsic factors, including the quality of the individual alloys and the smoothness of the heterointerface, affect the properties of InP-based quantum wells and superlattices grown by molecular beam epitaxy. The perfection of these prepared materials has steadily improved, although the growth of high quality AlInAs remains difficult.

REFERENCES

[1]   G.J. Davies, R. Heckingbottom, H. Ohno, C.E.C. Wood, A.R. Calawa [*Appl. Phys. Lett. (USA)* vol.37 (1980) p.290]

[2]   M.B. Panish [*J. Electrochem. Soc. (USA)* vol.127 (1980) p.2729]

[3]   A.Y. Cho [*J. Vac. Sci. Technol. (USA)* vol.16 (1989) p.276]

[4]   G.W. Wicks et al [*Appl. Phys. Lett. (USA)* vol.59 (1991) p.342]

[5]   J.N. Baillargeon, A.Y. Cho, R.J. Fisher [*J. Vac. Sci. Technol. B (USA)* vol.13 (1995) p.64]

[6]   G.W. Wicks, M.W. Koch, F.G. Johnson, J.A. Varriano, G.E. Kohnke, P. Colombo [*J. Vac. Sci. Technol. B (USA)* vol.12 (1994) p.1119]

[7]   J.N. Baillargeon, A.Y. Cho, R.J. Fisher, P.J. Pearah, K.Y. Cheng [*J. Vac. Sci. Technol. B (USA)* vol.13 (1994) p.1106]

[8]   J. Singh, K.K. Bajaj [*J. Appl. Phys. (USA)* vol.57 (1985) p.5433]

[9]   H. Ohno, C.E.C. Wood, L. Rathbun, D.V. Morgan, G.W. Wicks, L.F. Eastman [*J. Appl. Phys. (USA)* vol.52 (1981) p.4033-7]

[10]   J. Singh, S. Dudley, B. Davies, K.K. Bajaj [*J. Appl. Phys. (USA)* vol.60 (1986) p.3167-73]

[11]   S. Gupta, P.K. Bhattacharya, J. Pamulapati, G. Mourou [*J. Appl. Phys. (USA)* vol.69 (1991) p.3219-25]

[12]   G.W. Wicks, C.E.C. Wood, H. Ohno, L.F. Eastman [*J. Electron. Mater. (USA)* vol.11 (1981) p.435]

[13]   M. Popp, M. Schiefele, M. Hurich, J.M. Schneider, H. Heinecke [*MBE VIII* Osaka, August 1994 published in *J. Cryst. Growth (Netherlands)* vol.150 (1995) no.1-4, pt.1]

[14]   A.S. Brown, U.K. Mishra, J.A. Henige, M.J. Delaney [*J. Vac. Sci. Technol. B (USA)* vol.6 (1988) p.678]

[15]   J.P. Praseuth, L. Goldstein, P. Henoc, H. Primot, G. Danan [*J. Appl. Phys. (USA)* vol.61 (1987) p.215-9]

[16]   M. Allovan, J. Primot, Y. Gao, M. Quillec [*J. Electron. Mater. (USA)* vol.18 (1989) p.505]

[17]   E. Tournie, Y.-H. Zhang, N.J. Pulsford, K. Ploog [*J. Appl. Phys. (USA)* vol.70 (1991) p.7362-9]

[18]   A.S. Brown, M.J. Delaney, J. Singh [*J. Vac. Sci. Technol. B (USA)* vol.7 (1989) p.384-6]

[19]   A. Hase, H. Kunzel, D.R.T. Zahn, W. Richter [*J. Appl. Phys. (USA)* vol.76 (1994) p.2495]

[20]   J.E. Oh, P.K. Bhattacharya, Y.C. Chen, O. Aina, M. Mattingly [*J. Electron. Mater. (USA)* vol.19 (1990) p.435]

[21]   Yu-Peng Hu, P.M. Petroff, Xueyu Qian, A.S. Brown [*Appl. Phys. Lett. (USA)* vol.53 (1988) p.2194-6]

[22]   A.S. Brown, J.A. Henige, M.J. Delaney [*Appl. Phys. Lett. (USA)* vol.52 (1988) p.1142-3]

[23]   M. Quillec [*Proc. SPIE (USA)* vol.1361 (1990) p.34-46]

[24]   T. Fujii, Y. Nakata, Y. Sugiyama, S. Hiyamizu [*Jpn. J. Appl. Phys. (Japan)* vol.25 (1986) p.L254-6]

[25]   D.F. Welch, G.W. Wicks, L.F. Eastman [*Appl. Phys. Lett. (USA)* vol.43 (1983) p.762]

[26]   D.F. Welch, G.W. Wicks, L.F. Eastman [*Appl. Phys. Lett. (USA)* vol.46 (1985) p.991]

[27]   H. Sakaki, M. Tanaka, J. Yoshino [*Jpn. J. Appl. Phys. (Japan)* vol.24 (1985) p.L417-20]

[28]   F.-Y. Juang, P.K. Bhattacharya, J. Singh [*Appl. Phys. Lett. (USA)* vol.48 (1986) p.290-2]

[29]   J.-P. Reithmaier, S. Hausser, H.P. Meier, W. Walter [*J. Cryst. Growth (Netherlands)* vol.127 (1993) p.755-8]

[30]   J.W. Matthews, A.E. Blakeslee [*J. Cryst. Growth (Netherlands)* vol.27 (1974) p.118]

[31]   H. Asai, Y. Kawamura [*Appl. Phys. Lett. (USA)* vol.56 (1990) p.746-8]

[32]   A.S. Brown, J.A. Henige, A.E. Schmitz, L.E. Larson [*Appl. Phys. Lett. (USA)* vol.62 (1993) p.66-8]

[33]   M.-H. Meynadier, J.-L. de Miguel, M.C. Tamargo, R.E. Nahory [*Appl. Phys. Lett. (USA)* vol.302 (1988) p.302-4]

[34]   J.L. de Miguel, M.C. Tamargo, M.-H. Meynadier, R.E. Nahory, D.M. Hwang [*Appl. Phys. Lett. (USA)* vol.52 (1988) p.892-4]

[35]   E. Tournie, K.H. Ploog, M. Henstein [*J. Vac. Sci. Technol. B (USA)* vol.11 (1993) p.1388-91]

[36]   E. Tournie, O. Brandt, C. Giannini, K. Ploog, M. Hohenstein [*J. Cryst. Growth (Netherlands)* vol.127 (1993) p.765-9]

[37]   C.E.C. Wood [in *GaInAsP Alloys Semiconductors* Ed. T.P. Pearsall (Wiley, New York, 1982) p.87]

[38]   W.T. Tsang, E.F. Schubert [*Appl. Phys. Lett. (USA)* vol.49 (1986) p.220-2]

[39]   M.B. Panish, H. Temkin, R.A. Hamm, S.N.G. Chu [*Appl. Phys. Lett. (USA)* vol.49 (1986) p.164-6]

[40]   J.N. Baillargeon, A.Y. Cho, F.A. Thiel, P.J. Pearah, K.Y. Cheng [*Inst. Phys. Conf. Ser. (UK)* no.141 (1995) p.173]

[41]   J.M. Vandenberg, M.B. Panish, R.A. Hamm, H. Temkin [*Appl. Phys. Lett. (USA)* vol.56 (1990) p.910-2]

[42]   G.J. Shiau, C.P. Chao, P.E. Burrows, S.R. Forrest [*Appl. Phys. Lett. (USA)* vol.66 (1995) p.201]

[43]   M. Mukai, M. Sugawara, S. Yamazaki [*J. Cryst. Growth (Netherlands)* vol.137 (1994) p.388]

[44]   T. Mozume [*J. Appl. Phys. (USA)* vol.77 (1995) p.1492]

[45]   L.M. Woods et al [*J. Electron. Mater. (USA)* vol.23 (1994) p.1229]

[46]   K. Naniwae, S. Sugou, T. Anan [*Jpn. J. Appl. Phys. (Japan)* vol.33 (1994) p.L156]

[47]   H.Q. Hou, C.W. Tu, S.N.G. Chu [*Appl. Phys. Lett. (USA)* vol.58 (1991) p.2954]

[48]   W.-P. Hong, R. Bhat, J. Hayes, D. DeRosa, M. Leadbeater, M. Koza [*Appl. Phys. Lett. (USA)* vol.60 (1992) p.109]

3.3 Organometallic vapour phase epitaxial growth of GaAs- and InP-based quantum wells and superlattices

R. Bhat

June 1995

A INTRODUCTION

A wide variety of GaAs- and InP-based quantum wells and superlattices have been grown by organometallic vapour phase epitaxy (OMVPE). Some important examples are presented in this Datareview. Since the most commonly used tools to evaluate interface quality are photoluminescence and photoluminescence excitation spectroscopy, we concentrate mainly on the results of these evaluations.

B GaAs/AlGaAs QUANTUM WELLS AND SUPERLATTICES

GaAs/AlGaAs quantum wells are important for the fabrication of optoelectronic devices, particularly lasers and modulators, operating in the 0.68 - 0.89 μm wavelength range. In addition, these quantum wells have been used to fabricate detectors and modulators in the 7 - 14 μm region [1-4] by utilising the intersubband transitions. GaAs/AlGaAs quantum wells have been grown both by atmospheric [5-12] and low pressure [13,14] OMVPE with similar results. The FWHM of the photoluminescence (PL) emission of 100 Å wide GaAs wells is 4 - 5 meV at 2 K, with the FWHM increasing to 6 - 10 meV for 20 Å wide wells. Dupuis et al [5] have extensively studied GaAs/AlGaAs quantum wells by photoluminescence excitation spectroscopy (PLE) and PL. They conclude that the interface roughness is of the order of +/- one monolayer and an interface grading of less than three monolayers can be achieved. However, despite the achievements to date, the FWHM of OMVPE grown quantum wells remains considerably wider than that achieved (FWHM = 1 meV for 57 Å wide wells [15]) by molecular beam epitaxy (MBE). This is particularly surprising in view of the recent atomic force microscopy (AFM) study [16,17] of the surfaces of GaAs, AlAs, and AlGaAs grown by OMVPE and MBE, wherein 1 μm wide atomically smooth terraces were obtained for OMVPE but only 200 Å wide terraces for MBE. This would suggest that factors other than interface roughness are limiting the PL FWHM of OMVPE quantum wells.

$(AlAs)_n/(GaAs)_n$ superlattices with n as low as 2 have been demonstrated [7-9]. Electron transport in the direction perpendicular to the plane of 600 period GaAs/Al$_x$Ga$_{1-x}$As superlattices with 20 Å barriers and x varying from 0.1 to 0.3 have been investigated by cyclotron resonance [18]. It was found that the effective mass for tunnelling along the growth direction increased with the barrier height at a rate that is in rough agreement with a model that includes the change in band mass in the barrier.

C   InGaAs/GaAs AND InGaAs/GaAs/AlGaAs QUANTUM WELLS

The addition of In to a GaAs quantum well enables the operation of optoelectronic devices beyond 0.85 μm. Despite the technological importance of these quantum wells in the fabrication of 0.98 μm pump lasers for erbium doped fibre amplifiers, there appear to have been few investigations. Bertolet et al [19] have studied single wells of $In_{0.12}Ga_{0.88}As/GaAs$, with well thicknesses ranging from 11 to 99 Å, by low temperature photoluminescence. The PL FWHMs they observed were significantly narrower than those for GaAs/AlGaAs quantum wells. As an example, the FWHM at 80 K was 2.9 meV for a 55 Å $In_{0.12}Ga_{0.88}As$ well in contrast to 7.6 meV for a 58 Å $GaAs/Al_{0.4}Ga_{0.6}As$ well grown in the same reactor. While the exact reason for this difference in behaviour is unknown, it is possible that the presence of In, which has a high surface mobility, helps to smooth the interfaces. In the case of InGaAs/AlGaAs quantum wells, Wang et al [20] observed an improvement in the quantum well quality when a GaAs layer was introduced between the InGaAs well and the AlGaAs barrier layer.

D   GaAs/GaInP QUANTUM WELLS

These quantum wells cover a wavelength range similar to that covered by GaAs/AlGaAs quantum wells. The interest in GaAs/GaInP quantum wells stems from an expected higher reliability of lasers fabricated from these materials due to the absence of Al. In addition, the absence of Al makes regrowth in the fabrication of devices easier. High quality GaAs/GaInP single and multiple quantum wells have been grown by low pressure OMVPE [21,22], and their properties investigated by low temperature PL, PLE and double crystal X-ray diffraction. Indium incorporation in the quantum wells and InGaAsP interfacial layer formation are a problem in the growth of these structures [22]. The In in the well has been variously attributed to a memory effect [21,23], In diffusion from the barrier [24], and In transported from the susceptor during growth of the well [22]. The formation of the InGaAsP interfacial layer has been attributed to the reaction of arsine with underlying GaInP upon commencing GaAs growth [22]. This interfacial layer can result in lower than expected PL emission energy and anomalously large splitting between the energies associated with the light and heavy hole emissions [22]. Intentional growth of a thin (8 Å) GaP interfacial layer was shown to eliminate these problems [22,25]. In addition to the interfacial layer, As carry-over into the GaInP may exist when making the transition from GaAs to GaInP [22]. By choosing a low growth temperature (510°C) and optimising the switching sequence, Omnes et al [21] have grown high quality single and multiple quantum wells with a PL FWHM of 4 meV at 4 K for a 90 Å wide well. The low growth temperature is expected to minimise the formation of interfacial layers and reduce In contamination of the wells.

E   InGaAs/InGaAsP/InGaP QUANTUM WELLS

This system, like InGaAs/AlGaAs quantum wells, is important for 0.98 μm pump lasers for erbium doped fibre amplifiers. Groves [26] has shown that, as in the case of InGaAs/AlGaAs quantum wells, substantial improvement in quantum well quality can be obtained by introducing a thin GaAs layer between the InGaAs well and InGaAsP barrier.

F   GaInP/AlGaInP QUANTUM WELLS

Quantum wells emitting in the 0.59 - 0.68 μm wavelength range, at room temperature, can be fabricated from GaInP/AlGaInP on GaAs substrates. GaInP/AlGaInP quantum wells, with and without strain, have been studied by Ikeda et al [27] and Kondo et al [28], with wells as narrow as 10 Å having been grown. The PL FWHM at 4 K was 11.9 - 13.2, 9, and 13 meV for lattice matched, 1% compressive, and 0.9% tensile strained GaInP wells. Transitions to both the light and heavy hole levels were observed in the room temperature PL spectra of strained quantum well layers. From these results Kondo et al [28] have deduced the light and heavy hole splitting as a function of strain.

G   InGaAs/InP AND InGaAs/InGaAsP/InP QUANTUM WELLS AND SUPERLATTICES

Numerous studies of InGaAs/InP and InGaAs/InGaAsP quantum well and superlattice growth by OMVPE exist [29-58], reflecting their importance in the fabrication of optoelectronic devices operating at 1.3 and 1.55 μm wavelengths for telecommunications applications. As in the case of GaAs/GaInP quantum wells, switching from InGaAs to InP and vice versa is difficult due to the carry-over of As into the InP on the one hand, and the reaction of arsine with the underlying InP upon commencing InGaAs growth on InP (also referred to as As/P exchange) on the other. There also have been a number of reports of the observation of the splitting of low temperature PL emission peaks due to the existence of regions larger than the exciton diameter but whose thicknesses differ by a monolayer [29,42,47]. We believe that these splittings are caused by improper switching sequences resulting in layer thickness fluctuations and are not necessarily an indication of superior wells.

Despite these difficulties, high quality InGaAs/InP quantum wells with PL FWHMs of 3.2 and 10 meV at 4.2 K for 100 and 6 Å wide wells, respectively, have been obtained [46].

In the growth of multiple quantum wells and superlattices thickness fluctuations in the growth plane have been observed [54-56]. This is accentuated by improper switching sequences [54-56] and thin (<50 Å) InP layers [54].

Compressive and tensile strained InGaAs/InP quantum wells also have been studied [57,58]. The compressively strained quantum wells showed multiple low temperature emission peaks due to well width fluctuations [57]. At room temperature the tensile strained quantum wells show PL peaks due to recombinations associated with both light and heavy holes [58].

H   GaInAs/AlInAs QUANTUM WELLS ON InP

These quantum wells, like the GaInAs/InP quantum wells, are important for optoelectronic devices operating in the 1.3 - 1.55 μm wavelength range. Moreover, the higher conduction band offset available in this system as compared to the GaInAs/InP system has made this system attractive for fabricating lasers capable of high temperature operation [59]. In addition, these quantum wells are useful for fabricating high performance modulators due to the valence band offset being lower than in GaInAs/InP quantum wells [60]. GaInAs/AlInAs quantum wells are also of interest for optical detectors operating at 3 - 4 μm using intersubband transitions [61]. Kamei et al [62] have studied GaInAs/AlInAs quantum wells using low temperature PL. At 4.2 K PL FWHMs of 16 and 21 meV for 88 and 6 Å wide wells were obtained. The broad PL emission observed in wide wells was attributed to band filling due to the 3×10^{16} cm^{-3} n-type background level in the AlInAs barrier layers.

I   CONCLUSION

In this Datareview we have covered the progress in the fabrication of quantum wells and superlattices based on GaAs and InP. In the literature the accuracy of the stated thicknesses of quantum wells, particularly when they are thin, is often questionable since they are usually inferred from the growth rates of thick layers. This makes it difficult, in many cases, to assess the validity of the extent of energy shift obtained in thin quantum wells. In addition, a cautionary note by Miller et al [63] that both PL and PLE spectra need to be studied to make conclusive statements regarding the quality of quantum wells, is often ignored. Detailed studies of many of the quantum wells grown by OMVPE have not been carried out since the thrust in many laboratories has been towards optimisation of devices utilising these quantum wells rather than studying the wells themselves and the benefit to device performance in obtaining the ultimate in interface abruptness is yet to be proven.

REFERENCES

[1]   G. Hasnain, B.F. Levine, C.G. Bethea, R.A. Logan, J. Walker, R.J. Malik [*Appl. Phys. Lett. (USA)* vol.54 (1989) p.2515]

[2]   B.F. Levine, K.K. Choi, C.G. Bethea, J. Walker, R.J. Malik [*Appl. Phys. Lett. (USA)* vol.50 (1987) p.1092]

[3]   A. Harwitt, J.S. Harris [*Appl. Phys. Lett. (USA)* vol.50 (1987) p.685]

[4]   M.J. Kane, N. Apsley [*J. Phys. Colloq. (France)* vol.48 (1987) p.545]

[5]   R.D. Dupuis, R.C. Miller, P.M. Petroff [*J. Cryst. Growth (Netherlands)* vol.68 (1984) p.398-405]

[6]   P.M. Frijlink, J. Maluenda [*Jpn. J. Appl. Phys. (Japan)* vol.21 (1982) p.L574-6]

[7]   H. Kawai, K. Kaneko, N. Watanabe [*J. Appl. Phys. (USA)* vol.56 (1984) p.463-7]

[8]   K. Kajiwara, H. Kawai, K. Kaneko, N. Watanabe [*Jpn. J. Appl. Phys. (Japan)* vol.24 (1985) p.L85-8]

[9]   A. Ishibashi, Y. Mori, M. Itabashi, N. Watanabe [*J. Appl. Phys. (USA)* vol.58 (1985) p.2691-5]

[10]   D.C. Bertolet, J.-K. Hsu, K.M. Lau [*J. Appl. Phys. (USA)* vol.62 (1987) p.120-5]

[11]   M.R. Leys, C. Van Opdorp, M.P.A. Viegers, H.J. Talen-Van Der Mheen [*J. Cryst. Growth (Netherlands)* vol.68 (1984) p.431-6]

[12]   R.D. Dupuis, J.G. Neff, C.J. Pinzone [*J. Cryst. Growth (Netherlands)* vol.124 (1992) p.558-64]

[13]   P. Norris, J. Black, S. Zemon, G. Lambert [*J. Cryst. Growth (Netherlands)* vol.68 (1984) p.437-44]

[14]   P.M. Frijlink [*J. Cryst. Growth (Netherlands)* vol.93 (1988) p.207-15]

[15]   C.W. Tu et al [*J. Cryst. Growth (Netherlands)* vol.81 (1987) p.159-63]

[16]   M. Shinohara, M. Tanimoto, H. Yokoyama, N. Inoue [*J. Cryst. Growth (Netherlands)* vol.145 (1994) p.113-9]

[17]   M. Shinohara, M. Tanimoto, H. Yokoyama, N. Inoue [*Appl. Phys. Lett. (USA)* vol.65 (1994) p.1418-20]

[18]   T. Duffield et al [*Phys. Rev. Lett. (USA)* vol.56 (1986) p.2724-7]

[19]   D.C. Bertolet, J.-K. Hsu, S.H. Jones, K.M. Lau [*Appl. Phys. Lett. (USA)* vol.52 (1988) p.293-5]

[20]   C.A. Wang, H.K. Choi [*J. Electron. Mater. (USA)* vol.20 (1991) p.929-34]

[21]   F. Omnes, M. Razeghi [*Appl. Phys. Lett. (USA)* vol.59 (1991) p.1034-6]

[22]   R. Bhat, M.A. Koza, M.J.S.P. Brasil, R.E. Nahory, C.J. Palmstrom, B.J. Wilkens [*J. Cryst. Growth (Netherlands)* vol.124 (1992) p.576-82]

[23]   D.P. Bour, J.R. Shealy, S. McKenna [*J. Appl. Phys. (USA)* vol.63 (1988) p.1241-3]

[24]   C. Francis, M.A. Bradley, P. Boucaud, F.H. Julien, M. Razeghi [*Appl. Phys. Lett. (USA)* vol.62 (1993) p.178-80]

[25]   F.E.G. Guimaraes et al [*J. Cryst. Growth (Netherlands)* vol.93 (1992) p.199-206]

[26]   S.H. Groves [*J. Cryst. Growth (Netherlands)* vol.124 (1992) p.747-50]

[27]   M. Ikeda, K. Nakano, Y. Mori, K. Kaneko, N. Watanabe [*J. Cryst. Growth (Netherlands)* vol.77 (1986) p.380-5]

[28]   M. Kondo, K. Domen, C. Anayama, T. Tanahashi, K. Nakajima [*J. Cryst. Growth (Netherlands)* vol.107 (1991) p.578-82]

[29]   M. Razeghi, J. Nagle, C. Weisbuch [*Inst. Phys. Conf. Ser. (UK)* vol.74 (1985) p.379-84]

[30]   B.I. Miller, E.F. Schubert, U. Koren, A. Ourmazd, A.H. Dayem, R.J. Capik [*Appl. Phys. Lett. (USA)* vol.49 (1986) p.1384-6]

[31]   M.S. Skolnik et al [*Appl. Phys. Lett. (USA)* vol.48 (1986) p.1455-7]

[32]   K.W. Carey, R. Hull, J.E. Fouquet, F.G. Kellert, G.R. Trott [*Appl. Phys. Lett. (USA)* vol.51 (1987) p.910-2]

[33]   D. Moroni, J.P. Andre, E.P. Menu, P.H. Gentric, J.N. Patillon [*J. Appl. Phys. (USA)* vol.62 (1987) p.2003-8]

[34]   T.Y. Wang, K.L. Fry, A. Persson, E.H. Reihlen, G.B. Stringfellow [*J. Appl. Phys. (USA)* vol.63 (1988) p.2674-80]

[35]   K. Streubel, F. Scholz, G. Laube, R.J. Dieter, E. Zielinski, F. Keppler [*J. Cryst. Growth (Netherlands)* vol.93 (1988) p.347-52]

[36]   Y. Miyamoto, K. Uesaka, M. Takadou, K. Furuya, Y. Suematsu [*J. Cryst. Growth (Netherlands)* vol.93 (1988) p.353-8]

[37]   M. Engel, R.K. Bauer, D. Bimberg, D. Grutzmacher, H. Jurgensen [*J. Cryst. Growth (Netherlands)* vol.93 (1988) p.359-64]

[38]   M. Irikawa, I.J. Murgatroyd, T. Ijichi, N. Matsumoto, A. Nakai, S. Kashiwa [*J. Cryst. Growth (Netherlands)* vol.93 (1988) p.370-5]

[39]   M. Kondo, S. Yamazaki, M. Suguwara, H. Okuda, K. Kato, K. Nakajima [*J. Cryst. Growth (Netherlands)* vol.93 (1988) p.376-81]

[40]   D. Grutzmacher et al [*J. Cryst. Growth (Netherlands)* vol.93 (1988) p.382-8]

[41]   R. Schwedler, F. Reinhardt, D. Grutzmacher, K. Wolter [*J. Cryst. Growth (Netherlands)* vol.107 (1991) p.531-6]

[42]   J. Hergeth, D. Grutzmacher, F. Reinhardt, P. Balk [*J. Cryst. Growth (Netherlands)* vol.107 (1991) p.537-42]

[43]   J. Camassel et al [*J. Cryst. Growth (Netherlands)* vol.107 (1991) p.543-8]

[44]   C.G. Cureton, E.J. Thrush, A.T.R. Briggs [*J. Cryst. Growth (Netherlands)* vol.107 (1991) p.549-54]

[45]   P. Wiedemann et al [*J. Cryst. Growth (Netherlands)* vol.107 (1991) p.561-6]

[46]   H. Kamei, H. Hayashi [*J. Cryst. Growth (Netherlands)* vol.107 (1991) p.567-72]

[47]   G. Landgren, P. Ojala, O. Ekstrom [*J. Cryst. Growth (Netherlands)* vol.107 (1991) p.573-7]

[48]   W. Seifert et al [*J. Cryst. Growth (Netherlands)* vol.124 (1992) p.531-5]

[49]   A.R. Clawson, T.T. Vu, S.A. Pappert, C.M. Hanson [*J. Cryst. Growth (Netherlands)* vol.124 (1992) p.536-40]

[50]   K. Streubel, J. Wallin, M. Amiotti, G. Landgren [*J. Cryst. Growth (Netherlands)* vol.124 (1992) p.541-6]

[51]   X.S. Jiang, A.R. Clawson, P.K.L. Yu [*J. Cryst. Growth (Netherlands)* vol.124 (1992) p.547-52]

[52]   R. Meyer, M. Hollfelder, H. Hardtdegen, B. Lengler, H. Luth [*J. Cryst. Growth (Netherlands)* vol.124 (1992) p.583-8]

[53]   M.J. Yates, M.R. Aylett, S.D. Perrin, P.C. Spurdens [*J. Cryst. Growth (Netherlands)* vol.124 (1992) p.604-9]

[54]   R. Bhat, M.A. Koza, D.M. Hwang, K. Kash, C. Caneau, R.E. Nahory [*J. Cryst. Growth (Netherlands)* vol.110 (1991) p.353-62]

[55]   P.C. Spurdens, M.R. Taylor, M. Hockly, M.J. Yates [*J. Cryst. Growth (Netherlands)* vol.107 (1991) p.215-20]

[56]   N. Agrawal, D. Franke, N. Grote, F.W. Reier, H. Schroeter-Janssen [*J. Cryst. Growth (Netherlands)* vol.124 (1992) p.610-5]

[57]   P.J.A. Thijs, E.A. Montie, T. van Dongen [*J. Cryst. Growth (Netherlands)* vol.107 (1991) p.731-40]

[58]   C.E. Zah et al [*Electron. Lett. (UK)* vol.27 (1991) p.1414-5]

[59]   C.E. Zah et al [*IEEE J. Quantum Electron. (USA)* vol.30 (1994) p.511]

[60]   S. Kondo, K. Wakita, Y. Noguchi, N. Yoshimoto [*Proc. 7th Int. Conf. on Indium Phosphide and Related Materials* Sapporo, Hokaido, Japan, 9-13 May (1995) p.61]

[61]   B.F. Levine, A.Y. Cho, J. Walker, R.J. Malik, D.A. Kleinman, D.L. Sivco [*Appl. Phys. Lett. (USA)* vol.52 (1988) p.1481]

[62]   H. Kamei, K. Hashizume, M. Murata, N. Kuwata, K. Ono, K. Yoshida [*J. Cryst. Growth (Netherlands)* vol.93 (1988) p.329-35]

[63]   R.C. Miller, R. Bhat [*J. Appl. Phys. (USA)* vol.64 (1988) p.3647-9]

3.4 GaAs- and InP-based quantum wells and superlattices on non-(100) substrates

P.K. Bhattacharya

April 1996

A   INTRODUCTION

Due to the differences in the band structure of semiconductors for different crystallographic orientations, the electrical and optical properties and growth phenomena are expected to be different. The study of GaAs- and InP-based heterostructures on the three low-index planes, namely (100), (110) and (111), is therefore of considerable importance since most of the physical properties of the materials on other high-index planes will fall in between those of the three basic low index planes. Most device technology with GaAs- and InP-based heterostructures has been developed with (100)-oriented substrates, particularly due to ease of substrate availability and epitaxial growth and due to the vast amount of work done to characterise the substrates and epitaxial materials. Nonetheless, interesting doping characteristics and material properties open the doors for new device phenomena. Differences in properties such as breakdown field and photoluminescence are observed in crystals of different orientations. In particular growth on (111)-oriented substrates, and in general on (n11)-oriented substrates, is of importance for several reasons. The first is the existence of a strong internal piezoelectric field in strained quantum wells resulting in a red shift of the excitonic transitions [1,2]. Application of an external bias can modulate these fields so as to obtain a blue shift of the absorption edge [3-8]. Second, due to a larger bulk heavy-hole mass in the (111) direction, the heavy-hole - light-hole band splitting is larger. This results in smaller valence band intermixing and the threshold currents of quantum well lasers grown on (111)B substrates are smaller than those of lasers grown on (100) substrates [9]. Detectors grown on (111)B substrates can be tuned to both longer and shorter wavelengths. It is also expected that a larger built-in birefringence in (111)- or (110)-oriented multilayers will lead to large optical non-linearities and enhanced electro-optic effects. This Datareview will mainly focus on aspects related to quantum wells and superlattices.

B   GROWTH ON POLAR (n11) SURFACES

FIGURE 1 shows schematically the geometry of surface bonds for some selected orientations of III-V compounds. It is seen that the (001) surface has double dangling bond sites, while the (110) surface has single dangling bond sites originating from equal numbers of group III and group V atoms. Surface polarity (A and B surfaces) exists for the (n11) orientations. It is seen from FIGURE 1 that the (111) surface has single dangling bond sites. The (211) surface consists of twice as many single dangling bond sites as there are double dangling bond sites. The (311) surface consists of equal densities of single and double dangling bond sites.

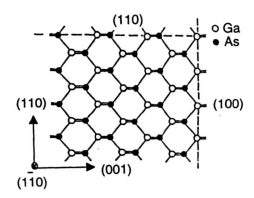

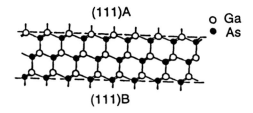

FIGURE 1. Surface bonding configurations for different crystallographic orientations in GaAs.

Quantum wells and superlattices on (111)-oriented GaAs and InP substrates have been grown using molecular beam epitaxy (MBE) and metal-organic vapour phase epitaxy (MOVPE) by a number of authors [10-13] and some general conclusions can be made. Although the apparent growth rate, compared to (100) growth, remains unchanged [10], the true growth rate is orientation dependent. The effect is masked by the limited 'surface' mobility of the group III species. The growth rate in the (111) direction varies significantly with growth temperature [10], while that in the (100) direction is almost independent of temperature. It is observed that misorientation improves the surface morphology and the electrical and optical properties of the crystals [8,9]. It is believed that such misorientation provides addition surface steps, or kink sites, necessary for growth [11]. In effect, the defect (vacancy) density is reduced. For growth on (111)-InP, a higher growth temperature than that for (100) growth is required. The necessity for high growth temperature indicates that the surface migration length on the (111) surface is probably much smaller than that on the (100) surface, as also reported for growth on the (311) surface [14].

C   DOPING CHARACTERISTICS

It can be expected that impurity and dopant incorporation behaviour will also be different on non-(100) substrates. In general, p-type doping is difficult for growth on (111)B GaAs and InP substrates [8,11]. The problem is slightly alleviated for growth on (211)B and (311)B substrates (FIGURE 2). On the other hand, it has been shown that Be (p-dopant) incorporation is dramatically improved on (311)A GaAs surfaces, compared to (100) growth [15]. Similarly, surface segregation of Sn as an n-type dopant was reduced and Sn incorporation was increased on (111)A substrates when MBE growth was carried out at a low growth temperature and/or high As_4:Ga flux ratio [16]. Carbon is usually an unintentional amphoteric dopant in crystals grown by MBE and MOVPE. It has been confirmed that the amount of C incorporated in GaAs and AlGaAs films depends on the orientation of crystals grown by MOVPE [17]. Carbon is usually a p-type dopant in (100)-oriented films. The concentration of electrons on (311)B substrates of unintentionally doped films was higher than on (100) and (311)A substrates. Films grown on (311)B substrates did not show p-type behaviour even when they were grown with a fairly low V/III ratio. The relative intensity of the free-to-carbon acceptor luminescence of the layers grown on (311)B substrates was smaller than that of layers grown on substrates of other orientations. Secondary ion mass spectrometry (SIMS) has confirmed that these results are consistent with unintentional C incorporation.

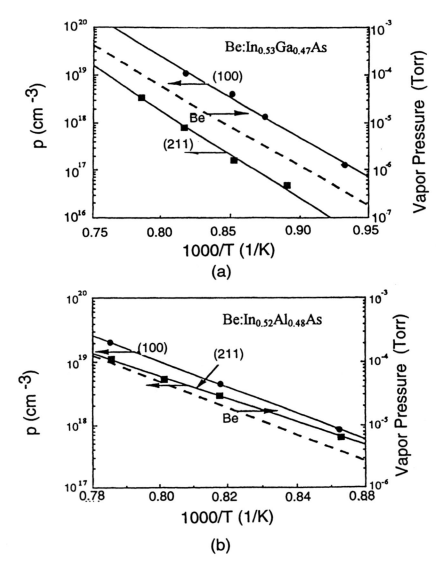

FIGURE 2. Be doping characteristics during molecular beam epitaxy on (100)- and (211)-oriented GaAs of (a) $In_{0.53}Ga_{0.47}As$/InP, and (b) $In_{0.52}Al_{0.48}As$/InP. The equilibrium vapour pressure curve for Be is indicated by the dashed line.

Silicon is widely used as an n-type dopant for most III-V semiconductors grown by MBE on (100) substrates. However, because Si is an amphoteric impurity, its doping behaviour depends on various factors including substrate orientation and morphology, growth temperature and the V/III ratio. It was first shown by Cho and Hayashi [18] that during MBE on the (111)A surface of GaAs, Si dopant atoms bond to surface Ga atoms, taking up As sites and acting as a p-type dopant, while on the (111)B surface, Si atoms bond to surface As atoms, taking up Ga sites and acting as an n-type dopant. This doping behaviour has been subsequently confirmed by many groups [19-24]. On high-index (n11) surfaces, the doping type depends on the competition between single and double dangling bond sites, the growth rate and temperature, and the V/III ratio. For example, it has been shown by Li et al [25] that reproducible and high levels of p-doping ($p = 2 \times 10^{19}$ cm^{-3}) can be achieved in GaAs for growth on (311)A oriented substrates, for which Si normally exhibits well-behaved n-type doping behaviour. The incorporation of Si atoms as n- and p-type dopants in (311)A GaAs, achieved by changing the growth parameters, has been confirmed by Raman spectroscopy [26] and high-performance all-Si doped n-p-n GaAs/AlGaAs heterojunction bipolar transistors have been grown and characterised [27]. Additional Si-related defect complexes have also been observed in material grown on (n11) substrates [26,28-30], and these may reduce the radiative efficiency of the material.

D   STRAINED-LAYER QUANTUM WELLS AND SUPERLATTICES ON (111) AND (110) SUBSTRATES

Interest in pseudomorphically strained quantum wells and superlattices grown on (n11) and (110) oriented substrates is primarily due to the built-in piezoelectric field, ~10^5 V/cm, which lends itself to the design of efficient electroabsorption or electro-optic modulators. Growth and characterisation of $In_{1-x}Ga_xAs/GaAs$ multiquantum well heterostructures have been undertaken by many authors [31-35]. Growth of $In_xGa_{1-x}As/In_{0.52}Al_{0.48}As$ quantum wells on (n11) InP substrates has been demonstrated by Sun et al [8]. As mentioned earlier, the growth morphology is improved and the photoluminescence intensity is enhanced by a slight (1 - 5°) misorientation. From the equilibrium theory of Matthews and Blakeslee [36] it was predicted that the critical thickness h_c in the (111) direction is about twice that in the (100) direction under the same growth conditions (FIGURE 3). This has been experimentally verified by Anan et al [32]. The dopant behaviour, effect of growth parameters and surface morphology of $In_xGa_{1-x}As/GaAs$ strained quantum well modulator structures grown on (110) GaAs substrates have been studied by Sun et al [33].

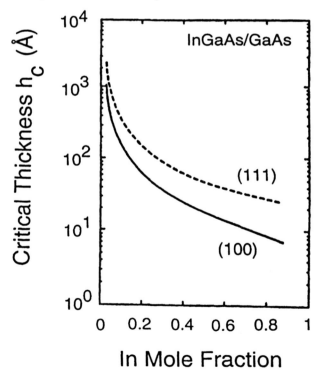

FIGURE 3. Calculated critical thickness for $In_xGa_{1-x}As$ on (111)- and (100)-oriented GaAs (from [32]).

E   MIGRATION ENHANCED EPITAXY ON (111)-ORIENTED SUBSTRATES

In migration enhanced epitaxy, group III and V atoms are deposited alternately by shuttering, for periods corresponding to the growth time of 1 - 4 monolayers. The group III atoms can migrate to longer distances than in conventional MBE which leads to two immediate advantages: surfaces are very smooth and growth is possible at lower substrate temperatures. MEE of single AlGaAs layers and GaAs/AlGaAs single quantum wells has been demonstrated by Imamoto et al [37], with in-situ reflection high-energy electron diffraction (RHEED) measurements. A photoluminescence intensity enhancement of 50 times, compared to photoluminescence from a (100)-oriented GaAs/AlGaAs SQW grown simultaneously, was observed for the SQW grown on (111)B GaAs. It is useful to note that similar luminescence enhancement has been reported by Hayakawa et al [9] for a GaAs/AlGaAs SQW grown on (111)B GaAs by conventional MBE.

F   CONCLUSION

Novel materials and device phenomena and the need to understand intrinsic growth phenomena have motivated growth of single and multilayer structures on non-(100) oriented GaAs and InP substrates [38]. The growth and doping behaviour for such growth has been summarised in this Datareview.

REFERENCES

[1]   D.L. Smith [*Solid State Commun. (USA)* vol.57 (1986) p.919]
[2]   C. Mailhot, D.L. Smith [*Phys. Rev. B (USA)* vol.35 (1987) p.1242]
[3]   T.S. Moise, L.J. Guido, J.C. Beggy, T.J. Cunningham, S. Seshadri, R.C. Barker [*J. Electron. Mater. (USA)* vol.21 (1992) p.119]
[4]   E.A. Caridi, T.Y. Chang, K.W. Goossen, L.F. Eastman [*Appl. Phys. Lett. (USA)* vol.56 (1990) p.659]
[5]   I.H. Campbell, D.E. Watkins, D.L. Smith, S. Subbanna, H. Kroemer [*Appl. Phys. Lett. (USA)* vol.59 (1991) p.1711]
[6]   K.W. Goossen, E.A. Caridi, T.Y. Chang, J.B. Stark, D.A.B. Miller, R.A. Morgan [*Appl. Phys. Lett. (USA)* vol.56 (1990) p.715]
[7]   A.S. Pabla et al [*Appl. Phys. Lett. (USA)* vol.63 (1993) p.752]
[8]   H.C. Sun, L. Davis, Y. Lam, P.K. Bhattacharya, J.P. Loehr [*Inst. Phys. Conf. Ser. (UK)* no.136 (1993) ch.4 p.197]
[9]   T. Hayakawa, M. Kondo, T. Suyama, K. Takahashi, S. Yamamoto, T. Hijikata [*Jpn. J. Appl. Phys. (Japan)* vol.26 (1987) p.L302]
[10]   S.D. Hersee, E. Barbier, R. Blondeau [*J. Cryst. Growth (Netherlands)* vol.77 (1986) p.310]
[11]   L. Vina, W.I. Wang [*Appl. Phys. Lett. (USA)* vol.48 (1986) p.36]
[12]   E. Morita, M. Ikeda, M. Inoue, K. Kaneko [*J. Cryst. Growth (Netherlands)* vol.106 (1990) p.197]
[13]   T. Takebe, M. Fujii, T. Yamamoto, K. Fujita, T. Watanabe [*Inst. Phys. Conf. Ser. (UK)* no.136 (1993) ch.9 p.577]
[14]   T. Fukunaga, H. Hakashima [*Jpn. J. Appl. Phys. (Japan)* vol.25 (1986) p.L856]
[15]   K. Mochizuki, S. Goto, C. Kusano [*Appl. Phys. Lett. (USA)* vol.58 (1991) p.2939]
[16]   S.J. Hu, M.R. Fahy, K. Sato, B. Joyce [*J. Electron. Mater. (USA)* vol.24 (1995) p.1003]
[17]   K. Tamamura, J. Ogawa, K. Akimoto, Y. Mori, C. Kojima [*Appl. Phys. Lett. (USA)* vol.50 (1987) p.1149]

[18]   A.Y. Cho, I. Hayashi [*Metall. Trans. (USA)* vol.2 (1971) p.777]

[19]   J.M. Ballingall, C.E.C. Wood [*Appl. Phys. Lett. (USA)* vol.41 (1982) p.947]

[20]   W.I. Wang, E.E. Mendez, T.S. Kuan, L. Esaki [*Appl. Phys. Lett. (USA)* vol.47 (1985) p.826]

[21]   D.L. Miller [*Appl. Phys. Lett. (USA)* vol.47 (1985) p.1309]

[22]   S. Subbanna, H. Kroemer, J.L. Merz [*J. Appl. Phys. (USA)* vol.59 (1986) p.488]

[23]   T. Takamori, T. Fukunaga, J. Kobayashi, K. Ishida, H. Nakashima [*Jpn. J. Appl. Phys. (Japan)* vol.26 (1987) p.1097]

[24]   T. Takamori, K. Watanabe, T. Fukunaga [*Electron. Lett. (UK)* vol.27 (1991) p.729]

[25]   W.-Q. Li, P.K. Bhattacharya, S.H. Kwok, R. Merlin [*J. Appl. Phys. (USA)* vol.72 (1992) p.3129]

[26]   S.H. Kwok, R. Merlin, W.-Q. Li, P.K. Bhattacharya [*J. Appl. Phys. (USA)* vol.72 (1992) p.285]

[27]   W.-Q. Li, P.K. Bhattacharya [*IEEE Electron Device Lett. (USA)* vol.13 (1992) p.29]

[28]   T. Takamori, T. Fukunaga, H. Nakashima [*Jpn. J. Appl. Phys. (Japan)* vol.26 (1987) p.L920]

[29]   R. Murray et al [*J. Appl. Phys. (USA)* vol.66 (1989) p.2589]

[30]   M. Vematsu, K. Maezawa [*Jpn. J. Appl. Phys. (Japan)* vol.29 (1990) p.L527]

[31]   J.G. Beery et al [*Appl. Phys. Lett. (USA)* vol.54 (1989) p.233]

[32]   T. Anan, K. Nishi, S. Sugou [*Appl. Phys. Lett. (USA)* vol.60 (1992) p.3159]

[33]   D. Sun, E. Towe, M.J. Hayduk, R.K. Boncek [*Appl. Phys. Lett. (USA)* vol.63 (1993) p.2881]

[34]   J.P.R. David et al [*J. Electron. Mater. (USA)* vol.23 (1994) p.975]

[35]   R.L. Tober, T.B. Bahder, J.D. Bruno [*J. Electron. Mater. (USA)* vol.24 (1995) p.793]

[36]   J.W. Matthews, A.E. Blakeslee [*J. Cryst. Growth (USA)* vol.27 (1974) p.118]

[37]   H. Imamoto, F. Sato, K. Imanaka, M. Shimura [*Appl. Phys. Lett. (USA)* vol.55 (1989) p.115]

[38]   P. Bhattacharya [*Microelectron. J. (UK)* vol.26 (1995) p.887]

CHAPTER 4

STRUCTURAL PROPERTIES

4.1 Transmission electron microscopy of quantum wells and superlattices

H. Okamoto and Y. Suzuki

May 1995

A INTRODUCTION

Transmission electron microscopy (TEM) of quantum wells and superlattices originated from an interest in directly observing the abruptness of the heterointerfaces. The feasibility of observing the local structure on an atomic scale has opened up the potential of TEM to study not only the interface sharpness but also regularity in atomic arrangement of the monolayer superlattice, lattice disordering of the superlattice suffered from ion implantation and annealing, and two or three dimensional superlattices (quantum wires and quantum boxes). It has been widely used also to study the epitaxial growth mechanism and various kinds of lattice defects. It is now quite common to provide ultra-thin film technology with TEM.

B SAMPLE PREPARATION FOR TEM

The success of TEM observation depends mostly on the preparation of the specimen. A detailed description is given here.

B1 Plan View

Selective chemical etching is used to remove layers other than the layer to be observed. The resultant layer should be as thin as possible, but handling the sample becomes difficult when the thickness is less than 50 nm.

For the example of a $GaAs/Al_xGa_{1-x}As$ $(x \sim 0.3)$ heterostructure, the PA-30 $(H_2O_2 : NH_4OH = 30:1)$ solution attacks only the GaAs layer, while the boiling HF solution attacks only the $Al_xGa_{1-x}As$ $(x \sim 0.3)$ layer.

B2 Cross-Sectional View

1. Two pieces of ~4 mm square cut GaAs wafer grown as hetero-epitaxial layers are contacted face-to-face with epoxy glue. Furthermore to both sides of the sample other dummy GaAs substrate chips of the same size are glued. In order to make the epoxy layer thin enough, the sample is pressed during heat treatment at ~80°C for 3 - 4 hours.

2. With a diamond cutter, the sandwich structure is cut along a <110> or a <100> direction to ~300 μm thickness.

3. By lapping both sides with 2500 Al_2O_3 abrasive the thickness is reduced to ~120 μm.

4.   One side is polished with 0.1 μm Al$_2$O$_3$ abrasive and then mechano-chemically etched by 5% bromine-methanol solution to 50 μm thickness. The other side is also polished and etched. A specimen ~30 μm thick is obtained.

5.   The specimen is glued with epoxy on to a stainless-steel one-hole mesh with a hole diameter of 1 mm. It is further thinned by Ar ion milling until a small hole is opened at the centre of the specimen. The typical acceleration voltage of the ion beam is 2 - 5 kV and the incidence angle is 5 - 15°.

The procedure described above is a standard one for cross-sectional TEM sample preparation. Modifications are sometimes introduced. For example, in order to reduce the ion thinning time, the sample is polished by a Dimpler (a machine which makes a small and shallow hole) to 10 μm thick before ion thinning. In order to remove ion damage, chemical etching is tried after Ar ion thinning.

C   OBSERVATION

C1   Superlattice Structure

The structure of GaAs/AlAs monolayer superlattices was analysed by TEM and shown as a high contrast bright field image clearly exhibiting each constituent layer. This showed that the resolution of TEM was enough to distinguish up to two monolayers [1]. The lattice image also showed that the lattice arrangement was quite regular across the GaAs/AlGaAs interfaces [2,3]. High resolution TEM (HRTEM) observation could also distinguish an array of isolated InAs clusters with subnanometre scale (quantum dots) which was embedded within the GaAs matrix [4]. InGaP mixed crystal grown on GaAs has been observed with a lattice image which is locally ordered or formed as natural superlattices, as well as the extra electron diffraction spots [5,6].

C2   Abruptness of (GaAs/AlGaAs) Semiconductor Interface

It is practically very important that the abruptness of heterointerfaces can be observed directly using HRTEM. Using a digital image processing technique the abruptness of the interfaces was estimated on a lattice image [7]. It was also proposed that, in contrast to the usual <110> projection lattice image, the <100> projection image gives an enhanced contrast in chemical compositional difference between GaAs and AlAs. By using this direction, the interface roughness was indeed estimated on an atomic scale [8-10] ([10] observed the InP/InGaAs interface). The reason for this was understood as follows. In the (100) transmission electron diffraction (TED) pattern, the spots nearest to the direct beam are {002} beams, which give the highest contrast between GaAs and AlAs. A lattice image with clear contrast was observed with the [100] electron beam incidence even from a GaAs/Al$_x$Ga$_{1-x}$As superlattice with relatively small compositional difference x = 0.28, which exhibited no clear contrast in the [110] incidence lattice image [8]. It has recently been reported that terraces elongating along the [1$\bar{1}$0] direction are formed on the GaAs surface as evidenced by scanning tunnelling microscopy (STM). When observing the interfacial steps running along this direction, however, the conventional [110] incidence lattice image is more advantageous. The GaAs/AlAs interface structure was observed and the effect of the growth interruption on the smoothness of the interfaces was presented using the [1$\bar{1}$0] incidence HRTEM [11]. The [1$\bar{1}$0] incidence HRTEM was also applied to a heterostructure grown on a vicinal (001) substrate tilted toward the <110> direction [12].

C3   Observation of Quantum-Wires and Quantum-Boxes

Fabrication of quantum confinement structures not only in the growth direction of the epitaxial growth but also in the lateral directions results in quantum-wires and quantum-boxes. Several methods have been tried to prepare these structures by making use of some particular epitaxial growth mechanism, focused ion beam implantation and so on. For example, a cross-sectional TEM observation showed the QW structure laterally confined by the intermixed region realised by focused ion beam implantation and annealing [13].

C4   Ion Implantation and Disordering of Superlattices

TEM observations were carried out on GaAs/AlAs superlattices which were disordered by ion implantation of various ion species (Ga$^+$, As$^+$, Si$^+$, Ar$^+$) and subsequent heat treatment [14].

C5   Epitaxial Growth Mechanism

Cross-sectional TEM clearly showed that different epitaxial growth mechanisms occur in GaSb/AlSb MBE growth with different V/III beam intensity ratios [15].

C6   Surface Defect Observation

TEM observation contributed greatly to the study of the origin of the surface defects in MBE grown layers [16,17].

D   SPECIALISED OBSERVATION METHOD

D1   Scanning-Transmission Electron Microscope (S-TEM) (Microbeam)

Observation of extremely small selected areas is possible mainly with microbeam diffraction in the S-TEM. Changes of diffraction patterns can be observed by changing the selected area irradiated by the microbeam with a diameter ranging from 0.3 to 1.5 nm. Compositional changes and strains around a heterointerface of a GaAs/AlAs superlattice were examined [18,19].

D2   Convergent Beam Electron Diffraction (CBED) Method

The principle of the method is described in detail in [20,21]. By analysing the CBED pattern, it is possible to detect on a nanometre scale the symmetry of the crystal, thickness of the specimen, lattice constant of a selected local area in the crystal, assignment of crystal defects, and lattice deformation (strain and atomic composition) around heterointerfaces.

D3 Composition Analysis by Thickness-Fringe (CAT) Method

This method is used to observe the thickness fringes realised when an electron beam is transmitted through a rectangle edge prepared by cleaving an epilayer crystal. Iso-thickness fringes reflect the Al composition in GaAs/AlGaAs heterostructures. The spatial resolution was demonstrated to be 0.4 nm. The relation between Al composition and thickness-fringes was calculated using the dynamical theory, the result of which agreed quite well with the experiment. The abruptness of the heterointerfaces could also be estimated by this method. This method has the great advantage of ease of specimen preparation, which overcomes the greatest difficulty in the conventional observation method [22,24].

D4 Reflection Electron Microscope (REM)

Recently ultra-high vacuum technology has been adopted in the TEM apparatus so that contamination problems have almost been overcome. The difficulty in specimen preparation, however, still remains unsolved. REM observation, on the other hand, is advantageous because it is possible to observe a clean cleaved surface without any need to prepare a special specimen. REM has the same spatial resolution as TEM in the direction perpendicular to the beam incidence, but image shrinking occurs in the direction parallel to it. Image shrinkage is proportional to $\sin \theta$, where θ is the beam incidence angle. REM showed clear contrast between GaAs and AlGaAs due to the difference in electron beam reflection yield [25].

E COMPUTER-SIMULATION

Simulation of image contrast has usually been carried out using the multi-slice method [26], which takes into account the dynamic effect. In this method, the specimen is cut into many thin slices. By treating each of the slices as a phase medium for the electron beam, the scattering intensity is calculated at the slice of the specimen. The fast Fourier transformation method has been developed in the process [27]. The large unit cell method is suitable to simulate a superlattice structure having many interfaces [28]. It is necessary to determine acceleration voltage, spherical aberration, and chromatic aberration prior to starting the simulation. A best-fit image is obtainable by changing the value of the thickness of the specimen and the defocus as adjustable parameters.

F CONCLUSION

We have discussed transmission electron microscopy applied as a diagnostic tool to superlattice and quantum well structures. At the present time, almost all epitaxial structures and modulation structures have been studied using TEM. Therefore many technical papers have been published, and referring to all of them is far beyond the scope of this Datareview.

REFERENCES

[1]   P.M. Petroff, A.C. Gossard, W. Wiegmann, A. Savage [*J. Crystal. Growth (Netherlands)* vol.44 (1978) p.5]

[2]   A. Olsen, J.C.H. Spence, P. Petroff [*Proc. 38th Annual Meeting of the Electron Microscopy Society of America* Ed. G.W. Baily (Claritons, Baton Rouge, LA, 1980) p.318]

[3]   H. Okamoto, M. Seki, Y. Horikoshi [*Jpn. J. Appl. Phys. (Japan)* vol.22 (1983) p.L367]

[4]   O. Brandt, K. Ploog, L. Tapfer, M. Hohenstein, R. Bierwolf, R. Phillipp [*Phys. Rev. B (USA)* vol.45 (1992) p.8443]

[5]   A. Gomyo, T. Suzuki, K. Kobayashi, S. Kawata, I. Hino, T. Yasuda [*Appl. Phys. Lett. (USA)* vol.50 (1987) p.673]

[6]   O. Ueda, M. Takikawa, J. Komeno, I. Umebu [*Jpn. J. Appl. Phys. (Japan)* vol.26 (1987) p.L1824]

[7]   T. Furuta, H. Sakaki, H. Ichinose, Y. Ishida, M. Sone, M. Onoe [*Jpn. J. Appl. Phys. (Japan)* vol.23 (1984) p.L265]

[8]   Y. Suzuki, H. Okamoto [*J. Appl. Phys. (USA)* vol.58 (1985) p.3456]

[9]   J.C.D. Hetherington, J.C. Barry, J.M. Bi, C.J. Humphreys, J. Grange, C. Wood [*Mater. Res. Soc. Symp. Proc. (USA)* (1985) p.41]

[10]   A. Ourmazd, W.T. Tsang, J.A. Rentschler, D.W. Taylor [*Appl. Phys. Lett. (USA)* vol.50 (1987) p.1447]

[11]   N. Ikarashi, M. Tanaka, H. Sakaki, K. Ishida [*Appl. Phys. Lett. (USA)* vol.60 (1992) p.1360]

[12]   N. Ikarashi, A. Sakai, T. Baba, K. Ishida, J. Motoshita, H. Sakaki [*Appl. Phys. Lett. (USA)* vol.57 (1990) p.1983]

[13]   Y. Hirayama, Y. Suzuki, S. Tarucha, H. Okamoto [*Jpn. J. Appl. Phys. (Japan)* vol.24 (1985) p.L516]

[14]   Y. Suzuki, Y. Hirayama, H. Okamoto [*Jpn. J. Appl. Phys. (Japan)* vol.25 (1985) p.L912]

[15]   Y. Suzuki, Y. Ohmori, H. Okamoto [*J. Appl. Phys. (USA)* vol.59 (1986) p.3760]

[16]   Y. Suzuki, M. Seki, Y. Horikoshi, H. Okamoto [*Jpn. J. Appl. Phys. (Japan)* vol.23 (1984) p.164]

[17]   H. Kakibayashi, F. Nagata, Y. Katayama, Y. Shiraki [*Jpn. J. Appl. Phys. (Japan)* vol.23 (1984) p.L846]

[18]   N. Tanaka [*Jpn. J. Appl. Phys. (Japan)* vol.27 (1988) p.L468]

[19]   N. Tanaka, K. Mihama, H. Kakibayashi [*Jpn. J. Appl. Phys. (Japan)* vol.30 (1991) p.L959]

[20]   J.W. Steeds [*Introduction to Analytical Electron Microscopy* Eds J.J. Hren, J.I. Goldstein, D. Joy (Plenum Press, New York, 1979) p.378]

[21]   M. Tanaka, M. Terauchi [*Convergent-Beam Electron Diffraction* (JEOL-Maruzen, Tokyo, 1985)]

[22]   H. Kakibayashi, F. Nagata [*Jpn. J. Appl. Phys. (Japan)* vol.24 (1985) p.L905]

[23]   H. Kakibayashi, F. Nagata [*Jpn. J. Appl. Phys. (Japan)* vol.25 (1986) p.1646]

[24]   H. Kakibayashi, F. Nagata, Y. Ono [*Jpn. J. Appl. Phys. (Japan)* vol.26 (1987) p.770]

[25]   N. Yamamoto, S. Muto [*Jpn. J. Appl. Phys. (Japan)* vol.23 (1984) p.L806]

[26]   P. Goodman, A.F. Moodie [*Acta Crystallogr. A (Denmark)* vol.30 (1974) p.280]

[27]   K. Ishizuka, N. Ueda [*Acta Crystallogr. A (Denmark)* vol.33 (1977) p.740]

[28]   P.M. Fields, J.M. Cowley [*Acta Crystallogr. A (Denmark)* vol.34 (1978) p.103]

4.2　X-ray investigation of superlattices and quantum wells

L. Tapfer

July 1995

A　INTRODUCTION

X-ray diffraction experiments are frequently used for the investigation of structural and geometrical parameters of multilayered semiconductor heterostructures [1-3]. The main advantages are that X-ray diffraction experiments (i) do not require any sample preparation, (ii) are nondestructive and (iii) are highly sensitive and accurate for the determination of strain fields, lattice mismatch, epilayer thickness and also chemical composition if appropriate diffraction models are used for the data analyses.

B　BASIC EXPERIMENTAL AND THEORETICAL CONCEPTS

B1　Experimental Data Acquisition

The X-ray diffraction measurements can be performed by using a double-crystal arrangement, i.e. monochromator-sample, or a triple-crystal configuration, i.e. monochromator-sample-analyser [4]. In the first case, the X-ray beam intensity scattered from the sample is recorded (rocking curve) while the sample is scanned through a reciprocal lattice point (rotation around a Bragg angle). The experimental rocking curves will show several features which are related to the particular structural properties of the heterostructures. Recording rocking curves in the vicinity of different reciprocal lattice points it is possible to characterise the strain fields completely and to determine all the strain tensor elements of the distorted unit cell of the heterostructure [5]. Measurements carried out in symmetrical diffraction geometries will monitor the strain fields along the growth direction, while rocking curves recorded in asymmetrical diffraction geometries are also sensitive to the in-plane strain and are, therefore, used for revealing lattice relaxation processes. In the second case, several rocking curves are recorded where the angle offset for the analyser-crystal changes for each rocking scan. In this way a two-dimensional intensity contour-map around a reciprocal lattice point is obtained (reciprocal space map), where one axis corresponds to a scan along the surface normal (out-of-plane momentum transfer) and the other axis to a scan along the surface (in-plane momentum transfer) [4]. Basically, one reciprocal space map may yield similar information to two or three rocking curves recorded at different reciprocal lattice points. As X-ray sources, mainly conventional X-ray tubes or rotating anodes with Cu- or Fe-targets are used, having an X-ray wavelength of 0.154 nm and 0.194 nm, respectively. For these wavelengths the X-ray extinction length in III-V semiconductor compounds of interest is comparable to the typical total thickness of whole heterostructures [6,7].

B2　Data Evaluation Procedures

Several pieces of information can be obtained directly from the experimental rocking curves by using analytical formulae which relate the diffraction data to structural and geometrical parameters of superlattices and quantum well structures (see Section C). However, in many cases a simulation of the experimental diffraction curves is necessary. A computer simulation of the experimental rocking curves can be performed by using kinematical [7], semi-kinematical [6,8] and dynamical [9,10] diffraction models. The kinematical and semi-kinematical models can be used if (i) the lattice mismatch between the heterostructure and the substrate is larger than 0.3%, (ii) the thickness of the whole epitaxial structure is greater than 0.1 μm and less than the X-ray extinction length (about 0.7 μm for III-V semiconductor compounds using CuKα₁ radiation), and (iii) if dynamical interference phenomena can be neglected [11,12]. In all other cases dynamical diffraction theories are required for a correct simulation of the experimental data. However, it should be noted that the dynamical models are valid for 'slightly' distorted crystal structures, where the contribution of structural defects such as dislocations, point defects and clusters to the scattered X-ray intensity (X-ray diffuse scattering) is negligible. In this case statistical diffraction theories, which are still under investigation and test, must be considered [13,14].

C BASIC RELATIONS FOR THE STRUCTURAL PARAMETERS

C1 Lattice Mismatch and Strain Tensor Components

X-ray diffraction experiments enable us to measure all the components of the strain tensor of a heterostructure. The angular splitting $\Delta\Theta^\pm$ between the main diffraction peak of the heterostructure and the substrate peak is related to the strain tensor components by [5]:

$$\Delta\Theta^\pm = \tan\theta_B(\varepsilon_{zz}\cos^2\varphi + \varepsilon_{xx}\sin^2\varphi) \pm \frac{1}{2}\sin 2\varphi(\varepsilon_{zz} - \varepsilon_{xx}) -$$

$$(2\sin^2\varphi \pm \tan\theta_B\sin 2\varphi)\varepsilon_{xz} + U_{zx} \qquad (1)$$

where the upper sign corresponds to the grazing incidence and the lower sign to the grazing exit geometry. Here, θ_B is the kinematic Bragg angle of the substrate crystal and φ is the angle between the diffraction planes and the crystal surface. The components ε_{zz}, ε_{xx} and ε_{xz} refer to the normal, in-plane and shear strain components of the elastic distortion tensor (X-ray strain), respectively, while U_{zx} is the component of the asymmetric part of the strain tensor and describes the rotation of the epilayer unit cell with respect to the substrate unit cell [5]. The strain tensor components can be determined with an accuracy of about 3×10^{-5}. However, it should be noted that the accuracy depends strongly on the reflection orders and the structural quality of the heterostructure.

EQN (1) is valid for arbitrarily-oriented surfaces. In the case of coherent growth (pseudomorphic structures) in EQN (1) we will have $\varepsilon_{xx} = 0$. If the superlattices and quantum well structures are grown on high-symmetry surfaces, i.e. (100), (110) and (111) surfaces, the shear strain component is $\varepsilon_{xz} = 0$ [15].

It should be noted that EQN (1) has been derived from the dynamical incidence parameter and by using a first-order approximation. This approximation strictly is valid only for heterostructures with a lattice mismatch smaller than 0.5%. This consideration is almost independent of the material systems under consideration and diffraction geometries employed for the measurements. For larger lattice mismatch values and highly asymmetric diffraction geometries the second- or third-order approximation must be used [16-18].

In the case of multilayered heterostructures the main diffraction peak gives an average value of the strain tensor components and the lattice mismatch. In order to obtain the values of the individual layers which constitute the multilayered heterostructure, a simulation of the experimental diffraction curves is necessary [19].

C2   Layer Thickness and Superlattice Periodicity

In epitaxial layers of good structural quality the X-ray wave scattered at the interfaces may interfere and produce interference fringes in the rocking curve. The angular spacing $\Delta\omega$ between the interference fringes is related to the thickness t_f of the epitaxial layer:

$$t_f = \frac{\lambda \cos|(\theta_B \mp \varphi)|}{\Delta\omega \sin 2\theta_B} \tag{2}$$

where λ is the X-ray wavelength and the upper and lower signs refer to the grazing incidence and grazing exit geometry, respectively. Superlattices, i.e. structures with a periodic stacking of epitaxial layers made of materials A and B of different strain fields and/or chemical composition, can be considered as a single epitaxial layer of a 'new' material system with a lattice constant that coincides with the superlattice periodicity. The X-ray interference in superlattices will produce satellite peaks in the diffraction spectra which are located symmetrically around the main superlattice peak, i.e. zeroth order satellite peak. The angular distance between the satellite peaks is related to the superlattice period and is identical to the relation given by EQN (2). In the case where the superlattice period length is smaller than 2 nm the correct equation should be used:

$$t_{SL} = \frac{|m - n|\lambda}{|\sin \theta_{\pm n} - \sin \theta_{\pm m}|} \tag{3}$$

where $\theta_{\pm n}$ and $\theta_{\pm m}$ are the angular positions of the satellite peaks of order $\pm m$ and $\pm n$, respectively. The accuracy for the determination of the average superlattice period is about a fraction of a monolayer.

In the case of perfect superlattices, i.e. with sharp heterointerfaces and without thickness fluctuation, the full width at half maximum (FWHM) is the same for all the satellite peaks independent of peak order and is related to the total thickness of the whole superlattice (EQN (3)). For superlattices with thickness fluctuation and rough interfaces a broadening of the satellite peak occurs. The linewidth broadening $\delta\omega$ will be more pronounced for higher-order satellite peaks and is related to the thickness fluctuation or interface roughness ρ_i by the approximation [12]:

$$\rho_i \approx \frac{t_{SL}^2 \, \delta\omega \sin 2\theta_B}{n\lambda|\cos (\theta_B \mp \varphi)|} \tag{4}$$

However, for a more correct evaluation of the interface roughness a computer simulation of the experimental rocking curve has to be performed using a statistical diffraction model [20].

It should also be noted that the correlation function describing the interface roughness in the statistical diffraction model is a very critical parameter which may influence the analysis itself considerably.

C3   Chemical Composition, Mole Fraction and Doping Concentration

The mole fraction x of an epitaxial layer $A_xB_{1-x}C$ can be easily determined if the strain tensor components of this layer are calculated by following the procedures described in Sections B2 and C1 and using EQN (1). Assuming the validity of Vegard's rule the strain tensor components are related linearly to the mole fraction x ($\varepsilon_{ij} = \varepsilon_{ij}(x)$), i.e. the chemical composition. The accuracy for the determination of the average mole fraction, assuming the validity of Vegard's rule, is about $\Delta x = 0.001$ or better. However, it should be noted that for some semiconductor compounds a possible deviation from Vegard's rule has been revealed [21,22]. The validity of Vegard's rule for many semiconductor compounds remains an open issue.

The lattice mismatch will not depend on the chemical composition (mole fraction) only, but also on the concentration of dopant elements on lattice sites (substitutional impurities). In particular at high doping levels, as may occur in modulation doped multiple quantum well structures and high electron mobility transistor (HEMT) structures, non-negligible strain fields may be introduced. Assuming a linear dependence of the dopant induced strain from the dopant concentration N_d, we have the following relation [23]:

$$N_d = \frac{c_{11}\varepsilon_{zz}}{\beta(c_{11} + 2c_{12})} \tag{5}$$

where β is the contraction coefficient and is given by:

$$\beta = \frac{r_d - r_A}{r_d} \frac{1}{N_0} \tag{6}$$

Here, N_0 is the number of atoms per cm^3 in the considered semiconductor compound, r_d is the covalent radius of the dopant or impurity atom and r_A is the covalent radius of the element A whose lattice site the impurity atom will occupy. If $\beta < 0$ the lattice will be contracted (compressive strain) and otherwise a dilatation (tensile strain) occurs. For example, if carbon is incorporated on Ga or As sites of the GaAs lattice, we have $\beta_C = -8.35 \times 10^{-24}$ cm^3/atoms, and for silicon and beryllium we obtain $\beta_{Si} = -2.24 \times 10^{-24}$ cm^3/atoms and $\beta_{Be} = -1.57 \times 10^{-24}$ cm^3/atoms.

D   ANALYSIS AND EXPERIMENTAL DATA EVALUATION

D1   Superlattices and Multiple Quantum Wells

X-ray diffraction measurements on superlattices reveal the periodicity of strain fields and/or chemical composition (structure factor). In the case of all binary superlattices of the type AC/BC grown onto BC substrates, the angular distance between the main superlattice peak (zeroth order satellite peak) and the substrate crystal gives the average lattice strain between the whole superlattice (SL) and the substrate crystal. The average lattice mismatch between SL and substrate can be determined by measuring the average strain tensor components of the superlattice, and is given by [18]:

$$m_{SL} = \left(\frac{\Delta a}{a_{SL}}\right)^r = \frac{a_S - a_{SL}}{a_{SL}} = \frac{c_{12}(\varepsilon_{xx} + \varepsilon_{yy}) + c_{11}\varepsilon_{zz}}{c_{11} + 2c_{12}} \tag{7}$$

where a_S and a_{SL} are the substrate lattice parameter and average lattice parameter of the superlattice, respectively. Since the mismatch between the bulk AC and BC materials is known, i.e. $m_S = (a_{AC} - a_{BC})/a_{BC}$, the thickness ratio between the individual superlattice layers AC and BC is given by:

$$\frac{t_{BC}}{t_{AC}} = \frac{m_S}{m_{SL}} - 1 \tag{8}$$

Using EQN (2) and EQN (8) the thickness of the individual superlattice layers can be determined with an accuracy of about one monolayer and in some cases to a fraction of a monolayer. If the thickness ratio is

$$\frac{t_{AC} + t_{BC}}{t_{AC}} = k \tag{9}$$

the satellite peaks of order nk (with $n = \pm 1, \pm 2, \dots$) are suppressed (forbidden order). If satellite peaks of a forbidden order are still revealed in the experimental rocking curves, this indicates that the heterointerfaces between the constituent superlattice layers are not sharp and well defined (interface roughness or interdiffusion of elements) [24,25].

In the case of superlattices made of binary/ternary, ternary/ternary, etc., materials or if the constituent materials of the binary superlattices are different from the substrate material (e.g. GaSb/AlSb on GaAs), only the superlattice period length and the average strain tensor components and lattice mismatch of the superlattice can be obtained directly from the rocking curves as described above. However, in many cases the thickness and lattice strain of the individual layers can be determined by simulating the experimental diffraction data.

In the following some particular features of the X-ray diffraction measurements on III-V superlattices are briefly summarised.

D1.1 $Al_xGa_{1-x}As/Al_yGa_{1-y}As$ superlattices

The lattice mismatch between AlAs and GaAs is rather small (0.14%). In addition, for reflections which are mostly used, such as (400) and (422), the structure factor is given by the sum of the Al (Ga) and As scattering factors and, therefore, the structure factors for AlAs and GaAs are not very different. The low strain and structure factor modulation results in weak satellite peaks for these reflections (very often only satellite peaks of positive order are detected). It is very likely that in the case of short period superlattices, no satellite peaks are revealed in the vicinity of the (400) reflection. For this reason the quasi-forbidden (200) reflection (the structure factor for this reflection is given by the difference between the Al (Ga) and As scattering factors) is favoured because the structure factors for AlAs and GaAs are very different. Even if the (200) reflection is rather insensitive to the strain modulation, pronounced satellite peaks of negative and positive order can be revealed. FIGURE 1 shows the double-crystal X-ray rocking curves recorded in the vicinity of the (400)GaAs (a) and (200)GaAs (b) reflections of a 25 period $Al_{0.79}Ga_{0.21}As/AlAs$ superlattice with t_{AlAs} = 5.8 nm and t_{AlGaAs} = 9.4 nm, respectively. The dotted lines indicate the measured rocking curve while the solid line is the simulated curve. The peak 'S' indicates the GaAs substrate peak and 'O' is the main superlattice peak. The satellite peaks are labelled by $\pm n$. The quasi-forbidden (200) reflection is particularly indicated to reveal the intermixing or interdiffusion of the group III elements at the heterointerfaces [26].

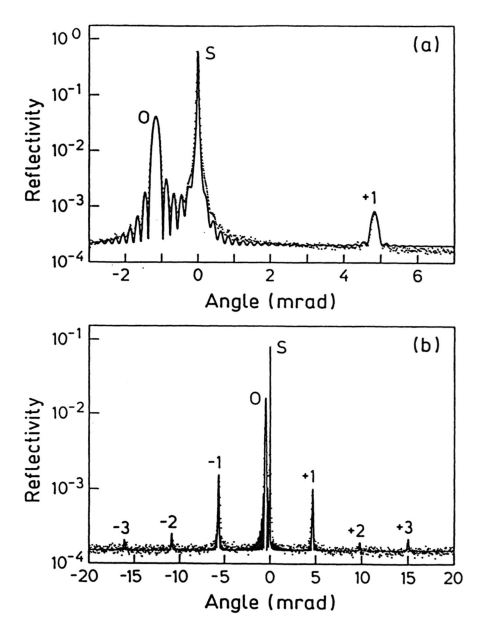

FIGURE 1. Double-crystal X-ray rocking curves recorded in the vicinity of the (400)GaAs (a) and the quasi-forbidden (200)GaAs (b) reflection of a 25 period $AlAs/Al_{0.79}Ga_{0.21}As$ superlattice grown by MBE onto (100)GaAs. The mole fraction and the AlAs and AlGaAs layer thicknesses (t_{AlAs} = 5.8 nm, t_{AlGaAs} = 9.4 nm) are determined by computer simulation (solid lines) of the experimental curves (dotted lines).

D1.2   In$_x$Ga$_{1-x}$As/GaAs superlattices

The strain modulation in these superlattices is rather strong due to the large lattice mismatch between InAs and GaAs (7%). The (400) and (422) reflections are very suitable for recording pronounced satellite peaks. The quasi-forbidden (200) reflection shows very weak satellite peaks because the structure factors F_{200}^{InAs} and F_{200}^{GaAs} are very similar. Interdiffusion of In and Ga atoms at the heterointerfaces will cause a broadening and intensity decrease of the satellite peaks.

D1.3   Ga$_x$In$_{1-x}$As/InP and Ga$_x$In$_{1-x}$As/Al$_y$In$_{1-y}$As superlattices

In general, these superlattice are grown lattice-matched onto InP substrates (x = 0.469 and y = 0.475). In the case of a perfectly lattice-matched superlattice or if there is no mismatch between the individual superlattice layers, no satellite peaks will be observed in the rocking curves. The structure factors of the constituent layers are very similar for all the reflections. If a small lattice strain between the constituent layers is present, the (400) and (422) reflections are the most appropriate to monitor the periodicity and to evaluate the amount of the strain modulation. Interdiffusion at the heterointerfaces will change the satellite peak intensities if the constituent layers still have coherent heterointerfaces [25]. If the interdiffusion zone exceeds the critical thickness, defects will be generated in the interface layer and the satellite peaks also will be broadened [24].

D1.4   Al$_x$Ga$_{1-x}$Sb/GaSb superlattices

Since the mismatch between AlSb and GaSb is about 0.65% all reflections will show pronounced satellite peaks for superlattices grown on GaSb as well as on GaAs substrate crystals. In particular, the (400), (422) and also the (511) reflections are strong, while the (200) reflection shows rather weak satellite peaks since the values for the structure factors of AlSb and GaSb are not very different.

D1.5   InAs/GaSb superlattices

The structure factors for InAs and GaSb are very similar for many reflections. However, the mismatch between InAs and GaSb is about 0.62% and the strain modulation is well pronounced independent of whether the superlattice is grown on GaSb or GaAs substrates. The reflections which are favoured for measuring the strain modulation along the growth direction are the (400) and (600) reflections.

D2   Single Quantum Wells

In non-periodic structures, like single quantum wells or laser and HEMT structures, more complicated interference phenomena may occur within the layered structure and give rise to an intensity modulation and phase shift of the interference fringes [19,27,28]. A simulation of the experimental rocking curve will yield the structural and geometrical parameters of the quantum well structure under investigation. The modulation of the interference fringes is basically related to the product of strain (lattice mismatch) and layer thickness. Also, ultrathin buried layers which are only a fraction of an atomic layer thick and single quantum wells were detected and characterised by this method [29,30]. It has been experimentally proven that structures with a strain-thickness product of about 0.005 nm could be detected by using a conventional double-crystal X-ray diffractometer [31]. FIGURE 2 shows the experimental (dotted line) and the simulated (solid line) rocking curve ($CuK\alpha_1$-radiation) of a 13 nm thick GaAs quantum well layer sandwiched between two $Al_{0.37}Ga_{0.63}As$ confinement (barrier) layers. The main heterostructure peak is labelled 'E' and the GaAs substrate 'S', respectively. The inset shows the layer structure and its structural parameters.

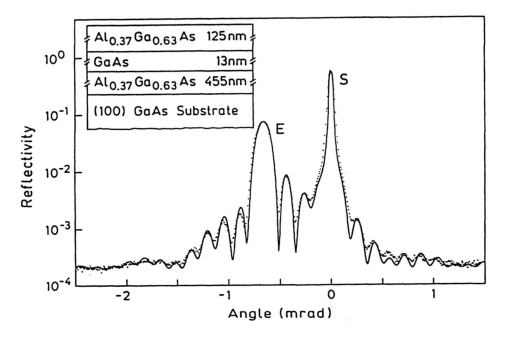

FIGURE 2. Experimental (a) and simulated (b) rocking curves (400 reflection) of a 13 nm thick single quantum well structure. The structural parameters (inset) of the heterostructure are determined by the simulation of the experimental curve.

Using this procedure the structural and geometrical parameters of more complicated structures such as HEMTs [19] and laser structures [19,28] could be determined and analysed. The modulation of interference fringes was also used for the analysis of the critical thickness [32] of highly strained buried layers, ultrathin single quantum wells and heterointerfaces.

E   CONCLUSION

High-resolution double- (or multiple-) crystal X-ray diffractometry is a powerful tool for the structural characterisation of multilayered semiconductor structures. Very accurate values for the structural parameters and chemical composition of quantum well structures and superlattices can be obtained, if the experimental rocking curves are analysed by using computer simulation procedures (line profile analysis). Also single quantum wells, ultrathin layers and interfaces can be revealed and characterised (determination of the layer thickness and/or strain).

REFERENCES

[1]   A. Segmuller, I.C. Noyan, V.S. Speriosu [*Prog. Cryst. Growth Charact. Mater. (UK)* vol.18 (1989) p.21-66]

[2]   A.T. Macrander [*Annu. Rev. Mater. Sci. (USA)* vol.18 (1988) p.283-302]

[3]   P.F. Fewster [*Appl. Surf. Sci. (Netherlands)* vol. 50 (1991) p.9-19]

[4]   P.F. Fewster [*J. Appl. Crystallogr. (Denmark)* vol.22 (1989) p.64-9]

[5]   Y.P. Khapachev, F.N. Chukhovskii [*Sov. Phys. Crystallogr. (USA)* vol.34 (1989) p.465-82]

[6]   L. Tapfer, K. Ploog [*Phys. Rev. B (USA)* vol.33 (1986) p.5565-74]

[7]   V.S. Speriosu, T. Vreeland Jr. [*J. Appl. Phys. (USA)* vol.56 (1984) p.1591-600]

[8]   C.R. Wie, H.M. Kim [*J. Appl. Phys. (USA)* vol.66 (1991) p.6406-12]

[9]   M.J. Hill, B.K. Tanner, M.A.G. Halliwell, M.H. Lyons [*J. Appl. Crystallogr. (Denmark)* vol.18 (1985) p.446-51]

[10]   W.J. Bartels, J. Hornstra, D.W.J. Loobeck [*Acta Crystallogr. A (Denmark)* vol.42 (1986) p.539-45]

[11]   P.F. Fewster, C.J. Curling [*J. Appl. Phys. (USA)* vol.62 (1987) p.4154]

[12]   L. Tapfer [*Phys. Scr. (Sweden)* vol.T25 (1989) p.45-50]

[13]   V. Holy, J. Kubena, E. Abramof, K. Lischka, A. Pesek, E. Koppensteiner [*J. Appl. Phys. (USA)* vol.74 (1993) p.136-1743]

[14]   T.J. Davis [*Acta Crystallogr. A (Denmark)* vol.49 (1993) p.755-62]

[15]   L. De Caro, L. Tapfer [*Phys. Rev. B (USA)* vol.48 (1993) p.2298-303]

[16]   M. Servidori, F. Cembali, R. Fabbri, A. Zani [*J. Appl. Crystallogr. (Denmark)* vol.25 (1992) p.46-51]

[17]   C. Bocchi, C. Ferrari, P. Franzosi, A. Bosacchi, S. Franchi [*J. Cryst. Growth (Netherlands)* vol.132 (1993) p.427-34]

[18]   C. Giannini, L. De Caro, L. Tapfer [*Solid State Commun. (USA)* vol.91 (1994) p.635-8]

[19]   L. Tapfer, K. Ploog [*Phys. Rev. B (USA)* vol.40 (1989) p.9802-10]

[20]   V. Holy, J. Kubena, J. Ohlidal, K. Ploog [*Superlattices Microstruct. (UK)* vol.12 (1992) p.25-35]

[21]   J.C. Mikkelsen Jr., J.B. Boyce [*Phys. Rev. B (USA)* vol.28 (1983) p.7130]

[22]   B.K. Tanner, A.G. Turnball, C.R. Stanley, A.H. Kean, M. McElhinney [*Appl. Phys. Lett. (USA)* vol.59 (1991) p.2272-4]

[23]   C. Giannini, C. Gerardi, L. Tapfer, A. Fischer, K.H. Ploog [*J. Appl. Phys. (USA)* vol.74 (1993) p.77-81]

[24]   L. Tapfer, W. Stolz, K.Ploog [*J. Appl. Phys. (USA)* vol.66 (1989) p.3217-9]

[25]   J.M. Vandenberg, S.N.G. Chu, R.A. Hamm, M.B. Panish, H. Temkin [*Appl. Phys. Lett. (USA)* vol.49 (1986) p.1302-4]

[26]   R.M. Fleming, D.B. McWhan, A.C. Gossard, W. Wiegmann, R.A. Logan [*J. Appl. Phys. (USA)* vol.51 (1980) p.357]

[27]   H. Holloway [*J. Appl. Phys. (USA)* vol.67 (1990) p.6229]

[28]   C.R. Wie, J.C. Chen, H.M. Kim, P.L. Liu, Y.-W. Choi, D.M. Hwang [*Appl. Phys. Lett. (USA)* vol.55 (1989) p.1774-6]

[29]   O. Brandt, L. Tapfer, R. Cingolani, K. Ploog, M. Hohenstein, F. Phillipp [*Phys. Rev. B (USA)* vol.41 (1990) p.12599-606]

[30]   C.R. Wie [*J. Appl. Phys. (USA)* vol.65 (1989) p.1036-8]

[31]   L. Tapfer, M. Ospelt, H. von Kanel [*J. Appl. Phys. (USA)* vol.67 (1990) p.1298-1301]

[32]   A. Mazuelas, L. Gonzales, F.A. Ponce, L. Tapfer, F. Briones [*J. Cryst. Growth (Netherlands)* vol.131 (1993) p.465-9]

4.3 X-ray diffraction from superlattices and quantum wells

H. Rhan

December 1995

A INTRODUCTION

Since the de Broglie wavelength of bound electrons - the most important quantity for quantum effects - typically lies in the range of lattice parameters (see TABLE 1) and the X-ray wavelengths, X-ray diffraction is a suitable method to investigate quantum well (QW) structures and related devices. Here, a review of different X-ray diffraction methods for characterising single and periodic QWs will be given.

TABLE 1. Dielectric susceptibilities along different reciprocal lattice points (0 0 0) (average), (2 0 0) (weak reflection), (4 0 0) (strong reflection): χ_0, χ_{200} and χ_{400}, the lattice parameter and the Poisson ratio for different III-V materials. All χ are given in units -1 x 10^{-6} and a is given in nm, $\lambda = 0.1540562$ nm [9,10,12,45].

Material	Re(χ_0)	Im(χ_0)	Re(χ_{200})	Im(χ_{200})	Re(χ_{400})	Im(χ_{400})	a	Poisson ratio
AlAs	21.0	0.63	7.7	0.40	10.8	0.57	0.56612	0.305
GaAs	30.0	0.89	1.2	0.14	15.3	0.80	0.565335	0.312
InAs	38.3	2.93	5.1	1.80	19.1	2.45	0.60584	0.352
InP	29.3	2.5	12.0	2.2	14.1	2.1	0.58687	0.360

Although the interference condition is strong, the expected signal intensity depends linearly on the number of scatterers, which is quite small in the QW. For this reason, it is difficult to observe a peak coming from the QW itself. However, this layer influences the scattered waves from other parts of the structure (substrate, top layer etc.) [1-6] and leads to interference, which makes it possible to characterise such thin layers.

Commonly, the thickness of periodic multilayers, multiquantum wells (MQW) or superlattices [7] consisting of many QWs and barriers is comparable with typical extinction lengths of X-rays [8], giving rise to at least one additional peak (or more for MQWs).

B REFLECTIVITY

Since the refractive index of condensed matter, n = 1 + χ_0/2, is less than unity for X-rays, the incoming beam is totally reflected as long as the incidence angle with respect to the surface (α_i) is less than the critical angle of total external reflection,

$$\alpha_c = \sqrt{|\chi_0|}$$

(see TABLE 1). The real part of the susceptibility

$$Re(\chi_0) \approx -\frac{N_A r_e Z}{2\pi A} \lambda^2 \rho$$

can be calculated knowing the averaged macroscopic density, ρ, (N_A = Avogadro number, r_e = electron radius, Z = atomic number, A = atomic weight) or the atomic form factors f [9,10]. For larger incidence angles the incoming beam penetrates into the crystal. The reflectivity (condition $\alpha_i = \alpha_f$) plotted versus the absolute value of the scattering vector

$$|\vec{q}| = \frac{2\pi (\sin \alpha_i + \sin \alpha_f)}{\lambda}$$

rapidly decreases as

$$R(q) \sim \left(\frac{q}{q(\alpha_c)}\right)^4$$

However, due to the interaction with the density jumps at the interfaces, an oscillatory behaviour is visible, coming from the interference of the waves reflected from the top and bottom of the layer.

The calculation of the X-ray reflectivity is based on Parrat's [11] recursive algorithm, which uses the Fresnel coefficients of every jth layer

$$f_j = \sqrt{\alpha_i + \chi_j}$$

The amplitude r_j of the reflectivity depends on that of the layer below (r_{j-1}):

$$r_j = \frac{r_{j-1} F_j}{1 + r_{j-1} F_j} e^{2ik f_j t_j}$$

$$F_j = \frac{f_{j-1} + f_j}{f_{j-1} - f_j} e^{-2k^2 \sigma^2 f_j f_{j-1}}$$

(1)

where $k = 2\pi/\lambda$ is the wave vector, λ is the wavelength, σ the roughness at the interface, and t_j is the thickness of the jth layer. The calculation starts in the substrate (j = 0) with r = 0 and F = 0 and goes up to j = N + 1, if N is the number of the layer, in order to consider the uppermost interface between the material and air. For small α_i values the reflectivity R(q) has to be corrected for a geometry factor [6].

The measurements shown in FIGURE 1 are performed using a double crystal diffractometer [5] and Cu Kα radiation from a rotating anode (FIGURE 2). FIGURE 1 shows the comparison of the measured and the simulated reflectivity profile. The model for the simulation consists of a 20 period GaAs/AlAs MQW with a well (GaAs) thickness of 6.5 nm and a barrier (AlAs) thickness of 15.0 nm. Besides the intensity decay above α_c, the first Bragg peaks coming from the MQW structure (period length τ) are clearly visible. The range of purely total reflection is cut off at the left-hand side. Above $\alpha_i = 1.1°$, the higher intensity in the tails is caused by the diffuse scattering, which is not included in the simulation. Note the destructive interference for the Bragg peaks at $\alpha_i \approx 2.1°$ and at $\alpha_i \approx 2.75°$. In terms of conventional X-ray diffraction, this may be interpreted as a (nearly) forbidden reflection and is caused by the thickness ratio of the well and the barrier ('structure factor'). To fit the experimental curve, an additional oxide layer of 2.1 nm and a small surface roughness of about 0.2 nm have to be taken into consideration.

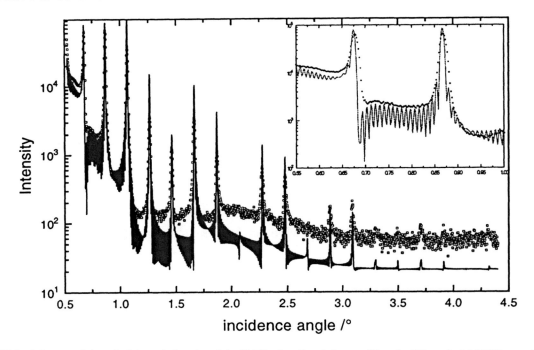

FIGURE 1. Measured (symbols) and simulated (solid line) reflectivity profile of a 20 period MQW consisting of 6.5 nm GaAs wells and 15 nm AlAs barriers. For larger incidence angles the diffuse scattering from interfaces leads to larger measured intensity, as can be seen particularly in the tails between the peaks. In the inset a part of both curves is enlarged. The fast oscillations due to the total thickness of the MQW are not visible in the experimental curve. All measured peaks are slightly broader than the simulated ones.

In the inset, the range around the two Bragg peaks at the left is enlarged. The (fast) oscillations due to the whole MQW thickness are not visible in the experimental curve and the peaks themselves are slightly broader than expected. However, this peak broadness does not depend on α_i. Therefore, the cause is probably the experimental resolution.

C   X-RAY DIFFRACTION

In contrast to reflectivity, X-ray diffraction (XRD) measurements are very sensitive to the crystalline structure. Since the lattice parameters, a, and the Poisson ratios of GaAs and the related materials (especially AlAs) agree very well (TABLE 1), the appropriate Bragg angles differ only slightly. Therefore, the required angular and spatial resolution of the diffractometer sometimes has to be higher than that for reflectivity, near the intrinsic width of the rocking curve (0.001° - 0.005°).

The scattering contrast between the layers is determined by the lattice mismatch perpendicular to the surface, which is the important quantity. In view of the operation of the QW, any mismatch parallel to the surface is not considered, although this could also be taken into account [4,5]. Because of the epitaxial growth, the strain relations (Poisson ratio) have to be considered [12,13]. These elastic moduli depend on both the materials and the growth direction.

Due to the small lattice mismatch, $\Delta a_\perp$, between GaAs and AlAs it is necessary to strengthen the contrast between the layers. By choice of a weak (e.g. 002) reflection, the contrast caused by the scattering factor can be enlarged by a factor of more than 10, so that it can be used for characterisation of QW structures [14-16]. The scattering factor $S(\vec{q})$ [9,10], which is roughly proportional to $S(q) \sim |f_A - f_B|^2 \approx |Z_{Ga} - Z_{As}|^2$, has a very small value in the zincblende structure. Any substitution of Ga by Al or In gives a much larger scattering factor (TABLE 1).

For symmetrical reflections, the amplitude ratio of diffracted, D_H, to incident, D_O, wave

$$X = \sqrt{\frac{\chi_{\bar{H}}}{\chi_H}} \frac{D_H}{D_0}$$

can be obtained by reduction of Takagi's dynamical differential equations [4,17,18] to:

$$-i\frac{\partial X}{\partial T} = X^2 - 2yX + 1 \tag{2}$$

where y and T are the normalised angular and thickness functions, respectively:

$$y = -\left(\left(\Delta\theta + \frac{\Delta a_\perp}{a \tan \theta_B} \sin 2\theta_B\right) - \chi_0\right) \times \left(C\sqrt{\chi_{\bar{H}}\chi_H}\right)^{-1} \tag{3}$$

$$T = \frac{\sqrt{\chi_{\bar{H}}\chi_H}}{\sin \theta_B} \times \frac{t}{\lambda} \tag{4}$$

$\chi_{\bar{H}}$, χ_H are the susceptibilities in the directions of the reciprocal lattice points $\bar{H}$ and H, and t is the layer thickness. Mostly, such a dynamical treatment is only necessary close to the substrate peak and for layer thicknesses above the extinction length. In the kinematical approximation, the square term in EQN (2) is neglected and the equation can easily be solved [3,4]. The amplitude ratio X_t at the top side of any layer is given in terms of that from its underside X_O:

$$X_t = X_0 e^{-2iyT} + ie^{-iyT} \frac{\sin y T}{y} \tag{5}$$

Approximating small lattice parameter differences, the period length τ of a multilayer structure results from the angular spacing (ω_S) of any MQW peak from its zero-order peak, θ_O:

$$\tau = n \times \lambda \sin \theta_O/(\sin 2\theta_O \omega_S) \tag{6}$$

where n is the order of the MQW reflection. If there are any fluctuations in the period, $\Delta\tau$, they directly affect the measured peaks. In the derivation of EQN (6) the enlargement of the full width at half maximum (FWHM) of the MQW peaks $\Delta\omega_S$ can be estimated from the fluctuations:

$$\Delta\omega_S/\omega_S = -\Delta\tau/\tau \tag{7}$$

For example, the slope of the FWHM of the MQW reflections is equal to the relative fluctuations of the periods. In this angular range, a change in instrumental resolution is not expected.

The experimental set-up of both the reflectivity and XRD measurements is given in FIGURE 2. The first Ge (Si) (220) crystal separates the $K\alpha_1$ line from the Cu doublet. The slits limit the beam size on the sample to 0.4 x 1 mm and reduce the incoherent background. To avoid an additional dispersion connected with a broadening of the FWHM, the monochromator and sample were arranged in a weakly-dispersive [19] set-up (reflections $0\bar{2}2$ and 002, respectively).

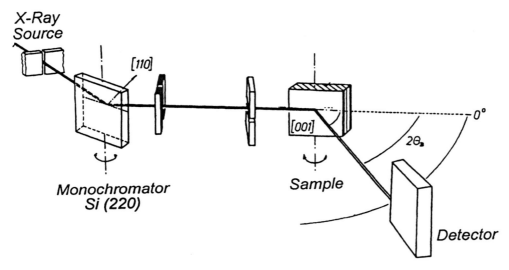

FIGURE 2. Schematic representation of the experimental set-up used for reflectivity and XRD measurements. The beam from the tube is made monochromatic and collimated by the Si (Ge) monochromator and the slit system.

FIGURES 3(a) and 3(b) show the measured diffraction curve of the MQW structure, which was used in the reflectivity, together with the simulation according to EQN (5), versus the deviation of the Bragg angle, $\Delta\theta = \theta - \theta_B$. Despite the small scattering factor, the highest and narrowest peak is that from the GaAs substrate and is superposed by the zero-order peak of the MQW (θ_0). It corresponds to a composition averaged over the whole MQW. In addition, at least six orders of MQW peaks can be seen. Their angular distance, ω_S, results directly from the MQW period, τ. The envelope of these peaks gives information about the inner structure of the MQW, e.g. the thickness ratio of the well to the barrier, as in the reflectivity. Using τ, the thickness ratio, and the average composition, the thicknesses of the single layers (well 6.5 nm, barrier 15.0 nm) in the MQW as well as their composition (assuming the well consists of pure GaAs) are determined [14-16].

FIGURE 3(b) is an enlargement of FIGURE 3(a) around the MQW peak of order -1 and shows that the expected oscillating behaviour caused by the total thickness of the MQW structure is also visible in the measured curve. The number of oscillations between two MQW peaks is the number of periods divided by 2. The experimental curve is smoothed by some real structure effects. Therefore the minima are not so pronounced as in the simulation.

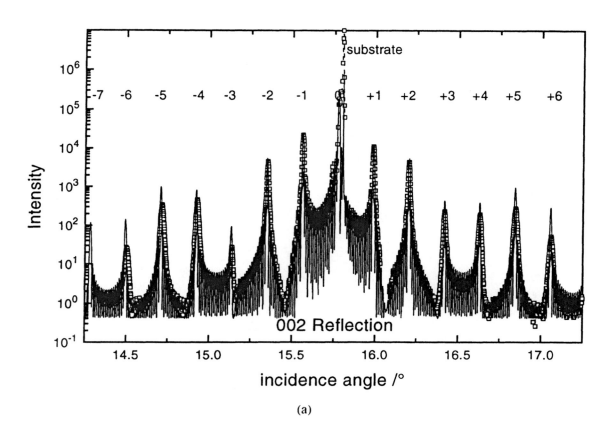

(a)

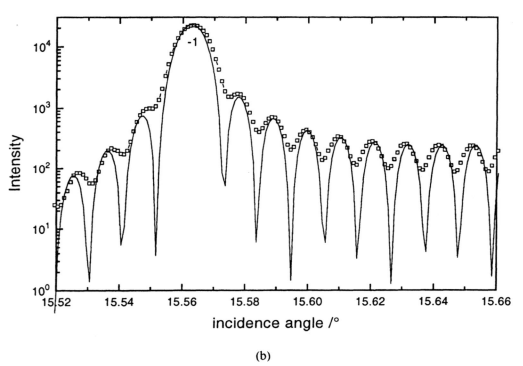

(b)

FIGURE 3. (a) Rocking curves of the same MQW structure as in FIGURE 1 (measured - symbols, simulated - solid line). In addition to the GaAs substrate peak in the centre many MQW peaks (numbered) are visible. (b) The inset shows the surrounding of the MQW peak -1. Due to the better resolution the fast oscillation can also be recognised in the measured curve; however, the weaker modulation is due to the real structure.

To obtain information on the disturbance of the periodicity, the FWHM $\Delta\omega_S$ of the MQW peaks is displayed as a function of the angular distance $\omega_S = \theta - \theta_0$ in FIGURE 4. The linear regression gives $\Delta\tau/\tau = 1\%$, which corresponds to $\Delta\tau \approx 0.22$ nm or ± 1 monolayer: according to EQN (7), τ varies by about ± 1 monolayer around its average yield. The oscillation dips seem to be averaged in FIGURE 4. These dips result from destructive interferences which are perturbed.

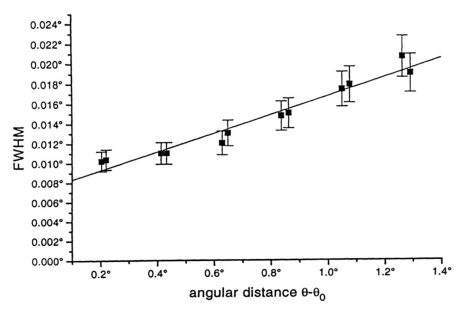

FIGURE 4. The widths of the MQW peaks are displayed against their angular distance from the zero-order peak, θ_0. The slope of about 1% gives the value of the relative period fluctuations (see text).

D   GRAZING INCIDENCE DIFFRACTION

For investigations of single quantum wells (SQW) which are located near the surface, the methods described above have a disadvantage due to the large penetration depth of X-rays. In the common energy range around 10 keV, this depth is some μm, depending on the material and possibly anomalous absorption. To overcome this, the incidence angle can be chosen to be as small as possible (asymmetric reflections). However, this is usually possible only down to $\alpha_i \approx 1°$ and it is necessary to consider some additional effects [20,21] such as the drastic change of the penetration depth over the rocking curve due to the variation of α_i.

By contrast, using diffraction at net planes, which stand perpendicular to the surface (grazing incidence diffraction - GID), the penetration depth can be varied (nearly) independently of the Bragg condition (FIGURE 5) [22,23]. Owing to the control of incidence and exit angle with respect to the surface (α_i, α_f), the penetration depth can be changed in the range 5 nm - 1 μm [24]. This advantage may be applied to relaxed layer structures or for investigations of interfacial layers within these structures [25]. For lattice-matched systems like GaAs/AlAs or very thin InAs layers on or sandwiched in GaAs, in GID the interesting signal is expected to be close to the lateral Bragg condition, θ_B, because of the unique lateral lattice parameter. However, at this point there is a strong difference in the scattering factor, $S(\vec{q})$, and the refractive index $n = 1 + \chi_0/2$.

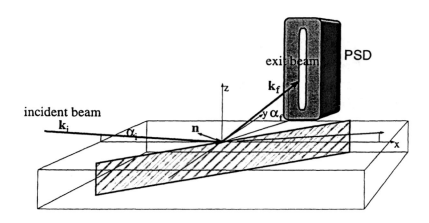

FIGURE 5. Schematic diffraction scheme in grazing incidence/exit geometry. The diffracting net plane stands perpendicular to the surface. The small divergence in the incident beam is transformed into a wide α_f fan.

For the calculation of the scattered intensity, the distorted wave Born approximation [28,29] is commonly used while the fully dynamical treatment [30-32] has rarely been used. Thus, the intensity $I(\vec{q})$ is given mainly by the transmission functions [24]:

$$I(q) = |T(\alpha_i)\,T(\alpha_f)\,S(q)|^2 e^{-\sigma^2 q_z^2} \tag{8}$$

The scattering function $S(q)$ may be split into $S(q) = S(q_z)\,S(q_\parallel)$, where

$$q_z = \frac{2\pi}{\lambda}\left(\sqrt{\alpha_i^2 + \chi_0} + \sqrt{\alpha_f^2 + \chi_0}\right)$$

is perpendicular to the surface and $q_\parallel$ is parallel to it. $S(q_z)$ is written as

$$S(q_z) = \chi_H \frac{1 - e^{-i q_z t}}{1 - e^{-i q_z a}} e^{-i q_z t_{top}} \tag{9}$$

where t_{top} is the depth of the layer. For an exact simulation of semiconductors, $S(q_\parallel)$ can become very important [27] and has to be considered.

The measurements were performed on the surface diffractometer at HASYLAB's beamline D4 [34] using a wavelength $\lambda = 0.142$ nm, which generally corresponds to FIGURE 2. The sample was a one monolayer thick InAs SQW embedded in GaAs. While no intensity modulation could be found in the strong (220) and (040) reflection, the weak (020) reflection was selected.

FIGURE 6 shows the intensity distribution in the position sensitive detector, e.g. vs. α_f, which is compared to a simulation of a one monolayer InAs layer covered by 38 nm GaAs and a 3 nm thick oxide. The slope on the left-hand side up to α_c is due to the transmission function and the top layer on the GaAs, that does not diffract. Above α_c an oscillating behaviour is visible. The cause is the interference of the waves scattered from the GaAs cap layer while the strength is determined by the thickness of the InAs layer. Near α_c there is a difference between the measured and the simulated curve because of the kinematical approximation, which is not strictly valid around α_C. The scattering contrast can be varied by changing the reflection order or by changing the incidence angle, α_i. By comparing the data sets obtained, it is possible to distinguish between layer thickness in the QW and the composition in a compound.

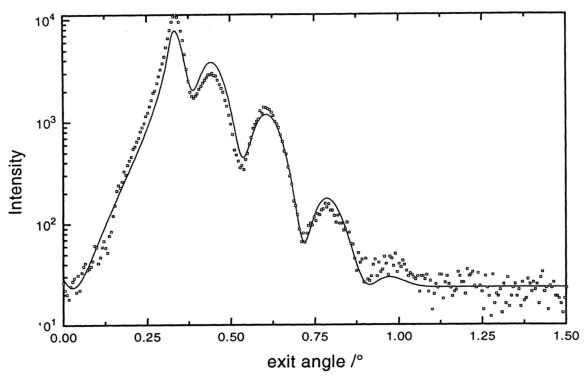

FIGURE 6. Measured (symbols) and simulated (line) intensity distribution over the exit angle α_f of a 1 monolayer InAs single quantum well sandwiched in GaAs. The oscillating behaviour is caused by the 38 nm thick GaAs cap layer.

E   CONCLUSION

X-ray diffraction methods have been discussed for the investigation of different quantum well structures. Due to the high scattering intensity in the forward direction, a large dynamic range is available and a fast characterisation is possible. For investigations of the interfaces [35], it is necessary to take into account the diffuse scattering ($\alpha_f \neq \alpha_i$). In this case, a lot of information about the disturbed interfaces and the surface (e.g. correlated roughness) can be obtained [36-38].

Systems such as (InGa)As on InP and on GaAs are not ideal for reflectivity measurements due to their small refractive index difference (TABLE 1). Well-periodic multilayers of such systems only may be characterised successfully by reflectivity.

Because of the perfect lattice of the semiconductors, XRD is a very useful tool to investigate QW structures. Periodic MQWs can be characterised with a high accuracy in length (e.g. part of a monolayer) and composition [39,40].

SQWs, which are capped or sandwiched by the much thicker barrier layer(s) like (InGa)As in GaAs or GaAs in (AlGa)As grown on GaAs, usually cannot be observed directly. Since the thickness is small and the number of scatterers is also small, only a very weak intensity distributed over a very large angular range is to be expected. Using the interference behaviour of the much larger barrier layers, characteristic values of the QW can be obtained [41,42]. Thus, in thinner layers the effects of thickness and strain (composition) of the QW are more difficult to determine [43]. A comparison of the data obtained by reflectivity and XRD shows that they agree well [6].

GID may be applied for both SQWs and MQWs. The remarkable advantage of GID is that it can distinguish between the influence of thickness and of composition on the diffraction pattern even for small QWs. For such a decision to be made it is necessary to vary α_i and/or the reflection order. Additionally, by changing α_i the penetration depth can be controlled. Because of the weak scattered intensity, this method is mostly performed in a synchrotron. Although the method is sensitive to the lateral lattice mismatch, GID also allows information on very thin interfacial layers in strained as well as in relaxed lattices to be obtained.

In conclusion, X-ray diffraction makes possible the nondestructive characterisation of quantum well structures down to monolayer resolution. However, if the structure consists of many layers with different thicknesses, the data analysis can be very difficult.

ACKNOWLEDGEMENT

The author wishes to thank Drs. D. Martin and F. Morier-Genaud of the Ecole Politechnique de Lausanne and Dr. V. Gottschalch of the Chemistry Department of the University of Leipzig for growth and preparation of the samples as well as for fruitful discussion. This work was supported by the Bundesministerium für Forschung und Technologie under contract number 055WMAXI.

REFERENCES

[1]   L. Tapfer, K. Ploog [*Phys. Rev. B (USA)* vol.40 (1989) p.9802]
[2]   E. Spiller, A.E. Rosenbluth [*Opt. Eng. (USA)* vol.25 (1986) p.954]
[3]   V.S. Speriosu, T. Vreeland Jr. [*J. Appl. Phys. (USA)* vol.56 (1984) p.1591]
[4]   W.J. Bartels, J. Hornstra, D.W.J. Lobeek [*Acta Crystallogr. A (Denmark)* vol.42 (1986) p.539]
[5]   T. Baumbach, H.-G. Bruhl, H. Rhan, H. Pietsch [*J. Appl. Crystallogr. (Denmark)* vol.21 (1988) p.386]
[6]   H. Rhan, U. Pietsch, S. Rugel, H. Metzger, J. Peisl [*J. Appl. Phys. (USA)* vol.74 (1993) p.146]
[7]   In X-ray diffraction there is no difference between SLs and QWs because in this energy range no quantum effects are expected and the treatment of both is classical.
[8]   V.M. Laue [*Rontgenstrahlinterferenzen* Frankfurt, 1932]
[9]   [*International Tables for Crystallography* Ed. T. Hahn (Dordrecht, Holland, 1987)]
[10]   C.T. Chantler [*J. Phys. Chem. Ref. Data (USA)* vol.24 (1995) p.71]
[11]   L. Parrat [*Phys. Rev. (USA)* vol.95 (1954) p.359]
[12]   J. Hornstra, W.J. Bartels [*J. Cryst. Growth (Netherlands)* vol.44 (1978) p.513]
[13]   W.J. Bartels, W. Nijman [*J. Cryst. Growth (Netherlands)* vol.44 (1978) p.518]
[14]   G. Oelgart et al [*Phys. Status Solidi B (Germany)* vol.49 (1992)]

[15]   M. Proctor, G. Oelgart, H. Rhan, F.-K. Reinhart [*Appl. Phys. Lett. (USA)* vol.64 (1994) p.3154]

[16]   G. Oelgart et al [*Phys. Rev. B (USA)* vol.49 (1994) p.10456]

[17]   S. Takagi [*J. Phys. Soc. Jpn. (Japan)* vol.26 (1969) p.1239]

[18]   C.R. Wie, T.A. Tombrello, T. Vreeland Jr. [*J. Appl. Phys. (USA)* vol.59 (1986) p.3743]

[19]   J.W.M. DuMond [*Phys. Rev. (USA)* vol.32 (1937) p.872]

[20]   F. Rustichelli [*Philos. Mag. (UK)* vol.31 (1974) p.1]

[21]   R. Zaus [*J. Appl. Crystallogr. (Denmark)* vol.26 (1993) p.801]

[22]   W.C. Marra, P. Eisenberger, A.Y. Cho [*J. Appl. Phys. (USA)* vol.50 (1979) p.6927]

[23]   R. Feidenhans'l, M. Nielsen, F. Gray, R.L. Johnson, I.K. Robinson [*Surf. Sci. (Netherlands)* vol.186 (1987) p.499]

[24]   H. Dosch [*Springer Tracts Mod. Phys. (Germany)* vol.126 (1992)]

[25]   H. Rhan [*Physica B (Netherlands)* submitted (1995)]

[26]   D. Rose, U. Pietsch, V. Gottschalch, H. Rhan [*J. Phys. D, Appl. Phys. (UK)* vol.28 (1995) p.A246]

[27]   H. Rhan, J. Peisl [*Z. Phys. B, Condens. Matter (Germany)* submitted]

[28]   G.H. Vineyard [*Phys. Rev. B (USA)* vol.26 (1982) p.4146]

[29]   S. Dietrich, H. Wagner [*Z. Phys. B, Condens. Matter (Germany)* vol.27 (1984) p.27]

[30]   A.M. Afanasiev, A.M. Melkonyan [*Acta Crystallogr. A (Denmark)* vol.39 (1983) p.207]

[31]   H. Rhan, U. Pietsch [*Z. Phys. B, Condens. Matter (Germany)* vol.80 (1990) p.347]

[32]   U. Pietsch, H. Rhan, A.L. Golovin, Y. Dmitriev [*Semicond. Sci. Technol. (UK)* vol.6 (1991) p.743]

[33]   J.W. Matthews, A.E. Blakeslee [*J. Cryst. Growth (Netherlands)* vol.27 (1974) p.118]

[34]   T. Salditt, H. Rhan, T.H. Metzger, J. Peisl, R. Schuster, J.P. Kotthaus [*Z. Phys. B, Condens. Matter (Germany)* (1994) p.227]

[35]   D.G. Stearns [*J. Appl. Phys. (USA)* vol.65 (1989) p.491]

[36]   S.K. Sinha, E.B. Sirota, S. Garoff, H.B. Stanley [*Phys. Rev. (USA)* vol.38 (1988) p.2297]

[37]   J.-P. Schlomka, M. Tolan, L. Schwalowski, O.H. Seek, J. Stettner, W. Press [*Phys. Rev. B (USA)* vol.51 (1995) p.2311]

[38]   V. Holy, T. Baumbach, M. Bessiere [*J. Phys. D, Appl. Phys. (UK)* (1995) p.A220]

[39]   J.M. Vandenberg, J.C. Bean, R.A. Hamm, R. Hull [*Appl. Phys. Lett. (USA)* vol.52 (1988) p.1152]

[40]   E.E. Fullerton, I.K. Schuller, H. Vanderstraeten, Y. Bruynseraede [*Phys. Rev. B (USA)* vol.45 (1992) p.9292]

[41]   L. Tapfer, K. Ploog [*Phys. Rev. B (USA)* vol.40 (1989) p.9802]

[42]   T. Baumbach, H. Rhan, U. Pietsch [*Phys. Status Solidi A (Germany)* vol.109 (1988) p.K7]

[43]   C. Bocchi, C. Ferrari [*J. Phys. D, Appl. Phys. (UK)* (1995) p.A164]

[44]   U. Pietsch et al [*Appl. Surf. Sci. (Netherlands)* vol.54 (1991) p.502]

[45]   K. Kamigaki, H. Sakashita, H. Kato, M. Nakayama, N. Sano, H. Terauchi [*Appl. Phys. Lett. (USA)* vol.49 (1986) p.1071]

4.4   Excitonic line broadening in quantum well structures due to interface roughness

K.K. Bajaj

September 1995

A   INTRODUCTION

A variety of modern electronic devices such as quantum well lasers, high electron mobility transistors, spatial light modulators, etc. involve interfaces between two semiconductors and thus the quality of the interface has an important effect on the performance of these devices. In quantum well structures, low temperature (4 K) photoluminescence (PL) studies are used extensively to estimate the quality of the interfaces forming the well. The linewidths of the excitonic spectra are generally used to characterise the quality of the interfaces since at low temperatures the principle mechanism for excitonic line broadening is the so-called interface roughness.

More than a decade ago Weisbuch et al [1] pointed out that in GaAs/AlGaAs quantum well structures, the topological disorder at the interfaces produces significant optical effects and that the optical methods such as PL and PL excitation spectroscopies provide useful means to characterise this disorder, complementary to the more microscopic methods of transmission electron microscopy (TEM) and X-ray diffraction. In the multiple quantum well structures where the low temperature PL is primarily due to intrinsic excitonic transitions, the width of these transitions is controlled by two effects: (i) layer-to-layer thickness variation in the sample which leads to different confinement energies for the different layers, and (ii) thickness fluctuation within each layer, all the layers having the same average thickness. This fluctuation arises due to the fact that after a layer of GaAs or AlGaAs has been grown, there exist a number of islands at the free interface as the number of atomic planes in a layer is never exactly an integral number. The next layer grown will freeze these islands, so that the microscopic interface position cannot be defined to better than one monolayer, although the macroscopic average position can be defined more accurately.

In this Datareview we will consider the effect of the thickness fluctuations within a given layer on the excitonic line broadening in quantum well structures. Though we will focus our attention on GaAs/AlGaAs structures the results discussed here are equally applicable to other similar systems. We will briefly review the results of various calculations of the excitonic linewidths in quantum wells.

B   THEORY OF EXCITONIC LINEWIDTH

After the early work of Weisbuch et al [1], Singh et al [2] developed a theory correlating the linewidths of the excitonic PL spectra in quantum wells with the microscopic details of the interface quality. This theory was based on the realisation that the optical probe, namely, the exciton, has effectively a finite extent (~250 Å in GaAs, for instance) and the energy of the emitted radiation reflects the average composition of the region seen by the exciton. Using arguments similar to those invoked by Lifshiftz [3] to understand the excitation spectra of disordered alloys, Singh et al calculated the probability distribution of the compositional fluctuations at the interface. These fluctuations were then related to the line shape of the PL spectra. The basic formalism used in their work is briefly outlined in the following.

To discuss the model for the interface we consider the most studied quantum well structure i.e. GaAs/Al$_x$Ga$_{1-x}$As. Depending on the growth conditions, there will be localised fluctuations in the well size around a mean value L_0. We assume that the non-ideal interface can be represented by an average interface at $z = 0$ with fluctuations extending a distance $\pm\delta_1$ from the interface. These fluctuations arise from the presence of islands of AlGaAs in the well and the islands of GaAs in the barrier region. We further assume that the correlated lateral radius of these islands (assumed circular for the sake of illustration) is δ_2, i.e. the smallest island has a radius δ_2. Larger islands have radii which are integral multiples of δ_2. Based on this model we view the interface as being described on a global scale by parameters C_a^0, C_b^0, C_c^0, representing the mean concentrations of the islands protruding into the well, islands projecting out of the well and regions that are flat.

The intrinsic PL spectrum arises as a result of radiative recombination of an exciton and the emission energy $E_{ex}(\underline{r},L)$ depends on the mean value of the width of the quantum well as seen by the exciton, namely

$$E_{ex}(\underline{r},L) = E_e(\underline{r},L) + E_h(\underline{r},L) - E_b(\underline{r},L) + E_g \qquad (1)$$

where $\underline{r}$ denotes the centre of mass coordinate of the exciton, E_e and E_h are the energies of the electron and hole subbands, respectively, E_b is the binding energy of the exciton and E_g is the bandgap of the well material (GaAs). In quantum well structures, the degeneracy of the valence band is removed leading to the formation of two types of excitons, namely the heavy-hole exciton and the light-hole exciton associated with the heavy-hole subband and light-hole subband respectively. Therefore in EQN (1) E_h and E_b refer to the appropriate exciton. The value of E_b changes very little due to small fluctuations in well size. However the values of E_e and E_h are quite sensitive to the changes in well size, particularly for narrow wells. The exciton wave function extends over an effective region of radius say, R, so that the relevant information regarding the quantum well corresponds to that part of the well with lateral extent R and centred around the position $\underline{r}$ of the exciton. The average width of the well around point $\underline{r}$ will be given by the average microscopic nature of the interface. If it is assumed that the excitons are created randomly in the well, the line shape is determined by the probability distribution P(R,L) of finding fluctuations in the well size extending over the effective exciton size.

To determine the probability distribution required to calculate the line shape, Singh et al employed the statistical thermodynamic arguments similar to those used by Lifshitz [3]. It was assumed that the islands, valleys and flat regions occur randomly at the interface and that the radii of the smallest clusters associated with islands, valleys and flat regions are δ_{2a}, δ_{2b}, δ_{2c}, respectively. The probability of finding fluctuation concentration C_a, C_b, and C_c over a region of effective radius R when the global concentrations are C_a^0, C_b^0, and C_c^0, is given as [3]

$$P(C_a,C_b,C_c,R) = \exp -\left[\frac{R^2}{\delta_{2a}^2} C_a \ln\left(\frac{C_a}{C_a^0}\right) + \frac{R^2}{\delta_{2b}^2} C_b \ln\left(\frac{C_b}{C_b^0}\right) + \frac{R^2}{\delta_{2c}^2} C_c \ln\left(\frac{C_c}{C_c^0}\right) \right] \qquad (2)$$

The above expression is valid in the case where we assume that the quality of the well is determined primarily by the interfacial quality of only one interface. The effective well size obtained in this case is

$$L = L_0 + \delta_1 \left[\left(C_a - C_a^0 \right) - \left(C_b - C_b^0 \right) \right] \qquad (3)$$

In the case where both the interfaces in a quantum well are of comparable quality R^2 in EQN (2) and δ_1 in EQN (3) are replaced by $2R^2$ and $2\delta_1$ respectively.

To evaluate the probability distribution (EQN (2)) R^2 was replaced by the expectation value of $\rho^2 = x^2 + y^2$ using the variational wave functions of Greene and Bajaj [4] who have calculated the binding energies of heavy- and light-hole excitons as a function of well width. The values of the electron and the hole subband energies E_e and E_h as a function of well size, are determined by solving the problem of a particle in a box. Combining EQNS (1), (2) and (3) one can determine the excitonic linewidth.

Singh et al calculated the full width at half maximum σ for the heavy-hole exciton as a function of well size in GaAs/Al$_{0.3}$Ga$_{0.7}$As quantum well structures. They chose, for the sake of illustration, a simple model in which it was assumed that the interface was made up of valleys and hills only with $C_a^0 = C_b^0 = 0.5$ and $\delta_{2a} = \delta_{2b} = \delta_2$ and took δ_1 equal to the thickness of one monolayer of GaAs (i.e. 2.83 Å). For a given value of δ_2 they calculated the value of σ using the following procedure. First they calculated C_{a1} and C_{a2} which are two points where probability distribution $P(C_a, R)$ is 0.5 at a given well size L_0. From EQN (3) they determined the effective well sizes, L_1 and L_2 corresponding to C_{a1} and C_{a2} where L_1 is less than L_0 and L_2 is larger than L_0. They then calculated the exciton transition energies $E_{ex}(L_1)$ and $E_{ex}(L_2)$ using EQN (1). The value of σ is given as a difference between $E_{ex}(L_2)$ and $E_{ex}(L_1)$.

In FIGURE 1 is displayed the variation of σ of the heavy-hole exciton as a function of well width for three different values of δ_2 assuming $\delta_1 = 2.83$ Å. We note that for a given well size the value of σ increases as a function of δ_2. In addition, for a given value of δ_2, the value of σ increases as the well size is reduced due to larger

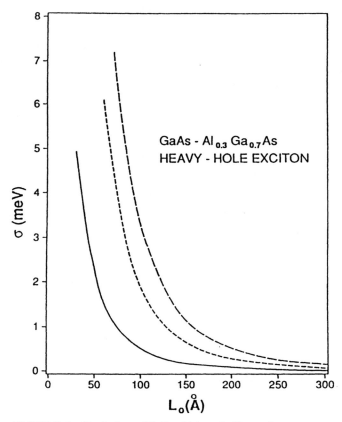

FIGURE 1. Variation of full width at half maximum (σ) of the heavy-hole exciton as a function of well width (L_0) for $\delta_2 = 20$ Å (solid line), $\delta_2 = 80$ Å (dotted line), and $\delta_2 = 160$ Å (dashed line), in GaAs/Al$_{0.3}$Ga$_{0.7}$As quantum wells.

changes in the values of E_e and E_h. These results correspond to the situation where the excitons are produced uniformly in the quantum well and do not migrate to regions of lower energy. As the value of δ_2 becomes considerably larger than the value of $<\rho^2>$, the excitons see flat surfaces and therefore the emission lines originating from different regions of the quantum well become sharp again. Therefore, in the limit of very small and very large values of δ_2 compared with the effective extent of the exciton, the emission lines become sharp. In the case of large values of δ_2 there will be more than one emission line associated with the free exciton.

It should be pointed out that in GaAs/AlGaAs quantum well structures, the quality of the alloy of the barrier layers also influences the excitonic linewidth in very narrow wells because a fraction of the wave function (depending on the well width and the Al concentration) resides in the barrier layers. Several groups [5-10] have calculated the excitonic linewidth in semiconductor alloys where the dominant mechanism for the line broadening is due to the statistical potential fluctuations caused by the components of the alloys. Singh and Bajaj [11] have developed a formalism in which they take into account the effects of both the microscopic interfacial quality and compositional disorder in alloys on the excitonic emission linewidth in quantum well structures. They consider the following three different cases:

Case I: barrier region is formed by an alloy while the well region is a single component material. This case would include the GaAs/AlGaAs quantum wells.

Case II: here the alloy forms the quantum well region while the barrier region is a single component material. This case would include InGaAs/InP quantum wells.

Case III: here both the barrier and the well regions are formed by the alloys. This would include InGaAs/InAlAs quantum wells.

In high quality cluster free semiconductor alloys such as $Al_{0.3}Ga_{0.7}As$, $In_{0.53}Ga_{0.47}As$ and $In_{0.52}Al_{0.48}As$ the typical values of the excitonic linewidth, at liquid He temperatures, are in the range of about 1 - 3 meV [8]. Thus, in most quantum well structures with well width ranging from about 20 - 200 Å, the excitonic linewidth is dominated by the interface roughness unless the interfaces are very sharp, namely dominated by very small island sizes or/and have alloy components of poor quality.

C   EXCITONIC LINEWIDTH IN ELECTRIC AND MAGNETIC FIELDS

Hong and Singh [12] have calculated the excitonic linewidth in quantum well structures in the presence of an electric field applied parallel to the growth direction assuming a model for the interface as described in the preceding section. The application of the electric field forces the electron and the hole of the exciton to move away from each other and closer to the two interfaces of the quantum well. This enhances the excitonic linewidth. They find that in a $GaAs/Al_{0.3}Ga_{0.7}As$ quantum well structure, for a 100 Å thick well, the linewidth increases from 1.2 to 1.9 meV when the electric field is increased from zero to 100 kV/cm, for $\delta_1 = 2.83$ Å and $\delta_2 = 150$ Å. However, for a very small island size, namely $\delta_2 = 20$ Å, there is practically no change in the linewidth as a function of the electric field, as expected.

As discussed in the previous section, changing the size of the well changes the size of the probe, namely, the exciton. It was pointed out by Singh et al [2] that a more convenient method of changing the exciton size would be the application of a magnetic field along the growth direction. It is well known that the magnetic field affects the size of the exciton so that a study of the linewidth as a function of the magnetic field for the same quantum well may yield useful information concerning the microscopic details of the interface.

Recently Lee et al [13] have calculated the excitonic linewidth in quantum wells in the presence of a magnetic field applied parallel to the growth direction using the model for the interface as described in Section B. The probability function (EQN (2)) is now obtained by replacing R^2 by $<\rho^2>$ calculated using exciton wave functions in a magnetic field [14]. In FIGURE 2, we display the variation of σ for the heavy-hole exciton as a function of well size for magnetic field $B = 75$ kG. We find that for a given value of L_0 and δ_2 the value of σ increases as a function of the magnetic field. This can be seen by comparing FIGURE 2 with FIGURE 1. This is to be expected since with increasing magnetic field the exciton wave function is compressed thus decreasing the value of $<\rho^2>$ and enhancing the value of σ. For instance, for $L_0 = 100$ Å and $\delta_2 = 80$ Å the value of σ at $B = 75$ kG is about 50% larger than its value at zero field. Similar results are also obtained in other lattice-matched quantum well structures such as InGaAs/InP.

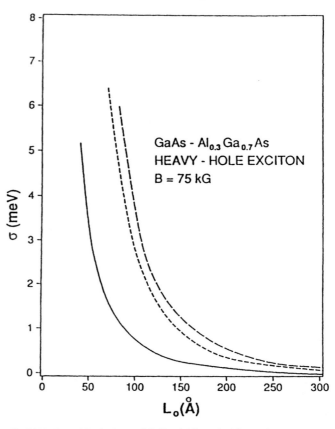

FIGURE 2. Variation of full width at half maximum (σ) of the heavy-hole exciton as a function of well width (L_0) for $\delta_2 = 20$ Å (solid line), $\delta_2 = 80$ Å (dotted line) and $\delta_2 = 160$ Å (dashed line), in GaAs/Al$_{0.3}$Ga$_{0.7}$As quantum wells at $B = 75$ kG.

Bajaj and Lee [15] have recently developed a formalism to calculate the linewidth of excitonic transitions in semiconductor quantum well structures with arbitrary potential profiles in the presence of electric and magnetic fields. They assume that at low temperatures the dominant mechanism responsible for line broadening is the interface roughness which is always present even in the so-called 'perfect' structures. For instance, in the case of GaAs/AlGaAs quantum well structures, the AlGaAs layers at the interfaces have a completely random distribution of Al and Ga in the ideal material and will always lead to excitonic line broadening. They have generalised the formalism which they recently developed [10] to calculate the excitonic linewidths in semiconductor alloys due to compositional disorder and have applied it to the calculation of the excitonic linewidths in quantum well structures. In particular, they have calculated the variation of the linewidth (σ) of a heavy-hole exciton as a function of well size and electric and magnetic fields in GaAs/$Al_{0.3}Ga_{0.7}As$ based quantum well structures with three different potential well profiles, namely, rectangular, parabolic and asymmetric triangular. Their formalism is general enough to be applicable to any quantum well structure where the quantum wells consist of binaries or alloys.

In the following we briefly review the main features of their formalism as applied to the calculation of the excitonic linewidth in semiconductor alloys due to compositional disordering. We will then outline the generalisation of this formalism for the calculation of excitonic linewidths in quantum well structures with arbitrary potential profiles in the presence of electric and magnetic fields.

A Wannier exciton in a semiconductor is described by its wave function which is determined by solving an appropriate Schrödinger wave equation. Since the exciton wave function, in principle, spreads throughout the semiconductor alloy, any definition of a so-called volume associated with the exciton becomes somewhat ambiguous. Thus the validity of the commonly used picture of an exciton subject to local potential fluctuations, whose range is much less than the exciton volume, becomes questionable.

Bajaj and Lee [15] have developed a quantum statistical mechanical approach to calculate the excitonic line broadening in semiconductor alloys due to compositional disordering. They treat the exciton as a quantum mechanical system with a well-defined wave function and therefore completely eliminate the need to use such ambiguous concepts as exciton volume. They calculate the exciton transition energy subject to the fluctuating bandgap of any particular alloy configuration using first-order perturbation theory. To obtain the excitonic lineshape, they sum over the contributions of all possible alloy configurations.

The exciton experiences a fluctuating potential, due to the difference in bond strengths between alloy components, which is a rapidly varying function of spatial coordinates compared to the exciton wave function. Therefore, the excitonic properties are generally obtained using the average material parameters in this region. However, excitons generated in different places in the crystal experience different bandgaps due to the random distributions of the alloy components. A detailed description of the fluctuating potential is clearly very complicated. They therefore estimate this fluctuating potential using a simple description, i.e. the virtual crystal approximation. There is either an atom of component A or B occupying each unit cell with a probability of x or (1-x), respectively. The difference in the bond strengths of A and B is the source of the fluctuating potential in the alloy. E_A and E_B are the bandgaps of A and B components, respectively; E_g is the average bandgap, and ΔE_g is the fluctuating potential.

In a random binary alloy, the average energy bandgap is given as

$$E_g(x) = E_A + C_A x + C_{AB} x^2 \qquad (4)$$

where x is the concentration of atom A in the alloy, and C_{AB} is the contribution from the lattice disorder generated by the random distributions of components A and B in the alloy. The radiative transition energy associated with the exciton ground state is defined as

$$E_{trans}(r) = E_g(r) + \frac{1}{2} \hbar \omega_c - E_b \qquad (5)$$

- 157 -

where E_b is the exciton binding energy, and ω_c is the cyclotron frequency. One must keep in mind that EQN (4) describes the average bandgap. On a microscopic scale, however, there is either an atom of component A or B occupying each unit cell with a probability of x or (1-x), respectively. Using the virtual crystal approximation, they approximate the fluctuating potential as the bandgap variation from its average value at the unit cell about the point $\underline{r} = r_{ijk}$ as

$$\Delta E_g(s_{ijk}) = E_g(s_{ijk}) - E_g(x) \tag{6}$$

where s_{ijk} is 0 for A or 1 for B residing at the site near r_{ijk}.

When an exciton is present in the alloy, it experiences fluctuating potentials from many unit cells. Although these local fluctuations are very large, they tend to cancel each other out. The excitonic transition energy is much larger than its linewidth. Therefore, we use a first-order perturbation theory for the excitonic transition energy in the presence of the fluctuating potentials, which should be accurate in calculating the line-shape function,

$$E = E_g(x) - E_b + \int \Delta E_g(\underline{r}) |\psi(\underline{r})|^2 d\underline{r} = E_g(x) - E_b + \sum_{ijk} \Delta E_g(s_{ijk}) |\psi_{ijk}|^2 \Delta V \tag{7}$$

where $\Delta V = a^3/4$ is the unit cell volume for a zincblende structure with a the lattice constant of the binary alloy and $\psi(\underline{r})$ is the excitonic wave function. The last term in EQN (7) sums over all unit cells in the crystal.

The details of the calculations of the values of the excitonic transition energies due to different alloy configurations and hence the line shape function are given in [11].

We now discuss the application of this formalism to calculate the excitonic linewidth in quantum well structures. In a rectangular $GaAs/Al_{0.3}Ga_{0.7}As$ quantum well structure, the effects of the broadening arise from the compositional disorder present at the interfaces and in the barrier alloy. We assume abrupt interfaces and calculate the line broadening using exciton wave functions appropriate for this quantum well structure. In graded quantum well structures such as parabolic and asymmetric triangular, we calculate the exciton linewidth by regarding them as inhomogeneous alloys. As pointed out earlier, in their formalism Bajaj and Lee [15] sum over the contributions from all possible alloy configurations and therefore the extension of their formalism to quantum well structures with arbitrary potential profiles is quite straightforward.

In FIGURE 3, we display the variation of the linewidth (σ) of a heavy-hole exciton as a function of well size in square, parabolic and asymmetric triangular $GaAs/Al_{0.3}Ga_{0.7}As$ quantum well structures. We find that the value of σ increases as the well size is reduced, in all three different structures, as expected. For well sizes larger than 100 Å, the value of σ is the highest in the asymmetric triangular wells and lowest in the rectangular wells. However, for well widths less than 100 Å, the value of σ in asymmetric triangular wells falls below those in rectangular or parabolic wells. This may be due to the fact that in narrow asymmetric triangular quantum wells the excitonic linewidth corresponds to an average alloy composition which is much less than 0.3, whereas in narrow rectangular wells much of the excitonic wave function resides in the barriers thus experiencing the effect of higher alloy composition.

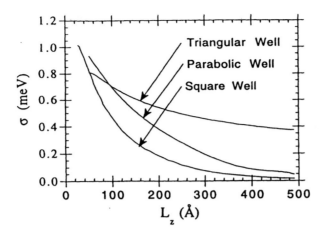

FIGURE 3. Variation of the linewidth (σ) of a heavy-hole exciton as a function of well size in square, parabolic and asymmetric triangular GaAs/Al$_{0.3}$Ga$_{0.7}$As quantum well structures.

In FIGURE 4, we display the variation of σ as a function of well size in rectangular quantum wells for several values of electric and magnetic fields applied parallel to the direction of growth. We find that for a given well size, the value of σ increases as a function of the magnetic field due to the shrinkage of the excitonic wave function. Also, for a given value of the well size and magnetic field, the value of σ increases as a function of the applied electric field. This is due to the fact that the application of the electric field pushes the excitonic wave function into the barrier. Similar results are also obtained for parabolic and asymmetric triangular quantum wells, and for light-hole excitons. It should be pointed out that Bajaj and Lee [15] assume a 'perfect' interface between GaAs and AlGaAs where the excitonic linewidth is caused by the random distribution of Al and Ga atoms. These linewidths are the narrowest one can achieve in these structures.

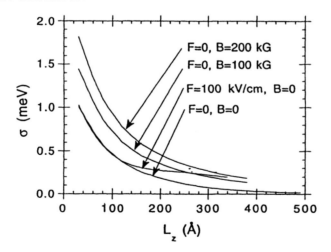

FIGURE 4. Variation of the linewidth (σ) of a heavy-hole exciton as a function of well size (L$_z$) in a square GaAs/Al$_{0.3}$Ga$_{0.7}$As quantum well for several values of the electric (F) and magnetic (B) fields.

It should be pointed out that in all the calculations described in the preceding sections, the effect of the relaxation of excitons into the lower potential energy regions of the well caused by the size fluctuations has not been considered. If such an effect were included, one would expect somewhat narrower and asymmetric emission lines.

D   CONCLUSION

We have reviewed the results of various calculations of the excitonic linewidths in a variety of quantum well structures such as rectangular, parabolic and asymmetric triangular due to interface roughness. We have considered the effects of applied electric and magnetic fields on the excitonic linewidth. We find that the presence of these external perturbations always enhances the excitonic linewidths.

REFERENCES

[1]   C. Weisbuch, R. Dingle, A.C. Gossard, W. Wiegmann [*Solid State Commun. (USA)* vol.38 (1981) p.709-12]

[2]   J. Singh, K.K. Bajaj, S. Chaudhri [*Appl. Phys. Lett. (USA)* vol.44 (1984) p.805-7]

[3]   I.M. Lifshitz [*Adv. Phys. (UK)* vol.13 (1965) p.483-536]

[4]   R.L. Greene, K.K. Bajaj [*Solid State Commun. (USA)* vol.45 (1983) p.831-5]

[5]   O. Goede, L. John, D.H. Hennig [*Phys. Status Solidi B (Germany)* vol.89 (1978) p.K183-6]

[6]   J. Singh, K.K. Bajaj [*Appl. Phys. Lett. (USA)* vol.44 (1984) p.1075-7]

[7]   E.F. Schubert, E.O. Gobel, Y. Horikoshi, K. Ploog, H.J. Queisser [*Phys. Rev. B (USA)* vol.30 (1984) p.813-20]

[8]   J. Singh, K.K. Bajaj [*Appl. Phys. Lett. (USA)* vol.48 (1986) p.1077-9]

[9]   S.M. Lee, K.K. Bajaj [*Appl. Phys. Lett. (USA)* vol.60 (1992) p.853-5]

[10]   S.M. Lee, K.K. Bajaj [*J. Appl. Phys. (USA)* vol.73 (1993) p.1788-96]

[11]   J. Singh, K.K. Bajaj [*J. Appl. Phys. (USA)* vol.57 (1985) p.5433-7]

[12]   S. Hong, J. Singh [*Appl. Phys. Lett. (USA)* vol.49 (1986) p.331-3]

[13]   J. Lee, G.D. Sanders, K.K. Bajaj [*J. Appl. Phys. (USA)* vol.69 (1991) p.4056-60]

[14]   R.L. Greene, K.K. Bajaj [*Phys. Rev. B (USA)* vol.31 (1985) p.6498-502]

[15]   K.K. Bajaj, S.M. Lee [*J. Phys. IV, Colloq. (France)* vol.3 (1993) p.449-51]

4.5 Impurity and vacancy disordering of quantum wells and superlattices

J.H. Marsh and A.C. Bryce

August 1995

A INTRODUCTION

Intermixing the wells and barriers of quantum well structures generally results in an increase in the bandgap and is accompanied by changes in the refractive index. A range of techniques, based on impurity diffusion, dielectric capping and laser annealing, have been developed to enhance the quantum well intermixing (QWI) rate in selected areas of a wafer - such processes offer the prospect of a powerful and relatively simple fabrication route for integrating optoelectronic devices and for forming photonic integrated circuits (PICs). The thermal stability of certain, mainly lattice-matched, material systems and recent progress in QWI techniques is described.

B GENERAL CHARACTERISTICS OF QWI PROCESSES

Quantum well intermixing (QWI) is emerging as a powerful technique for fabricating PICs and optoelectronic integrated circuits (OEICs). In intermixing processes, the bandgap of quantum well (QW) structures is modified in selected regions, after growth, by intermixing the wells with the barriers to form an alloy semiconductor. The bandgap of the intermixed alloy is usually larger than that of the original QW structure thus providing a route to form low-loss optical waveguides, and bandgap shifted QCSE (quantum confined Stark effect) modulators, lasers and detectors using only post-epitaxy processing. It is therefore possible to fabricate PICs using only one epitaxial step. In addition, because the bandgap is increased, the refractive index is modified. In structures containing only a few QWs this refractive index change will have only a small effect on the optical propagation constant, but in MQW structures the optical overlap between the intermixed well and the optical wave can be large enough to give useful changes in the refractive index. The most important intermixing techniques are impurity induced disordering (IID), impurity-free vacancy disordering (IFVD) using dielectric caps, and laser induced disordering (LID).

QWI takes place when the matrix elements of a semiconductor interdiffuse - in certain material systems this need only take place on one lattice site (e.g. the group III site in the GaAs/AlGaAs system) or interdiffusion may need to take place on both lattice sites (e.g. the group III and group V sites in the GaInAs/GaInAsP system) in order to prevent or minimise strain. Because diffusion of matrix elements proceeds via native crystal defects, the self-diffusion rate is determined by the diffusion rate of the defects and their concentration. The equilibrium number of defects is determined by the temperature of the wafer, the partial pressure of matrix elements (in particular the group V elements) and the position of the crystal Fermi energy. It is also possible to introduce an excess (nonequilibrium) concentration of defects, for example, by introducing ion implantation damage or by using IFVD.

C THERMAL STABILITY OF DOPED AND UNDOPED ALLOY MATERIAL SYSTEMS

Self-diffusion has been much studied in the GaAs/AlGaAs system. There is considerable evidence that interdiffusion of Al and Ga can be a concentration dependent process [1,2], and the effects of concentration dependent interdiffusion on QWI have been modelled [3]. However, at least two experimental studies [4,5] suggest that, under the conditions used in QWI, no concentration dependence is observed. Although strained material systems are not considered in depth in this Datareview, it is worth noting that the strain energies are generally very small compared to the activation energy for diffusion. Strain is therefore neither expected nor observed to play a major role in determining QWI rates [6].

The thermal stability of the AlGaAs system and the dependence of the QWI rate on the As vapour pressure has been studied by Guido et al [7]. In their experiments undoped single QW (SQW) samples were annealed in sealed quartz ampoules at 825°C for 25 h. The As_4 vapour pressure over the crystal was varied in the range 0.1-10 atm by enclosing measured quantities of excess elemental As in the ampoule. The Al-Ga self diffusion coefficient lay in the range $7\text{-}13 \times 10^{-19}$ cm^2 s^{-1} and showed a clear dependence on the As_4 vapour pressure, with a minimum occurring at a vapour pressure of around 0.5 atm. This behaviour has been explained by Deppe and Holonyak [8] as follows: the self-diffusion rate is controlled by the concentration of native defects which, in turn, is sensitive to the stoichiometry of the crystal. As the As vapour pressure is increased, so defects associated with an excess of As control the group III self-diffusion. These defects are the As interstitial As_I, the As anti-site defect As_{III} and the group III vacancy V_{III}. In contrast, under a low As overpressure, the As-poor defects - interstitial Al or Ga , the group III anti-site defect and the As vacancy - increase in concentration and control the self-diffusion. Of these six defects, the group III self-diffusion rate is determined principally by the group III vacancy (As-rich conditions), or by the group III interstitial (As-poor conditions).

Many workers have shown that the presence of a large concentration of impurities in the crystal can increase the group III self-diffusion rate by orders of magnitude. The model of Deppe and Holonyak [8] explains the effect of the crystal Fermi energy on the QWI rate. In their model, the equilibrium vacancy concentration is found from simple thermodynamic considerations. Of the six point defects, only two - the group III vacancy V_{III} and the group III anti-site defect - behave like acceptors, whilst the other four are expected to behave as donors. Assuming the native defects are either neutral or singly ionised, the group III self-diffusion rate, D_{III} is given by:

$$D_{III} = f_1^{''} P_{As_4}^{1/4} \left\{ D_{V_{III}^x} + D_{V_{III}} \exp\left[(E_F - E_A)/(k_B T)\right] \right\}$$
$$+ f_2^{''} P_{As_4}^{-1/4} \left\{ D_{I_{III}^x} + D_{I_{III}} \exp\left[(E_D - E_F)/(k_B T)\right] \right\} \tag{1}$$

Here P_{As_4} is the partial pressure of As_4, $D_{V_{III}^x}$, $D_{V_{III}}$, $D_{I_{III}^x}$ and $D_{I_{III}}$ are the diffusion coefficients of neutral and charged group III vacancies, and neutral and charged group III interstitials respectively, $f_1^{''}$ and $f_2^{''}$ are functions which depend on the crystal structure, E_A is the vacancy acceptor energy level and E_D the interstitial donor level. From EQN (1) it can be seen that the IID rate depends on the type of doping (n-type or p-type) and the As overpressure. Intermixing takes place in p-type QW structures under As-poor conditions and in n-type QW structures under As-rich conditions because the concentrations of group III interstitials and group III vacancies are raised in the respective cases. Although different degrees of ionisation have been suggested for the defects [9], the behaviour is qualitatively similar.

In practice, ampoule annealing is rarely used. Furnace annealing or rapid thermal processing (RTP) is usually carried out with a flowing gas, and a proximity cap of GaAs (or InP) is used to protect the sample and provide a group V overpressure. In addition to the use of a large As overpressure, the use of an SiO_2 cap favours group III vacancy generation because Ga dissolves in the SiO_2 layer. SiO_2 caps can be used to promote QWI with or without active dopants. For example, it has been shown that Si-doped material will intermix beneath an SiO_2 cap but not beneath an Si_3N_4 cap [10]. Equally, undoped material beneath dielectric caps will intermix through the IFVD mechanism if the dielectric cap creates point defects.

By contrast, the QW material systems used at 1.5 μm are of much more limited thermal stability, and the GaInAs/InGaAsP (P-quaternary) system exhibits considerable intermixing even at temperatures normally encountered in epitaxial growth. Marsh et al [11] have demonstrated that unimplanted P-quaternary samples (capped with either Si_3N_4 or SiO_2 and in a P-ambient provided by pieces of InP) disorder at annealing temperatures above 500°C, with a blue shift always observed. It is believed that the blue shift is caused by the diffusion of phosphorus into and arsenic out of the wells [12]. This is likely to be due to group V vacancies, generated by phosphorus desorption from the surface, diffusing through the structure. To investigate this further, unimplanted samples were annealed in silica ampoules loaded with red phosphorus for 30 min at 650°C. It was found that the exciton energy shift could be reduced to 15 meV, compared to 25 meV in the conventional furnace anneal. This result is similar to that found by Komiya et al [13] who found that the thermal degradation of InP/InGaAsP/InP DH structures was less when their samples were annealed in an ampoule containing red phosphorus rather than in an LPE furnace containing InP loaded with tin. They proposed that the degradation of the structures was due to vacancies caused by the desorption of phosphorus from the surface, and concluded that the phosphorus overpressure generated by InP is insufficient to prevent this. Given that most QWI studies of the P-quaternary system have not used sealed ampoules, it is probable that intermixing takes place in samples which always contain a large concentration of group V vacancies. It is therefore not possible to apply the model of Deppe and Holonyak directly to this material system.

Bryce et al [14] have demonstrated that GaInAs/AlGaInAs (Al-quaternary) samples are very much more stable than the P-quaternary, with negligible shifts being observed for annealing temperatures up to 650°C. Even above this temperature only small red shifts in the low temperature PL measurements were observed, probably due to Ga diffusing out of the wells and being replaced by In from the barriers. This is unexpected since the In concentration is initially constant but this effect has also been observed in GaInAs/AlInAs QW structures [15,16]. The effect is probably caused by the low mobility of the Al compared to the Ga and In and the requirement of the material to remain stoichiometric. Since the aluminium does not diffuse significantly the In diffuses to maintain the structure of the material. Some evidence was, however, found for increased interface roughness after annealing.

D   IMPURITY INDUCED DISORDERING

In IID processes, an impurity is introduced into an epitaxial wafer and the wafer is then annealed. During the annealing step the layers intermix and ion-implantation damage, if present, is to a large extent removed. Current understanding [8] suggests that the role of impurities is to induce the disordering process through the generation of free-carriers which, in turn, increase the equilibrium number of vacancies at the annealing temperature. A number of species have been demonstrated to disorder the GaAs/AlGaAs system, the most important of which are Zn (p-type) and Si (n-type). Impurities need to be present in concentrations greater than around 10^{18} cm^{-3} in order to enhance the interdiffusion rates of the lattice elements. As a consequence, the lowest reported absorption coefficients in material intermixed using electrically active dopants are around 43 dB cm^{-1} (10 cm^{-1}), which is a consequence of free-carrier absorption, and the material is of low resistivity.

Electrically neutral dopants have also been studied. These include O, Ar, F and B. However, traps associated with these species and residual damage from implantation make IID an unsuitable technique for forming bandgap-tuned modulators or lasers. The mechanism by which neutral species disorder III-V semiconductor QW structures is not completely clear. In the case of boron and fluorine it has been noted that a variety of deep levels are associated with these species [17-21]. During intermixing it is suggested that these deep levels become ionised so influencing the position of the Fermi level [22].

D1   IID of GaAs/AlGaAs

Impurity induced disordering of GaAs/AlGaAs superlattices was first observed by Laidig et al [23] who showed that Zn diffusion into a GaAs/AlGaAs structure caused interdiffusion of the Ga and Al at temperatures much lower than for impurity free structures. It has also been shown that the diffusion of Si [24,25], S [26], Sn [27] and Ge [25] cause intermixing. Investigation of implanted Si [28] followed by annealing also caused intermixing. Implantation increased the range of impurities that can be used [29-31] to include elements such as As, Ar, B and F which cannot be diffused into the structure. Some impurities were better at inducing intermixing than others: for example Ar and Be [29] were reported as producing no enhancement of the interdiffusion process, while Si can cause complete intermixing.

The impurities boron and fluorine are electrically neutral at room temperature in the GaAs/AlGaAs system. O'Neill et al [32,33] have demonstrated the potential of these impurities as disordering species in PIC applications, with losses as low as those measured in bandgap widened waveguides. The excess propagation loss associated with boron intermixing is also very small. Hansen et al [34] have carried out systematic studies of the effect of disordering on the refractive index. The structure investigated was an MQW waveguide where the MQW consisted of 54 periods of 6.0 nm GaAs wells and 6.0 nm $Al_{0.26}Ga_{0.74}As$ barriers. The largest changes in the refractive index occurred at the exciton resonances in the starting material. At 840 nm, the limiting wavelength for TE measurements on the control sample, the TE refractive index change was 0.027 and 0.036, and for the TM polarisation at the corresponding wavelength of 835 nm, 0.014 and 0.021 for boron and fluorine respectively. After annealing for 4 hr following fluorine implantation, the material refractive index in the waveguide core was virtually identical for the two polarisations, confirming that the MQW was completely disordered.

Early work used long anneals, of several hours, in conventional furnaces; more recent work has made use of rapid thermal annealing (RTA) on a time scale of tens of seconds to a few minutes. RTA has the advantage over conventional furnace annealing in that the processing is carried out quickly and unwanted diffusion effects can be minimised because of the much shorter annealing times. The implant species which have been reported are Si [35], Al [35], Ga [35,36] and As [37]. The mechanism postulated for causing the intermixing is the recombination of the vacancies and interstitials introduced during the implantation process [36,38]. Further work by Kahen and Rajeswaran [35] has shown that, above an implant generated vacancy concentration of 6×10^{19} cm^{-3}, the shift in the bandgap saturates. They attribute this saturation to the formation of extended defects (vacancies coalesce) during the annealing process. Such extended defects have been observed in B implanted and annealed structures using transmission electron microscopy (TEM) [39]. So there exists an optimum dose for each implanted impurity which is determined by the concentration of the vacancies generated during implantation.

A variation of the implantation technique is to use a focused ion beam [40]. This can be used to write patterns into the structure instead of using masks to define the areas to be intermixed.

D2   IID of GaInAs/GaInAsP and GaInAs/AlGaInAs

QWI of MQW layers in the P- and Al-quaternary systems has been very much less widely studied than in the GaAs/AlGaAs system. In the P-quaternary system it has been shown that in-diffusion of zinc [41-43] results in preferential mixing on the group III lattice sites. SIMS studies [42,43] have shown that the wells become increasingly InAs-like which, in turn, causes a narrowing of the bandgap and a corresponding increase in the refractive index. Large changes in the bandgap wavelength can be induced, for example from 1.4 μm to 1.9 μm [43]. The effects of the resulting strain have recently been modelled [44]. Sulphur [45] (at a concentration of ~10^{18} cm^{-3}) and high concentration proton [46] implants (5×10^{18} cm^{-3} to 10^{21} cm^{-3}) have been demonstrated to give increases in the bandgap energy, causing intermixing on both the group III and group V lattice sites so the wells become increasingly GaInAsP-like. The amphoteric impurities, germanium [47] and silicon [48], and the isoelectronic impurities, gallium [49] and phosphorus [50], have also been demonstrated to give bandgap increases if implanted at high doses (~10^{19} cm^{-3}). These processes, however, either involve the use of active dopants or high implantation doses - free-carrier absorption losses are significant in the former case, and residual damage in the latter.

Using AlInAs as an alternative to InP and AlGaInAs as an alternative to GaInAsP simplifies the QWI process in that only the group III sites need to be intermixed. The improved superior stability of this material system compared to the P-quaternary has already been discussed. IID again using the active dopants silicon [16] and zinc [16,51] has been reported, in processes which give rise to the associated free-carrier absorption losses. Moreover, the work using silicon demonstrated that a very high concentration was required to induce intermixing, >3×10^{18} cm^{-3}.

As in the GaAs/AlGaAs system, the use of either neutral impurity IID or IFVD offers the prospect of low propagation loss in intermixed material. Fluorine and boron induced disordering of both the material systems used at 1.5 μm have been investigated -

GaInAs/AlGaInAs (Al-quaternary) by Bryce et al [14] and GaInAs/GaInAsP (P-quaternary) by Marsh et al [11], in both cases lattice-matched to InP.

D3   Rapid Thermal Annealing of P-Quaternary MQW Structures

As discussed above, there are serious problems in using conventional furnace annealing to process the P-quaternary, but Marsh et al [11], and Bradshaw et al [52] have demonstrated that annealing in a rapid thermal processing (RTP) system can overcome some of the limitations associated with furnace annealing. Studies of fluorine implanted samples have been made by Bradshaw et al [52]. It was found that, by annealing at 700°C for around 30 s, it was possible to obtain a bandgap shift of 40 meV in implanted material whilst unimplanted samples showed a shift of less than 4 meV. The authors also reported that there is an optimum concentration of fluorine implant (around 10^{18} cm^{-3}) which gives the maximum energy shift. Implanting with fluorine concentrations greater than this inhibits the intermixing processes.

Zucker et al [53] have reported losses of 14.7 dB cm^{-1} in an MQW waveguide in which phosphorus was implanted into the MQW at 200°C and the sample was then overgrown with InP at 650°C for 15 min. This loss figure is encouragingly low, although photocurrent measurements showed residual damage was still present in the structure. Using fluorine intermixing and rapid thermal annealing, Marsh et al [11] have reported significantly lower losses in the range 0.6 to 8.5 dB cm^{-1}. Whitehead et al [54] used phosphorus implantation to form an undisordered waveguide between two intermixed regions in an MQW structure; they reported losses of 4 dB cm^{-1}.

E   IMPURITY-FREE VACANCY DISORDERING

QWI has mainly been achieved by diffusion or ion-implantation followed by annealing. Although neutral IID can circumvent the large optical propagation losses associated with IID techniques, neutral impurities still introduce substantial changes in the material resistivity and trap concentrations and the residual damage associated with implantation raises concerns about device lifetimes. Impurity-free vacancy disordering (IFVD) can create large bandgap energy shifts without these disadvantages. IFVD is based on creating vacancies in the III-V semiconductor. Gallium, for example, is known to dissolve in certain dielectric caps, particularly SiO_2. If a wafer is coated with SiO_2 then heated, Ga migrates into the dielectric cap, creating group III vacancies which then diffuse through the QW structure resulting in intermixing. IFVD can produce low-loss waveguides in both the GaAs/AlGaAs [55] and in the GaInAsP/InP [56] systems, and bandgap shifted QCSE modulators [57] and lasers [58]. In the GaAs/AlGaAs system, SiO_2 is generally used to promote disordering whilst Si_3N_4 is used to prevent disordering. In the P-quaternary system, however, Si_3N_4 has been used to promote disordering, and an integrated extended cavity laser has been formed in this way [56]. Unfortunately, IFVD has poor reproducibility, especially from run-to-run [59]. In particular, 'silicon nitride' films are rarely pure Si_3N_4 and usually contain a substantial fraction of SiO_2. Also, films of Si_3N_4 are usually highly strained and, although it is possible to control the residual strain at room temperature by changing the deposition conditions or deposition technique, when the samples are heated large differential strain effects occur. It is believed that both the presence of SiO_2 in the silicon nitride and high interface strain can effect disordering beneath the cap. Further complications occur due to chemical reactions between the dielectric cap and the semiconductor. This is a particularly serious problem with Al-containing semiconductor alloys because Al is a very effective reducing agent. When an SiO_2 cap is in direct contact with an Al-containing alloy, reactions of the type

$$3SiO_2 + 4Al \rightarrow 2Al_2O_3 + 3Si \qquad (2)$$

occur, and the resulting free Si can then act as an IID species. Any excess Si present in the dielectric film itself resulting from nonstoichiometric deposition can also act as an IID species.

GaAs based QW material usually shows large shifts for SiO_2 caps [60] and smaller shifts for Si_3N_4 capped samples. Beauvais et al [61] have studied the use of SrF_2 as a dielectric cap which is inert, well-matched to GaAs in its thermal and mechanical properties, and which protects the wafer during annealing cycles. By comparing IFVD in the GaAs/AlGaAs system using three different dielectric caps, namely SiO_2, Si_3N_4, and SrF_2, they have shown that the use of SrF_2 as a cap almost completely inhibits QWI under conditions where SiO_2, and to a lesser extent Si_3N_4, promote rapid intermixing.

A variation of the IFVD technique allows different bandgap changes to be realised across a wafer in a single annealing stage. Beauvais et al [62] have patterned SrF_2 into small squares (using electron beam lithography) before depositing the SiO_2. If the dimensions of the squares are sufficiently small, uniform intermixing takes place, with the degree of intermixing controlled by the fraction of the area masked by the SrF_2. This process, called selective intermixing in selected areas (SISA), allows a complete range of bandgaps to be created across a wafer in a single processing stage. Bandgap tuned semiconductor lasers have recently been demonstrated.

Another approach, employed by Poole et al [63], is not to implant the impurity into the QW region, but rather to implant above the QWs and let the vacancies generated by the implantation process diffuse down through the QWs, intermixing them as they pass through. Arsenic was used in this study and the samples underwent repeated implant and RTA cycles. As with IID, there was found to be an optimum dose for which the bandgap shift is a maximum, the authors attributing this effect to surface damage during implantation. This damage being annealed out during the RTA step, in order to achieve useful amounts of intermixing it is necessary to use repeated implant and anneal stages.

F   LASER INDUCED DISORDERING

LID of QW structures was first demonstrated using CW radiation from an Ar laser. The beam was focused and scanned across the semiconductor surface. The maximum transient temperature reached by the semiconductor was sufficient to melt the surface layer, and the molten material then recrystallised as an alloy semiconductor. Variations of the same technique using pulsed laser irradiation have also been reported. LID therefore differs from the other techniques in that it does not necessarily rely on creating point defects to enhance intermixing rates. Because melting needs a high temperature, dielectric capping is vital to preserve the stoichiometry of the semiconductor, but such caps can also act as sources of impurities or of vacancies. The quality of recrystallised material is also inferior to that of the original semiconductor. An all solid-state variation of the LID process has been developed and will be described below.

Like IFVD, LID processes do not necessarily change the impurity concentrations in a semiconductor structure. Both annealing of implanted samples [64] and transient melting of multiple layers by pulsed laser irradiation [65] have been shown to be effective. Transient melting using scanning of CW lasers has been utilised to introduce encapsulant Si into the epitaxial layers as a source for impurity induced disordering which takes place during thermal post-annealing [66]. These techniques and their derivatives, although clearly effective, tend to be complicated by one or more of a number of additional processes, such as the introduction of impurities, various masking steps and high temperature annealing, and can suffer from the increased free-carrier population introduced, as well as from implantation induced damage. LID processes, although having the potential of being impurity-free and offering the possibility of direct write capability, require high power densities to melt the material, can introduce thermal shock damage if used in a pulsed mode and can cause a potentially undesirable redistribution of dopants into the active region of the device. Furthermore, melting of the semiconductor results in complete intermixing and the process cannot be controlled to give partial bandgap shifts.

McLean et al [67] have developed an alternative laser induced disordering process - photo-absorption induced disordering (PAID) - which takes advantage of the limited thermal stabilities of the GaInAs/GaInAsP and GaInAs/AlGaInAs systems. The method is impurity free, requires only a fraction of the power densities of existing CW techniques (~1-20 W mm^{-2} compared [68] with 10^5 W mm^{-2}) and does not involve a melt phase in the semiconductor processing. A variation of the technique, using a pulsed laser to create point defects followed by an annealing stage, has spatial resolution capabilities of better than 20 μm [69].

G   DEVICE APPLICATIONS

Many applications of QWI techniques have been identified: e.g. bandgap-tuned modulators [57], bandgap-tuned lasers [58,70], low-loss waveguides for interconnecting components on an OEIC, integrated extended cavities for line-narrowed lasers, single-frequency DBR lasers and mode-locked lasers, non-absorbing mirrors and either gain- or phase-gratings for DFB lasers. Taking the examples which utilise low-loss waveguides [71], three parameters are of particular importance: the absorption coefficient, the material resistivity, and the refractive index change induced by intermixing. An ideal loss target within a PIC is <1 dB cm^{-1} but 10 dB cm^{-1} would be acceptable in many applications. In integrated extended cavity lasers, losses can be 40 dB cm^{-1} in the passive section and still give a useful reduction in laser linewidth. A further requirement is that the electrical resistance of waveguides should be sufficiently high to isolate individual components and studies of an integrated laser/modulator structure [72] have demonstrated that the required isolation resistance between the laser and the modulator might need to be as large as 100 kΩ. The refractive index changes can be used to provide optical confinement [73,74], gratings [75], or even laser reflectors [76,77].

Pioneering work by Thornton et al [78] has demonstrated that QWI can be used to fabricate moderately complex PICs - in their case consisting of a gain region, an interconnecting waveguide and a electro-absorption modulator. The QWI technique used was IID using silicon. Although the resulting propagation losses were high (17 cm^{-1}), the integrated devices all functioned individually as expected. Silicon disordering has also been used to provide optical and electrical confinement in low threshold current lasers [79]. To illustrate the use of one QWI technique in a photonic integrated circuit (PIC) application, Andrew et al [80] have fabricated and studied extended cavity GaAs/AlGaAs lasers containing an integrated passive optical waveguide formed using neutral IID. A passive loss of 19 dB cm^{-1} was deduced. The lowest loss reported for comparable DQW devices fabricated by silicon disordering is 50.4 dB cm^{-1} (25.2 dB cm^{-1} per well) [81].

IFVD has been used in several device applications including bandgap tuned modulators, particularly planar modulators [57], low-loss waveguide devices [55] and buried heterostructure lasers [82], in demonstrations of bandgap tuned lasers [58,70], and to fabricate simple spatial mode filters within lasers [83]. Using the single step SISA intermixing process, described above, five distinguishable lasing wavelengths have been observed from five selected intermixed regions on a single chip [84].

PAID has been used to fabricate bandgap tuned lasers [85] and modulators [86]. QWI, using focused ion beams, has been used to make reduced dimensional structures [87]. The use of laser induced intermixing to fabricate wires and dots has also been reported [88,89].

H CONCLUSION

QWI processes are still the subject of much investigation and further studies are needed before they are adequately understood. They are, however, being used in device applications and are showing considerable promise as technologies for photonic integration.

REFERENCES

[1]   L.L. Chang, A. Koma [*Appl. Phys. Lett. (USA)* vol.29 (1976) p.138-41]

[2]   R.M. Fleming, D.B. McWhan, A.C. Gossard, W. Weigmann, R.A. Logan [*J. Appl. Phys. (USA)* vol.51 (1980) p.357-63]

[3]   I. Harrison [*Semicond. Sci. Technol. (UK)* vol.9 (1994) p.2053-5]

[4]   H. Peyre, J. Camassel, W.P. Gillin, K.P. Homewood, R. Grey [*Mater. Sci. Eng. B (Switzerland)* vol.28 (1994) p.332-6]

[5]   I. Gontijo et al [*J. Appl. Phys. (USA)* vol.76 (1994) p.5434-8]

[6]   W.P. Gillin, D.J. Dunstan [*Phys. Rev. B (USA)* vol.50 (1994) p.7495-8]

[7]   L.J. Guido et al [*J. Appl. Phys. (USA)* vol.61 (1987) p.1372]

[8]   D.G. Deppe, N. Holonyak Jr. [*J. Appl. Phys. (USA)* vol.64 (1988) p.R93-113]

[9]   T.Y. Tan, U. Gosele [*Appl. Phys. Lett. (USA)* vol.52 (1988) p.1240]

[10]   L.J. Guido et al [*J. Appl. Phys. (USA)* vol.61 (1987) p.1329]

[11]   J.H. Marsh, S.A. Bradshaw, A.C. Bryce, R.M. Gwilliam, R.W. Glew [*J. Electron. Mater. (USA)* vol.20 (1991) p.973-8]

[12]   K. Nakashima, Y. Kawaguchi, Y. Kawamura, H. Ashasi, Y. Imamura [*Jpn. J. Appl. Phys. (Japan)* vol.26 (1987) p.L1620]

[13]   S. Komiya, T. Tanahashi, I. Umebu [*Jpn. J. Appl. Phys. (Japan)* vol.24 (1985) p.1053]

[14]   A.C. Bryce, J.H. Marsh, R.M. Gwilliam, R.W. Glew [*IEE Proc.. Optoelectron. (UK)* vol.138 (1991) p.87-90]

[15]   R.J. Baird, T.J. Potter, G.P. Kothiyal, P.K. Bhattacharya [*Appl. Phys. Lett. (USA)* vol.52 (1988) p.2055]

[16]   R.J. Baird, T.J. Potter, R. Lai, G.P. Kothiyal, P.K. Bhattacharya [*Appl. Phys. Lett. (USA)* vol.53 (1988) p.2302]

[17]   P. Dansas [*J. Appl. Phys. (USA)* vol.58 (1985) p.2212]

[18]   W.J. Moore, R.L. Hawkins, B.V. Shanabrook [*Physica B (Netherlands)* vol.146 (1987) p.65]

[19]   D.W. Fischer, P.W. Yu [*J. Appl. Phys. (USA)* vol.59 (1986) p.1952]

[20]   Y. Makita, S. Gonda [*Appl. Phys. Lett. (USA)* vol.17 (1976) p.333]

[21]   B. Tell, K.F. Brown-Goebeler [*J. Appl. Phys. (USA)* vol.62 (1987) p.813]

[22]   J.H. Marsh [*Semicond. Sci. Technol. (UK)* vol.8 (1993) p.1136-55]

[23]   W.D. Laidig et al [*Appl. Phys. Lett. (USA)* vol.38 (1981) p.776]

[24]   K. Meehan, N. Holonyak Jr., J.M. Brown, M.A. Nixon, P. Gavrilovic, R.D. Burnham [*Appl. Phys. Lett. (USA)* vol.45 (1984) p.549]

[25]   R.W. Kaliski et al [*J. Appl. Phys. (USA)* vol.58 (1985) p.101]

[26]   E.V.K. Rao, H. Thibierge, F. Brillouet, F. Alexandre, R. Azoulay [*Appl. Phys. Lett. (USA)* vol.46 (1985) p.867]

[27]   E.V.K. Rao, P. Ossart, F. Alexandre, H. Thibierge [*Appl. Phys. Lett. (USA)* vol.50 (1987) p.588]

[28]   J.J. Coleman, P.D. Dapkus, C.G. Kirkpatrick, M.D. Camras, N. Holonyak Jr. [*Appl. Phys. Lett. (USA)* vol.40 (1982) p.904]

[29]   Y. Hirayama, Y. Suzuki, H. Okamoto [*Jpn. J. Appl. Phys. (Japan)* vol.24 (1985) p.1498]

[30]   Y. Hirayama, Y. Suzuki, S. Tarucha, H. Okamoto [*Jpn. J. Appl. Phys. (Japan)* vol.24 (1985) p.L516]

[31]   P. Gavrilovic, D.G. Deppe, K. Meehan, N. Holonyak Jr., J.J. Coleman, R.D. Burnham [*Appl. Phys. Lett. (USA)* vol.47 (1985) p.130]

[32]   M. O'Neill, A.C. Bryce, J.H. Marsh, R.M. De La Rue, J.S. Roberts, C. Jeynes [*Appl. Phys. Lett. (USA)* vol.55 (1989) p.1373]

[33]   M. O'Neill, J.H. Marsh, R.M. De La Rue, J.S. Roberts, R.M. Gwilliam [*Electron. Lett. (UK)* vol.26 (1990) p.1613-5]

[34]   S.I. Hansen, J.H. Marsh, J.S. Roberts, R.M. Gwilliam [*Appl. Phys. Lett. (USA)* vol.58 (1991) p.1398-400]

[35]   K.B. Kahen, G. Rajeswaran [*J. Appl. Phys. (USA)* vol.66 (1989) p.545]

[36]   J. Cibert, P.M. Petroff, D.J. Werder, S.J. Pearton, A.C. Gossard, J.H. English [*Appl. Phys. Lett. (USA)* vol.49 (1986) p.223]

[37]   B. Elman, E.S. Koteles, P. Melman, C.A. Armiento [*J. Appl. Phys. (USA)* vol.66 (1989) p.2104]

[38]   S.T. Lee, G. Braunstein, P. Fellinger, K.B. Kahen, G. Rajeswaran [*Appl. Phys. Lett. (USA)* vol.53 (1988) p.2531]

[39]   B.S. Ooi, A.C. Bryce, J.H. Marsh, J. Martin [*Appl. Phys. Lett. (USA)* vol.65 (1994) p.85]

[40]   W. Beinstingl, Y.J. Li, H. Weman, J. Merz, P.M. Petroff [*J. Vac. Sci. Technol. B (USA)* vol.9 (1991) p.3479]

[41]   M. Razeghi, O. Archer, F. Launay [*Semicond. Sci. Technol. (UK)* vol.2 (1987) p.793]

[42]   K. Nakashima, Y. Kawaguchi, Y. Kawamura, Y. Imamura [*Appl. Phys. Lett. (USA)* vol.52 (1988) p.1383-5]

[43]   I.J. Pape, P. Li Kam Wa, J.P.R. David, P.A. Claxton, P.N. Robson, D. Sykes [*Electron. Lett. (UK)* vol.24 (1988) p.910-1]

[44]   J. Micallef, E.H. Li, B.L. Weiss [*Appl. Phys. Lett. (USA)* vol.61 (1992) p.435-7]

[45]   I.J. Pape, P. Li Kam Wa, J.P.R. David, P.A. Claxton, P.N. Robson [*Electron. Lett. (UK)* vol.24 (1988) p.1217-8]

[46]   I.J. Pape, P. Li Kam Wa, D.A. Roberts, J.P.R. David, P.A. Claxton, P.N. Robson [*Inst. Phys. Conf. Ser. (UK)* no.96 (1988) p.397]

[47]   M.A. Bradley et al [*Electron. Lett. (UK)* vol.26 (1990) p.209]

[48]	B. Tell et al [*Appl. Phys. Lett. (USA)* vol.52 (1988) p.1428-30]
[49]	H. Sumida, H. Asahi, S. Jae Yu, K. Asami, S. Gonda, H. Tanoue [*Appl. Phys. Lett. (USA)* vol.54 (1989) p.520-2]
[50]	B. Tell et al [*Appl. Phys. Lett. (USA)* vol.54 (1989) p.1570]
[51]	Y. Kawamura, H. Asahi, A. Kohyen, K. Wakita [*Electron. Lett. (UK)* vol.21 (1985) p.219]
[52]	S.A. Bradshaw, J.H. Marsh, R.W. Glew [*Proc. 4th Int. Conf. on InP and Related Materials* Newport, Rhode Island, 1992 (IEEE Publishing Services, New York, 1992) p.604-7]
[53]	J.E. Zucker et al [*Appl. Phys. Lett. (USA)* vol.60 (1992) p.3036]
[54]	N.J. Whitehead, W.P. Gillin, I.V. Bradley, B.L. Weiss, P.A. Claxton [*Semicond. Sci. Technol. (UK)* vol.5 (1990) p.1063-6]
[55]	Y. Suzuki, H. Iwamura, O. Mikami [*Appl. Phys. Lett. (USA)* vol.56 (1990) p.19-20]
[56]	T. Miyazawa, H. Iwamura, M. Naganuma [*IEEE Photonics Technol. Lett. (USA)* vol.3 (1991) p.421]
[57]	J.D. Ralston, W.J. Schaff, D.P. Bour, L.F. Eastman [*Appl. Phys. Lett. (USA)* vol.54 (1990) p.534]
[58]	D.G. Deppe et al [*Appl. Phys. Lett. (USA)* vol.49 (1986) p.510-2]
[59]	E.S. Koteles, B. Elman, P. Melman, J.Y. Chi, C.A. Armiento [*Opt. Quantum Electron. (UK)* vol.23 (1991) p.S779-87]
[60]	J.D. Ralston, S. O'Brien, G.W. Wicks, L.S. Eastman [*Appl. Phys. Lett. (USA)* vol.52 (1988) p.1511-3]
[61]	J. Beauvais, J.H. Marsh, A.H. Kean, A.C. Bryce, C. Button [*Electron. Lett. (UK)* vol.28 (1992) p.1670-2]
[62]	J. Beauvais, S.G. Ayling, J.H. Marsh [*Electron. Lett. (UK)* vol.28 (1992) p.2240]
[63]	P.J. Poole et al [*Semicond. Sci. Technol. (UK)* vol.9 (1994) p.2134]
[64]	A. Rys, Y. Shieh, A. Compaan, H. Yao, A. Bhat [*Opt. Eng. (USA)* vol.29 (1990) p.329]
[65]	J.D. Ralston, A.L. Moretti, R.K. Jain, F.A. Chambers [*Appl. Phys. Lett. (USA)* vol.50 (1987) p.1817]
[66]	J.E. Epler, R.D. Burnham, R.L. Thornton, T.L. Paoli, M.C. Bashaw [*Appl. Phys. Lett. (USA)* vol.49 (1986) p.1447]
[67]	C.J. McLean, J.H. Marsh, R.M. De La Rue, A.C. Bryce, B. Garrett, R.W. Glew [*Electron. Lett. (UK)* vol.28 (1992) p.1117-9]
[68]	J.E. Epler, R.D. Burnham, R.L. Thornton, T.L. Paoli [*Appl. Phys. Lett. (USA)* vol.51 (1987) p.731]
[69]	C.J. McLean, A. McKee, G. Lullo, A.C. Bryce, R.M. De La Rue, J.H. Marsh [*Electron. Lett. (UK)* vol.31 (1995) p.1285-6]
[70]	S.G. Ayling, A.C. Bryce, I. Gontijo, J.H. Marsh, J.S. Roberts [*Semicond. Sci. Technol. (UK)* vol.9 (1994) p.2149-51]
[71]	J.H. Marsh, S.I. Hansen, A.C. Bryce, R.M. De La Rue [*Opt. Quantum Electron. (UK)* vol.23 (1991) p.S941]
[72]	M. Suzuki, H. Tanaka, S. Akiba, Y. Kushiro [*J. Lightwave Technol. (USA)* vol.6 (1988) p.779]
[73]	E. Kapon, N.G. Stoffel, E.A. Dobisz, R. Bhat [*Appl. Phys. Lett. (USA)* vol.52 (1988) p.351-3]
[74]	T. Wolf, C.-L. Shieh, R. Englemann, K. Alavi, J. Mattz [*Appl. Phys. Lett. (USA)* vol.55 (1989) p.1412-4]
[75]	J.D. Ralston, L.H. Camnitz, G.W. Wicks, L.F. Eastman [*Inst. Phys. Conf. Ser. (UK)* no.83 (1987) p.367-72]
[76]	N. Holonyak Jr., W.D. Laidig, M.D. Camras, J.J. Coleman, P.D. Dapkus [*Appl. Phys. Lett. (USA)* vol.39 (1981) p.102-4]
[77]	W.D. Laidig, J.W. Lee, P.J. Caldwell [*Appl. Phys. Lett. (USA)* vol.45 (1984) p.485-7]
[78]	R.L. Thornton, W.J. Mosby, T.L. Paoli [*J. Lightwave Technol. (USA)* vol.6 (1988) p.786]

[79] P.D. Floyd, C.P. Chao, K.K. Law, J.L. Merz [*IEEE Photonics Technol. Lett. (USA)* vol.5 (1993) p.1261-3]

[80] S.R. Andrew, J.H. Marsh, M.C. Holland, A.H. Kean [*IEEE Photonics Technol. Lett. (USA)* vol.4 (1992) p.426-8]

[81] J. Werner et al [*Appl. Phys. Lett. (USA)* vol.57 (1990) p.810]

[82] J.S. Major et al [*Appl. Phys. Lett. (USA)* vol.54 (1989) p.913]

[83] K. McIlvaney, J. Carson, A.C. Bryce, J.H. Marsh, R. Nicklin [*Electron. Lett. (UK)* vol.31 (1995) p.553-4]

[84] B.S. Ooi, S.G. Ayling, A.C. Bryce, J.H. Marsh [*Conf. on Lasers and Electro-Optics Technical Digest* Baltimore, USA, 1995, vol.15 (1995) p.152-3]

[85] A. McKee, C.J. McLean, A.C. Bryce, R.M. De La Rue, C. Button, J.H. Marsh [*Appl. Phys. Lett. (USA)* vol.65 (1994) p.2263]

[86] G. Lullo, A. McKee, C.J. McLean, A.C. Bryce, C. Button, J.H. Marsh [*Electron. Lett. (UK)* vol.30 (1994) p.1923]

[87] W. Beinstingl, Y.J. Li, H. Weman, J. Merz, P.M. Petroff [*J. Vac. Sci. Technol. B (USA)* vol.9 (1991) p.3479-82]

[88] K. Brunner, G. Abstreiter, M. Walther, G. Bohm, G. Trankle [*Surf. Sci. (Netherlands)* vol.267 (1992) p.218-22]

[89] K. Brunner et al [*Phys. Rev. Lett. (USA)* vol.69 (1992) p.3216-9]

CHAPTER 5

ELECTRONIC PROPERTIES

5.1 Carrier effective masses and mobilities in GaAs/AlGaAs quantum wells and superlattices

B. Vinter

April 1995

A INTRODUCTION

We review low-field mobilities in quantum wells (QWs) and superlattices (SLs) made of GaAs for the QWs and $Al_xGa_{1-x}As$ for the barriers. Transport properties depend on the direction of the electric field applied to the QW structure. When the field is applied parallel to the QW interfaces the carriers remain in the same quantum well. In vertical transport perpendicular to the interfaces there is transport between different QWs, either by tunnelling through the barriers or by thermal excitation above the barriers. This Datareview is restricted to genuine QWs with at least two interfaces.

B EFFECTIVE MASSES

The effective masses which are relevant in quantum wells and superlattices are the same as those of single heterojunctions. They have been reviewed in earlier Datareviews [1,2].

C TRANSPORT PARALLEL TO INTERFACES

All measurements of mobility parallel to the interfaces of QWs and SLs have been obtained by the Hall effect. The efforts have concentrated on determining the limiting scattering processes and on designing structures of the highest possible mobilities. For finite temperatures, phonon scattering determines the ultimate limit. For low temperatures in wells that are not too narrow the limit is due to scattering on ionised donor impurities in the barrier material, and in narrow wells the fact that the heterojunction interfaces are not perfectly planar sets the upper limit.

C1 Impurity Limited Mobility

The seminal idea of using modulation doping in quantum well structures has made it possible to increase the overall mobility and conductivity above that of bulk material for a given average concentration of carriers. This is due to the strong reduction of carrier scattering on the ionised donors which furnish the carriers but are spatially separated from them. FIGURE 1 shows measurements [3] of the Hall mobility in several multiple quantum well samples, in which the undoped AlGaAs spacer layer between the heterojunction interfaces and the dopants is varied. The average densities and mobilities at T = 4.2 K of the 4 samples are given in TABLE 1. Even higher mobilities ($>10^5$ cm^2/V s at T = 77 K) have been obtained [4] in very large (45 nm) wells but wells of this size strongly resemble double separate heterojunctions and are not strictly quantum wells. In simple cases analytical theoretical results in closed form may be obtained for the Coulomb impurity scattering [5] but in general numerical methods involving the self-consistent QW wavefunctions must be employed [6].

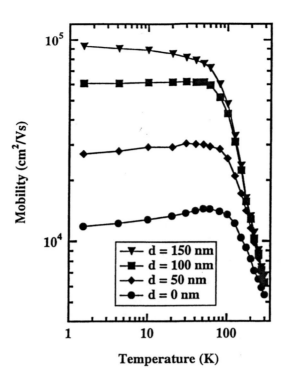

FIGURE 1. Measured mobility vs. temperature [3] for the 4 samples described in TABLE 1.
d is the thickness of the undoped barrier layer (spacer).

TABLE 1. Low-temperature (4.2 K) results for modulation doped multi-quantum wells [3].

GaAs QW thickness (nm)	Doped $Al_xGa_{1-x}As$ layer thickness (nm)	Undoped $Al_xGa_{1-x}As$ spacer thickness (nm)	Average free electron density (cm^{-3})	Mobility (cm^2/V s)	Resistivity (Ω cm)
25.5	30.2	0	16×10^{16}	12500	3.13×10^{-3}
24.5	29.0	5.1	12×10^{16}	28000	1.86×10^{-3}
24.4	28.7	9.9	7.7×10^{16}	62000	1.31×10^{-3}
25.0	29.3	15.1	4.9×10^{16}	93000	1.37×10^{-3}

C2   Interface Roughness Limited Mobility

For narrow quantum wells the thickness fluctuations due to interface roughness play a decisive
role. The confinement energy of the lowest subband in an infinitely deep QW of width L is
proportional to L^{-3} so for a fixed interface fluctuation one expects a very strong dependence of
the effective potential of the order of L^{-6}. The standard description of interface roughness as a
mean average fluctuation Δ in barrier position with a Gaussian autocorrelation function in the
interface plane with correlation length Λ leads to a low temperature mobility of the form [7-9]:

$$\mu \propto \frac{L^6}{\Delta^2 \Lambda^2} g(\Lambda, V, N_s) \qquad (1)$$

where g is a function of the correlation length, the barrier height V, and the electron density in
the well.

Experimental studies have been made by two groups [8-12]. Their results for densities in the dark in the 2 to 6 x 10^{11} cm^{-2} range are shown in FIGURE 2. In these cases the mobility is expected to be entirely dominated by interface roughness scattering. Fitting the results to the expression of EQN (1) leads to values of Δ of the order of one monolayer and Λ between 6 and 25 nm. However, caution is required in attaching too much quantitative importance to these values, since the function g depends on different microscopic models used for the effective scattering potential associated with the interface roughness.

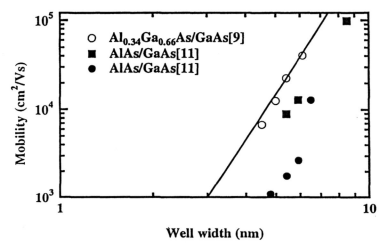

FIGURE 2. Measured interface roughness mobilities vs. well width.
The line represents an L^6-dependence.

C3   Phonon-Limited Mobility

As can be seen in FIGURE 1, the mobility at higher temperatures decreases owing to interactions with phonons and reaches values of the same order of magnitude as those of bulk GaAs. In quantum wells, the only study of the mobilities above 77 K with a systematic comparison to theoretical calculations of all the relevant scattering mechanisms has been that of Inoue and Matsuno [13], whose experimental results are displayed in FIGURE 3. The modification of the theory for phonon scattering due to the electronic structure of quasi-2D systems has been developed in several papers [14-17]. More refined treatments of the fact

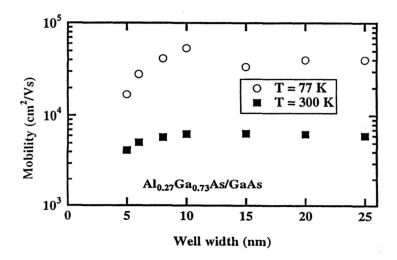

FIGURE 3. Finite-temperature mobilities in quantum wells of different width [13].

that the phonon spectra are influenced by the presence of the well [17-19] have shown that for the mobility, these details of the phonon spectra have negligible quantitative influence. Indeed, as has been found also in single heterostructures [20-22], the phonon-limited mobility is not strongly dependent on the two-dimensionality of the system, i.e. the well width.

D   TRANSPORT PERPENDICULAR TO INTERFACES

The low-field transport perpendicular to the interfaces can be classified according to the thickness of the barriers between the quantum wells. If the barriers are thin, tunnelling through the barriers leads to the formation of minibands [23,24] in which mobility is determined by the effective mass in the miniband; for thicker barriers phonon-assisted hopping from well to well becomes predominant.

D1   Miniband Effective Masses

In a tight-binding description of a miniband the effective mass in the bottom of the miniband is given by $m_z = \hbar^2/(d^2\Delta)$, where d is the SL period and Δ is half the miniband width, or twice the tunnelling matrix element between two neighbouring wells. Above the top of the miniband the Fermi surface contains open orbits in the z-direction. To our knowledge the only systematic measurement of the Fermi surface has been performed by Yoshino et al [25] in 2.3 nm $Al_{0.25}Ga_{0.75}As$/6.0 nm GaAs samples of different doping. A miniband mass anisotropy $m_z/m_\parallel$ of 1.45 at the bottom of the miniband was found in this SL structure. Furthermore, cyclotron resonance measurements [26,27] on non-intentionally doped thin-barrier structures (2 nm $Al_xGa_{1-x}As$/10 nm GaAs) for varying barrier heights (x = 0.12 - 0.29) have shown a mass for motion across the barriers which increases with x to around 0.11 m_0, corresponding to the average effective mass the electron experiences in a cyclotron orbit.

D2   Miniband Mobilities

The interest in vertical transport in superlattices has mainly centred on high-field transport, so the reported low-field mobilities are scarce and rather unsystematic. Furthermore, a strong dependence is expected from the exponential reduction of bandwidth with the barrier thickness. TABLE 2 shows values in the case when the miniband is still rather large. Calculations of mobilities in this regime for a large number of samples can be found in [28].

TABLE 2. Miniband transport mobilities.

Ref	d(barrier) (nm)	d(well) (nm)	x	T (K)	μ (cm^2/V s)
[29]	1.4	4.5	1	77	150
[30]	2.5	3.4	0.33	4.2	4100
[31]	2.8	2.8	0.33-0.17	15	500 (ambipolar)
[31]	5.6	5.6	0.33-0.17	15	1800 (ambipolar)
[31]	8.5	8.5	0.33-0.17	15	50 (ambipolar)

D3   Phonon-Assisted Mobilities

When the tunnelling probability through the barriers decreases to quite low values, vertical transport is best described as a phonon-assisted hopping from well to well. Theoretical work has been carried out for a number of years [32-25]. The few measurements reported show very low mobilities of around 1 - 10 cm^2/V s at room temperature for barriers ($x = 0.3$) up to 3.5 nm and decreasing rapidly to below 0.1 cm^2/V s for 5 nm barriers [36]. Similar results have been found in a detailed study of the temperature dependence of the mobility [37].

E   CONCLUSION

This Datareview shows that in transport parallel to the interfaces very high mobilities can be obtained, especially at low temperatures because of the reduction of impurity scattering. Even higher mobilities can, however, be produced in single heterostructures [22,38,39]. In vertical transport perpendicular to the interfaces, mobility studies are scarce and lead to highly differing values because of the exponential dependence on barrier width.

REFERENCES

[1]   S. Adachi [in *Properties of Aluminium Gallium Arsenide* Ed. S. Adachi (INSPEC, IEE, London, UK, 1993) p.58-65]

[2]   S. Adachi [in *Properties of Aluminium Gallium Arsenide* Ed. S. Adachi (INSPEC, IEE, London, UK, 1993) p.66-72]

[3]   H.L. Störmer, A. Pinczuk, A.C. Gossard, W. Wiegmann [*Appl. Phys. Lett. (USA)* vol.38 (1980) p.691-3]

[4]   K. Inoue, H. Sakaki, J. Yoshino [*Jpn. J. Appl. Phys. (Japan)* vol.23 (1984) p.L767-9]

[5]   G. Fishman, D. Calecki [*Phys. Rev. B (USA)* vol.29 (1984) p.5778-87]

[6]   S. Mori, T. Ando [*J. Phys. Soc. Jpn. (Japan)* vol.48 (1980) p.865-73]

[7]   T. Ando, A.B. Fowler, F. Stern [*Rev. Mod. Phys. (USA)* vol.54 (1982) p.437-672]

[8]   H. Sakaki, T. Noda, K. Hirakawa, M. Tanaka, T. Matsusue [*Appl. Phys. Lett. (USA)* vol.51 (1987) p.1934-6]

[9]   R. Gottinger, A. Gold, G. Abstreiter, G. Weimann, W. Schlapp [*Europhys. Lett. (Switzerland)* vol.6 (1988) p.183-8]

[10]   U. Bockelmann, G. Abstreiter, G. Weimann, W. Schlapp [*Phys. Rev. B (USA)* vol.41 (1990) p.7864-7]

[11]   T. Noda, M. Tanaka, H. Sakaki [*Appl. Phys. Lett. (USA)* vol.57 (1990) p.1651-3]

[12]   T. Noda, M. Tanaka, H. Sakaki [*J. Cryst. Growth (Netherlands)* vol.111 (1991) p.348-52]

[13]   K. Inoue, T. Matsuno [*Phys. Rev. B (USA)* vol.47 (1993) p.3771-8]

[14]   P.J. Price [*Ann. Phys. (USA)* vol.133 (1981) p.217-39]

[15]   F.A. Riddoch, B.K. Ridley [*J. Phys. C, Solid State Phys. (UK)* vol.16 (1983) p.6971-82]

[16]   B. Vinter [in *Heterojunctions and Semiconductor Superlattices* Eds G. Allan, G. Bastard, N. Boccara, M. Lannoo, M. Voos (Springer-Verlag, Berlin, Heidelberg, 1986) p.238-51]

[17]   N. Mori, T. Ando [*Phys. Rev. B (USA)* vol.40 (1990) p.6175-88]

[18]   N. Mori, K. Taniguchi, C. Hamaguchi [*Semicond. Sci. Technol. (UK)* vol.7 (1992) p.B83-7]

[19]   H. Rücker, E. Molinari, P. Lugli [*Phys. Rev. B (USA)* vol.45 (1992) p.6747-56]

[20]   B. Vinter [*Appl. Phys. Lett. (USA)* vol.45 (1984) p.581-2]

[21]   B. Vinter [*Surf. Sci. (Netherlands)* vol.170 (1986) p.445-8]

[22]   K. Hirakawa, H. Sakaki [*Phys. Rev. B (USA)* vol.33 (1986) p.8291-303]

[23]   G. Bastard [*Wave Mechanics Applied to Semiconductor Heterostructures* (Les Editions de Physique, Les Ulis, 1988)]

[24]   C. Weisbuch, B. Vinter [*Quantum Semiconductor Structures: Fundamentals and Applications* (Academic Press, Boston, 1991)]

[25]   J. Yoshino, H. Sakaki, T. Furuta [in *Proc. 17th Int. Conf. on the Physics of Semiconductors* San Francisco, 1984, Eds J.D. Chadi, W.A. Harrison (Springer-Verlag, New York, 1985) p.519-23]

[26]   T. Duffield et al [*Phys. Rev. Lett. (USA)* vol.56 (1986) p.2724-7]

[27]   T. Duffield et al [*Solid State Commun. (USA)* vol.65 (1987) p.1483-7]

[28]   G. Etemadi, J.F. Palmier [*Solid State Commun. (USA)* vol.86 (1993) p.739-43]

[29]   C. Minot, H. Le Person, J.F. Palmier, F. Mollot [*Phys. Rev. B (USA)* vol.47 (1993) p.10024-7]

[30]   H.J. Hutchinson, A.W. Higgs, D.C. Herbert, G.W. Smith [*J. Appl. Phys. (USA)* vol.75 (1994) p.320-4]

[31]   B. Deveaud, J. Shah, T.C. Damen, B. Lambert, A. Regreny [*Phys. Rev. Lett. (USA)* vol.58 (1987) p.2582]

[32]   R. Tsu, G.H. Döhler [*Phys. Rev. B (USA)* vol.12 (1975) p.680-6]

[33]   J.F. Palmier, A. Chomette [*J. Phys. (France)* vol.43 (1982) p.381]

[34]   D. Calecki, J.F. Palmier, A. Chomette [*J. Phys. C. Solid State Phys. (UK)* vol.17 (1984) p.5017-30]

[35]   I. Dharssi, P.N. Butcher [*J. Phys., Condens. Matter (UK)* vol.2 (1990) p.119-25]

[36]   J.F. Palmier [in *Heterojunctions and Semiconductor Superlattices* Eds G. Allan, G. Bastard, N. Boccara, M. Lannoo, M. Voos (Springer-Verlag, Berlin, Heidelberg, 1986) p.127-45]

[37]   H.T. Grahn, K. von Klitzing, K. Ploog, G.H. Döhler [*Phys. Rev. B (USA)* vol.43 (1991) p.12094-7]

[38]   L. Pfeiffer, K.W. West, H.L. Störmer, K.W. Baldwin [*Appl. Phys. Lett. (USA)* vol.55 (1989) p.1888-90]

[39]   C.T. Foxon, J.J. Harris, D. Hilton, J. Hewett, C. Roberts [*Semicond. Sci. Technol. (UK)* vol.4 (1989) p.582-5]

5.2 Carrier dynamics in quantum wells

T.B. Norris

September 1995

A INTRODUCTION

The dynamics of electrons and holes determine many of the properties of devices based on semiconductor quantum wells and superlattices. Furthermore, carrier dynamics in these systems are of basic interest since they provide insight into the dimensionality dependence of fundamental electronic processes in semiconductors. Not surprisingly, the body of work on carrier dynamics is quite extensive. This Datareview outlines the most basic results on carrier dynamics in quantum wells and superlattices, with sufficient references that the reader will have a useful introduction to the literature.

The field of carrier dynamics is generally concerned with two kinds of processes: (1) the relaxation of carriers when they are out of equilibrium with the lattice (e.g. when they have been electrically or optically injected with excess kinetic energy), and (2) nonequilibrium transport. Relaxation includes hot-electron thermalisation and energy relaxation, intersubband scattering, capture into the well, and recombination. Transport can occur in the quantum-well plane (parallel transport), or perpendicular to the quantum-well plane (including tunnelling and Bloch oscillations).

At room temperature or in high electric fields, the dynamics of free electrons and holes must be considered; free-carrier relaxation is discussed in Section B, and free-carrier transport is covered in Section C. At low temperatures and low fields, carriers relax into exciton states; the dynamics of excitons is considered in Section D.

B FREE-CARRIER RELAXATION

B1 Intrasubband Hot-Carrier Relaxation

When carriers are injected into a quantum-well subband with excess kinetic energy larger than the lattice temperature, the carriers are said to be hot. They will generally thermalise within the subband via carrier-carrier scattering, and lose their excess energy to the lattice via phonon emission until the carriers and lattice have the same temperature. Carrier-phonon interactions are discussed in detail in Datareview 5.3 of this volume. Here, we briefly describe the dynamical aspects of hot carrier relaxation. Extensive reviews may be found in [1-4].

The most commonly used techniques for directly time-resolving the dynamics of hot-carrier relaxation are transient luminescence and absorption spectroscopies. In both cases, a picosecond or femtosecond optical pulse injects carriers into the semiconductor with some excess kinetic energy. Luminescence experiments then measure (approximately) the product of the dynamic electron and hole distribution functions: $I \propto f_e f_h$. Transient absorption experiments measure the sum $f_e + f_h$. Electron distributions f_e have been measured by observing the recombination of hot electrons with neutral acceptors.

The first stage of hot-carrier relaxation is the evolution of the initially excited nonthermal distribution into a thermal one within a subband via carrier-carrier scattering. The thermalisation process has been directly time-resolved in undoped and modulation-doped GaAs/AlGaAs quantum wells in a series of experiments by Knox and co-workers (these experiments are summarised in [5]). Transient absorption spectroscopy (with ≈ 100-fs resolution) was used to observe the nonthermal distribution that occurred when electrons were photoinjected at an energy 20 meV above the gap (leading to a 'spectral hole' in the absorption spectrum), and the approach of the nonthermal distribution to a Boltzmann-like one. The data for undoped quantum wells could be reasonably well understood within a relaxation-time approximation, with a thermalisation time of 100 fs at a photoinjected carrier density of 2×10^{10} cm^{-2} and 30 fs at 5×10^{11} cm^{-2} [6]. In n-type modulation-doped quantum wells [7] (doping density 3.5×10^{11} cm^{-2}), and similar excitation conditions (20 meV excess energy and an optically generated density around 5×10^{11} cm^{-2}), no spectral hole was observed, indicating thermalisation on a time scale less than 10 fs. For p-type modulation-doped wells (and similar experimental conditions), nonthermal distributions were observed, with a thermalisation time constant of about 60 fs. These experiments [7] indicated that thermalisation is faster in a Fermi sea of electrons, but not in a Fermi sea of holes, relative to undoped quantum wells. This indicates that even in undoped wells, electron-electron scattering is the dominant mechanism responsible for thermalisation.

The density dependence of electron scattering from an electron-hole plasma has been investigated using hot electron - neutral acceptor luminescence [8]. Those experiments measured the energy loss rate of a nonthermal electron to a thermalised electron-hole plasma. At the highest density (10^{11} cm^{-2}), the loss rate was 200 meV/ps (essentially the same as for coupling to optical phonons), and the density dependence was found to be essentially linear.

After the rapid thermalisation of the plasma, the electrons and holes lose energy to the lattice via emission of phonons. Although the details of the phonon scattering are somewhat different for energy relaxation in the bulk and within a subband of a quantum well, the net energy loss rate is found not to depend substantially on dimensionality [1]. For large initial excess carrier energies, in III-V quantum wells the energy loss is dominated by emission of polar optical (LO) phonons. In GaAs, the LO phonon energy is 36 meV, and LO-phonon emission dominates the energy loss for carrier temperatures above about 30 K [4]. For GaAs the electron energy relaxation time is calculated to be 130 fs, and the net heavy-hole relaxation time less than about 80 fs [4]. In many experiments which have measured the energy relaxation of hot thermal distributions, much longer time constants are obtained. For low lattice temperatures, this discrepancy is understood to arise from hot phonon effects [1,4]: at the large injected carrier densities of most optical experiments the cooling of the hot carriers results in the emission of a large number of optical phonons. The LO-phonon lifetime is of the order of 8 ps [4], so that the optical phonon population can become significantly larger than that given by the lattice temperature. Hot carriers can then reabsorb phonons, thus slowing down the cooling process. An example of the measured energy loss rate for electrons and holes is given in FIGURE 1 [9], showing clearly the effect of hot phonons.

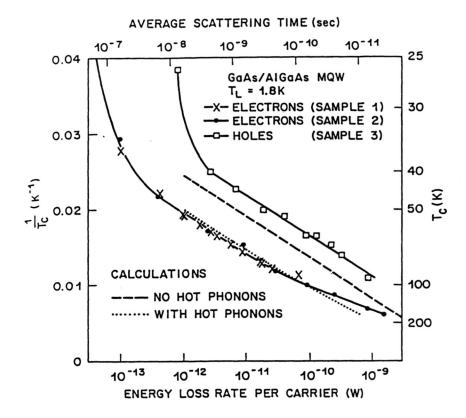

FIGURE 1. Energy loss rates for hot electrons and holes as a function of carrier temperature; the dashed and dotted curves are theoretical calculations. (From [9].)

The final stage of carrier cooling for low lattice temperatures is via acoustic phonon emission. This occurs on a nanosecond time scale for lattice temperatures around 10 K [4].

B2   Intersubband Relaxation and Carrier Capture

Although intrasubband relaxation in quantum wells is not substantially different from bulk energy relaxation, there are two relaxation processes in quantum wells that have no analogue in bulk material: scattering between subbands, and capture from barrier states into quantum-well subbands. Although these processes are still subjects of active study, the basic mechanisms and timescales have been established. Intersubband scattering and capture are important factors determining the performance of quantum-well lasers, far-infrared detectors, and some tunnelling devices.

The dynamics of scattering between the $n = 1$ and $n = 2$ subbands of quantum wells has been studied by several techniques, including transient absorption [10,11], luminescence [12,13], and Raman [14,15] spectroscopies. When the subband spacing is larger than the LO-phonon energy, the intersubband relaxation is dominated by LO-phonon emission. Hunsche et al [11] have measured this relaxation time in a 150 Å quantum well to be 160 fs. When the subband spacing is smaller than the LO-phonon energy, carriers at the $n = 2$ subband edge can no longer emit LO phonons and the relaxation time increases significantly. It has been noted, however [10], that the net intersubband scattering rate will be a rate averaged over the thermal carrier distribution, and carriers in the high-energy thermal tail may be able to emit LO phonons. At sufficiently low carrier temperature, the intersubband relaxation will be mediated by acoustic phonons. At low carrier density, relaxation times of the order of several hundred picoseconds have been measured [13,14], while at high density, 20 - 40 ps has been observed [10].

When carriers are injected into the barrier region of a quantum well structure (e.g. in a quantum-well laser), they must scatter from the three-dimensional barrier states into the two-dimensional subbands of the well. This important problem has been extensively studied, as the carrier-capture rate into quantum wells is a fundamental limitation to the modulation bandwidth of high-speed quantum-well lasers; a complete set of reviews on the subject and references to the original literature can be found in [16]. The capture time for separate-confinement heterostructures or in multiple quantum wells with thick barriers can be understood as the sum of a classical diffusion time (transport to the quantum well) and a local capture time. For thick barriers the diffusion dominates, whereas for thin barriers (smaller than the mean free path) the local capture dominates. The local capture time must be understood as an LO-phonon-mediated scattering process between the quantum-mechanical states above the barrier and in the quantum well [17,18]. Capture via LO-phonon emission results in resonances in the capture rate as the well width is varied, since the rate will be a minimum when one of the quantum well subbands is exactly one LO-phonon energy below the barrier, or when one subband is very close to the top of the well. Capture into a variety of multiple-quantum-well and separate-confinement-heterostructure quantum well laser structures has been investigated by transient luminescence [18-21] and absorption [22] spectroscopies. Resonant local capture has been observed, with the precise rates determined by the details of the quantum well and/or separate-confinement heterostructure design; the fastest observed room-temperature resonant capture time to date is 650 fs in a 58 Å GaAs quantum well [22].

B3   Recombination

The final stage of relaxation of free carriers injected into a semiconductor quantum well is electron-hole recombination. We review here only radiative recombination. At low temperatures and densities, free carriers relax into excitonic states; the recombination of excitons is discussed in Section D. For temperatures above approximately 70 K, excitons ionise, and free-carrier recombination becomes important [23]. At high densities (such as commonly occur in quantum-well lasers), excitons do not exist, and free-carrier recombination dominates. A number of in-depth studies of free-carrier recombination in GaAs quantum wells have been reported in recent years [23-26]; a summary of the basic results is as follows.

The electron-hole pair radiative lifetime is defined as

$$\tau_r^{-1} \equiv n^{-1} \int_0^\infty r_s(h\nu) dh\nu$$

where r_s is the total spontaneous emission rate and n the pair density. At low densities and high temperatures, radiative recombination of electrons with free holes results in a radiative lifetime which varies as $\tau_r \approx BT/n$. B has been measured to be 7.74 s/(K cm^2) in a GaAs/AlGaAs quantum well [25]. Hence lifetimes of typically a few ns are found at densities $\approx 10^{12}$ cm^{-2}. The relation $\tau_r \propto T/n$ near room temperature holds for electron-hole pair densities in the range of about 10^9 to 10^{12} cm^{-2}. It should be noted that, due to the presence of a background acceptor concentration of 10^{13}-10^{16} cm^{-3} in most quantum well samples, at even lower densities electron recombination with holes on neutral acceptors may dominate;

this results in a density-independent lifetime which is sample-dependent [24]. At high densities ($>3 \times 10^{12}$ cm^{-2} at room temperature), the electron and hole plasmas are degenerate, and the spontaneous emission rate saturates to a constant value [26]. In the degenerate limit, a saturated pair lifetime of 400 ps has been calculated and observed [25].

B4   Summary of Relaxation Rates

A useful summary of the time scales involved in the various relaxation processes discussed in this section is given in FIGURE 2 [4].

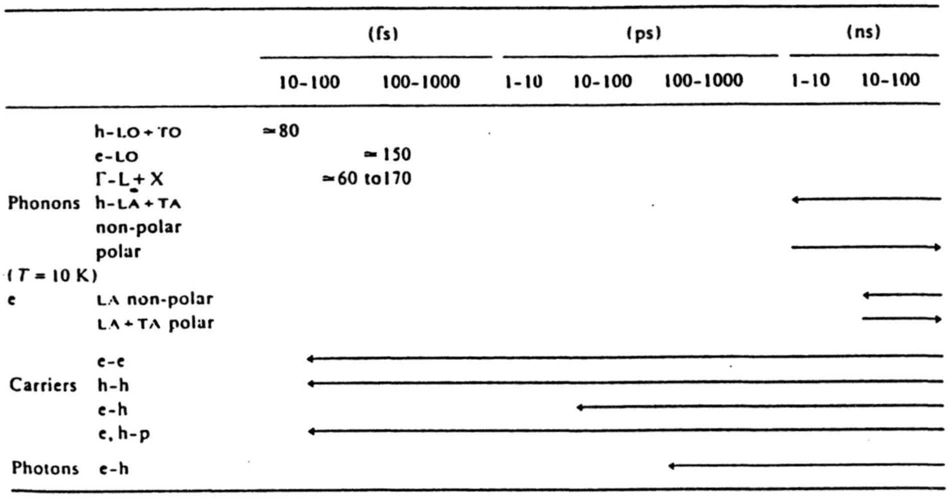

		(fs)		(ps)			(ns)	
		10-100	100-1000	1-10	10-100	100-1000	1-10	10-100
Phonons	h-LO+TO	≈80						
	e-LO		≈150					
	Γ-L+X	≈60 to170						
	h-LA+TA							
	non-polar						←	
	polar						→	
(T = 10 K)								
e	LA non-polar							←
	LA+TA polar							→
	e-e		←					
Carriers	h-h		←					
	e-h				←			
	e,h-p		←					
Photons	e-h					←		

Key: e = electron, h = hole, p = plasmon.
The arrows denote rough ranges of time constants traversed as the carrier density increases.

FIGURE 2. Summary of energy relaxation times for various relaxation processes in quantum wells. (From [4].)

C   TRANSPORT DYNAMICS

C1   Parallel Transport

Nonequilibrium transport in the plane of quantum wells is important for the operation of devices such as MODFETs or high electron mobility transistors (HEMTs). When a high electric field is applied in the plane of a quantum well, an electron will be accelerated by the field. As the electron kinetic energy increases, its momentum relaxation rate may also increase due to the possibility of intervalley scattering or real-space transfer. As a result, the electron average velocity is expected to display an overshoot as a function of time, with a peak value several times that of the saturated velocity [27].

Evidence for this picture of parallel transport dynamics has been obtained in several time-resolved optical experiments. The earliest stage of ballistic acceleration of electrons was directly observed by femtosecond transient-absorption spectroscopy, in which a 100 fs optical pulse generated carriers at the band edge in the presence of a strong electric field. The time-dependent distribution functions obtained from the differential spectra directly showed the presence of electrons which gained energy as $E = e^2F^2t^2/2m^*$, as expected for ballistic acceleration [28]. Evidence for velocity overshoot was obtained in measurements of photoconductive transients on coplanar transmission lines on multiple-quantum-well samples [29]. Again, a short optical pulse generated carriers with no initial kinetic energy in the presence of a parallel field. The time-dependent photocurrent (measured via femtosecond electro-absorption sampling) displayed an overshoot which matched qualitatively a Monte Carlo simulation. The details of velocity overshoot in parallel transport are not yet complete, since the presence of space-charge effects complicates the interpretation of the experiments performed to date.

C2   Perpendicular Transport

Transport perpendicular to the quantum-well planes has been a topic of fundamental importance since the earliest days of semiconductor-heterostructure physics, when Esaki and Tsu [30] proposed the superlattice to enable the observation of resonant tunnelling, negative differential conductance, and Bloch oscillations. Since then, a large number of structures have been developed for the investigation of perpendicular transport. In this section, the main features of the dynamics of perpendicular transport are outlined. As the fundamental physical process governing perpendicular transport is tunnelling between quantum wells, the main issue is the dynamics of tunnelling.

There are several distinct tunnelling processes which have been investigated: (i) tunnelling from a discrete quantum-well state into a continuum, (ii) incoherent tunnelling from one discrete state (subband) to another in coupled quantum wells, and (iii) coherent tunnelling of electron wavepackets between quantum wells; these will be considered in turn. Detailed reviews and extensive literature references may be found in [31-33].

Case (i) is particularly relevant to the operation of double-barrier structures, or resonant tunnelling diodes (RTDs; see Datareview 8.4). The speed of RTDs is fundamentally limited by the lifetime of the quasi-bound state in the quantum well formed by the double-barrier structure [34]. The quasi-bound-state lifetime can be expressed as $\tau = h/\Delta E$, where ΔE is the width of the transmission resonance through the double-barrier structure. The width must be calculated by solving the Schrödinger equation for the structure (usually via a transfer matrix approach). A useful way to calculate the lifetime is to consider the electron as a particle oscillating back and forth in the well, so that it hits a barrier with a frequency ν (the attempt frequency). Each time it is incident on a barrier, it has a probability T of being transmitted through it, so the tunnelling lifetime is then given by $\tau_t = 1/\nu T$. For a double-barrier structure

$$\nu = \sqrt{2E_r/m^*w^2}$$

and

$$T = \exp\left\{-2L\sqrt{2m^*(V-E_1)}\right\}$$

is the single-barrier transmission coefficient, where E_r is the energy in the well, w is the well width, V is the barrier height, and L the barrier thickness. Due to the exponential dependence on the barrier height and width, the tunnelling time can vary by orders of magnitude for typical RTD structures. This basic model of tunnelling escape rates was verified by time-resolved photoluminescence for thin barriers by Tsuchiya et al [35]; for example a 62 Å GaAs quantum well bounded by two 28 Å AlAs barriers will have a tunnelling escape time of 60 ps under flat-band conditions (see FIGURE 3). As a bias voltage is applied, the barriers are tilted, resulting in a reduction in the tunnelling escape time [36].

In case (ii), an asymmetric double quantum well is typically used so that the tunnelling of a carrier in a subband of one well into a subband of the second well may be observed. The relative energies of the two subbands may be controlled by situating the double-well structure in a diode and varying the bias voltage. The carriers are nonresonantly injected via a short optical pulse into the structure, where they rapidly thermalise in the subband of the first well (i.e. no phase information is retained in the initial state). The tunnelling process may then be thought of as an intersubband relaxation process, where the relaxation is mediated by scattering (e.g. from interface defects or phonons) [37,38]. The incoherent tunnelling is fastest when the subbands of the two wells are resonant, as has been observed in time-resolved photoluminescence by Oberli et al [39] (see FIGURE 4). They observed a resonant tunnelling time of $\tau_t = 7$ ps for a 50 Å tunnel barrier. Incoherent

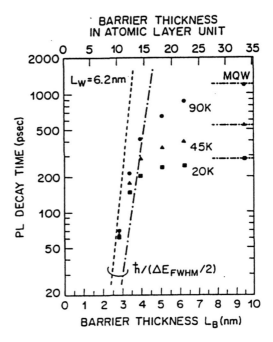

FIGURE 3. Photoluminescence decay times in a double-barrier resonant tunnelling structure vs. barrier thickness. For thin barriers, the decay rates correspond to the electron tunnelling escape rates. (From [35].)

tunnelling between heavy-hole subbands in asymmetric double wells has also been observed with $\tau_t = 1.3$ ns at resonance [40]. Resonant tunnelling between the heavy-hole subband of one well and a light-hole subband of the second well has also been observed ($\tau_t < 75$ ps) [41]; this surprisingly fast rate was attributed to tunnelling of holes with nonzero in-plane momentum, where the hole wavefunction has some light-hole character [42]. In general, incoherent tunnelling rates are determined by scattering and interface quality in double-well structures, so a wide range of tunnelling rates has been observed [31,33].

When the subbands of an asymmetric double-quantum-well structure are brought into resonance (by tuning the bias voltage), the eigenstates of the coupled-well system are no longer localised in each well, but are the 'bonding' (symmetric-like) and 'antibonding' (antisymmetric-like) states delocalised over the two wells. As the structure is tuned through resonance, the energy levels display an anticrossing with a minimum separation ΔE (which is determined by the strength of the coupling between the wells, i.e. the barrier height and width). In case (iii), a coherent short optical pulse is used to generate a wavepacket which is a superposition of the double-well eigenstates, thus localising the electron in one of the wells. When such a coherent wavepacket is formed, the time-dependent Schrödinger equation shows that the electron will oscillate back and forth between the two wells. Thus coherent tunnelling between discrete subbands is oscillatory, in contrast to incoherent tunnelling (which is a rate process). Time-resolved optical studies of coherent tunnelling are reviewed in Leo [43]. For a structure consisting of a 150 Å GaAs well coupled to a 100 Å GaAs well through a 25 Å $Al_{0.2}Ga_{0.8}As$ barrier, a tunnelling oscillation period of 800 fs is observed, which corresponds to the observed eigenstate splitting at resonance of 5 meV. The spatial oscillation of the electron wavepacket has been directly observed through the terahertz radiation emitted by the spatially oscillating dipole by Roskos et al [44]. The frequency of the tunnelling oscillations observed was 1.5 THz.

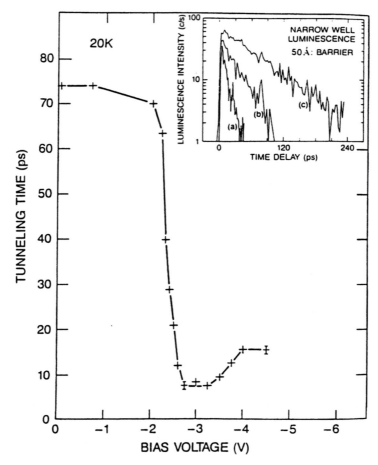

FIGURE 4. Time-resolved photoluminescence measurement of rate at which electrons tunnel from the lowest (n = 1) subband in a narrow quantum well into the n = 2 level of a wide well closely coupled to the first by a thin tunnel barrier. The two subbands are tuned in and out of resonance by an electric field applied perpendicular to the quantum wells; the tunnelling time is minimum at resonance. (From [39].)

Finally, we consider perpendicular transport in superlattices. When many identical quantum wells are coupled together through thin tunnel barriers, a superlattice is formed, and the discrete envelope wavefunctions of the wells overlap forming a miniband. Time-resolved photoluminescence spectroscopy has been used by Deveaud and co-workers to observe directly drift and diffusive carrier transport in superlattice minibands [45,46]; for a 20 Å/20 Å GaAs/AlGaAs superlattice, they find an electron mobility of 6300 cm^2/V s.

As discussed in Datareview 2.5 in this volume, when an electric field is applied along the growth direction of a superlattice, the miniband breaks up into a set of discrete levels known as the Wannier-Stark ladder. The energy separation between levels is eFd, where d is the superlattice period. The eigenstates are localised over a length $L = \Delta/2eF$. In experiments very similar to those discussed above for coherent tunnelling (case iii), a short optical pulse has been used to excite a coherent superposition of Wannier-Stark states, creating an electronic wavepacket which undergoes a spatial oscillation of length L and with period $\tau_B = h/eFd$. These oscillations are referred to as Bloch oscillations, due to their close connection to the 'classical' Bloch oscillations caused by Bragg reflection of ballistically accelerated electrons at the Brillouin zone boundary [47]. The oscillation period and amplitude are the same for the coherent wavepacket of Wannier-Stark states and for the classical Bloch electron moving in a miniband. However, the observations of Bloch oscillations must be understood as quantum interference of Wannier-Stark states [48,49]. Evidence for Bloch oscillations was first obtained in four-wave mixing experiments [50,51]; the spatially oscillating dipole associated with the Bloch oscillation was observed via the emitted THz radiation [52]. For typical GaAs/AlGaAs superlattices, oscillation frequencies tunable (by varying the field F) from 0.5 to over 2 THz can be obtained.

D   EXCITON DYNAMICS

At low temperatures, the lowest-energy photoexcited electron-hole pairs are the Coulombically bound excitons. Excitonic effects dominate the radiative properties of QWs at low temperature. Thus the dynamics of carriers at low temperature consist of the formation, scattering, and recombination (radiative decay) of excitons.

D1   Intrinsic Radiative Lifetime

Photoluminescence experiments to observe the dynamics of radiative decay have been performed by a number of groups. Although not all the details are settled, a basic picture has emerged. Exciton radiative dynamics must be understood in terms of the fundamental two-dimensional exciton-polariton states formed by the coupling of the quantum-well exciton with the electromagnetic field. Because of the translational symmetry of the QW, momentum in the plane of the QW ($k_\parallel$) must be conserved when a free exciton absorbs or emits light (i.e. $k_\parallel$ is a good quantum number for the exciton-polariton). Thus, if k_o is the wavevector of light in the medium, an exciton with $k_\parallel < k_o$ is coupled to a continuum of propagating (real-k_z) photon modes with wavevector $k = (k_\parallel, k_z)$ having the same value of $k_\parallel$, and can therefore undergo radiative decay; however, excitons with $k_\parallel > k_o$ cannot emit light [53,54]. At $k_\parallel = 0$, the intrinsic exciton-polariton population radiative decay rate is given by

$$\Gamma_o = \frac{\pi}{n} \frac{e^2}{m_o c} f \tag{1}$$

where f is the oscillator strength per unit area. For a typical GaAs QW of width 100 Å, $f = 50 \times 10^{-5}$ Å^{-2}, so that $\Gamma_o^{-1} = 25$ ps.

In general, the intrinsic exciton-polariton decay rate varies with $k_\parallel$; however, this is in practice not important, for two reasons. First, momentum relaxation of the excitons gives rise to an approximately thermal distribution of excitons over $k_\parallel$ states, so an average radiative decay results. (Note that as the temperature is increased so that $k_B T > E_0 = \hbar^2 k^2 / 2M$ where M is the exciton mass, the radiative lifetime will increase since a smaller fraction of excitons are within $k_\parallel < k_o$.) Second, excitons within a homogeneous linewidth Γ_h of $k_\parallel = 0$ can emit light. The radiative population decay rate for free excitons thermalised with temperature T and with a homogeneous linewidth Γ_h has been calculated to be [54,55]

$$\Gamma = \frac{E_0}{3\hbar\Gamma_h} \left(1 - e^{-\hbar\Gamma_h / k_B T}\right) \Gamma_o \tag{2}$$

Thus it is clear that the observed exciton decay will depend on the distribution of the population, and hence the excitation conditions. The two principal cases are resonant and nonresonant excitation.

D2   Resonant Excitation

Experimental investigations of excitonic luminescence dynamics under resonant excitation were first reported by Deveaud et al [56]. When excitons are resonantly created by a short (picosecond) optical pulse (usually near normal incidence), excitons with $k_\parallel \approx 0$ are created directly, and the initial decay of the luminescence is determined by a combination of radiative recombination, scattering to different $k_\parallel$ states, and spin relaxation (in the case of polarised excitation and detection). Vinattieri et al [57,58] have found that at temperatures for typical experimental conditions (10 - 30 K), all these processes have similar rates, and thus an extensive series of experiments and curve fitting is necessary to elucidate the different rates. They find for the intrinsic radiative lifetime $\Gamma_o^{-1} = 40$ ps, roughly twice the theoretical value. Sermage et al [59] have reported an experimental value $\Gamma_o^{-1} = 18$ ps. Thus, although the intrinsic lifetime is semi-quantitatively understood, open questions remain regarding the details of the radiative decay. Vinattieri et al [58] have also determined the scattering rate for initially excited excitons to large-$k_\parallel$ nonradiative states to be 6×10^{10} s^{-1} for a 150 Å well at 12 K, depending on the well width as $1/L$ as expected for excitons interacting with acoustic phonons [60]. Damen et al [61] have also reported an exciton-acoustic-phonon scattering rate from the T-dependent initial decay of the luminescence, with a slope of 2×10^9 s^{-1} K^{-1} for T < 25 K.

Following the fast initial decay of the luminescence, the exciton population thermalises over the exciton band. Since only those excitons within a homogeneous linewidth of $k_\parallel = 0$ can radiate, the lifetime becomes much longer (of the order of 100 ps) [59]. The decay dynamics of the thermalised excitons are essentially the same as for nonresonantly excited excitons.

D3   Nonresonant Excitation

When a quantum well is optically excited above the band edge, free electron-hole pairs are generated. Nevertheless, at low temperatures (T < 100 K), the luminescence is dominated by excitonic recombination [62,63]. Damen et al [63] have shown that for T $\approx$ 10 K, the free carriers form large-$k_\parallel$ excitons in about 20 ps. Because these large-$k_\parallel$ excitons cannot radiate, the luminescence shows a long rise time reflecting the slow relaxation of the excitons to $k_\parallel \approx 0$ via exciton-exciton and acoustic-phonon scattering, the latter (former) dominating below (above) a density of 4×10^9 cm^{-2}. At low densities and T $\approx$ 10 K, the rise time is as long as 400 ps. As the density or temperature is increased, the rise time is shortened. At 60 K, it is about 20 ps.

The decay of the thermalised exciton distribution depends on the temperature of the exciton population. Feldman et al [62] have observed (at 5 K) a lifetime increasing from 300 ps to 1.6 ns as the well width is increased from 25 to 150 Å, with a lifetime proportional to T. The latter may be expected from EQN (2), which shows that if the temperature is high compared to the homogeneous width, the free exciton radiative decay rate will be given by $(E_o/3k_BT)\Gamma_o$.

It is important to note that in many samples the lowest exciton states are localised by interface roughness. Their contribution to the luminescence decay has been considered by Citrin

[54,55]. The radiative lifetime for localised states is shown to be longer than that for free excitons, since the effect of localisation is to mix nonradiative large-$k_\parallel$ states into the exciton wavefunction. The calculation shows lifetimes of the order of 100 ps, dependent on the degree of localisation (i.e. on interface roughness). This is consistent with the wide range of long-time-scale lifetimes reported in the literature (see references reported in [54,55]). Typically lifetimes in the several-hundred-ps range are found, with sample-to-sample variations probably arising from the variable interface roughness.

D4   Exciton Ionisation

When excitons are created by resonant optical absorption at room temperature, the excitons are ionised by optical phonon scattering to form a free-carrier plasma. The exciton ionisation time in GaAs/AlGaAs quantum wells has been directly measured to be 300 fs [64].

E   CONCLUSION

Free-carrier and excitonic relaxation and transport processes occur on timescales from nanoseconds to femtoseconds. Direct time-resolved studies of these processes have established the basic physical mechanisms and timescales in quantum wells, especially in the GaAs/AlGaAs system. Nevertheless, as semiconductor technology continues to progress rapidly, and new low-dimensional structures are continuously being developed, the investigation of the dynamics of carriers in low-dimensional semiconductors continues to be a very active field of research.

REFERENCES

[1]   J. Shah (Ed.) [*Hot Carriers in Semiconductor Nanostructures* (Academic Press, Boston, 1992)]

[2]   J. Shah [*IEEE J. Quantum Electron. (USA)* vol.22 (1986) p.1728]

[3]   S.M. Goodnick [in *Spectroscopy of Semiconductor Nanostructures* Eds G. Fasol, A. Fasolino, P. Lugli (Plenum, New York, 1989) p.561-84]

[4]   B.K. Ridley [*Rep. Prog. Phys. (UK)* vol.54 (1991) p.169]

[5]   W.H. Knox [in *Hot Carriers in Semiconductor Nanostructures* Ed. J. Shah (Academic Press, Boston, 1992) ch.IV.2]

[6]   W.H. Knox, C.A. Hirlimann, D.A.B. Miller, J. Shah, D.S. Chemla, C.V. Shank [*Phys. Rev. Lett. (USA)* vol.56 (1986) p.1191]

[7]   W.H. Knox, D.S. Chelma, G. Livescu, J. Cunningham, J.E. Henry [*Phys. Rev. Lett. (USA)* vol.61 (1988) p.1290]

[8]   J.A. Kash [*Phys. Rev. B (USA)* vol.48 (1993) p.18336]

[9]   J. Shah, A. Pinczuk, A.C. Gossard, W. Wiegmann [*Phys. Rev. Lett. (USA)* vol.54 (1985) p.2045]

[10]   J.A. Levenson, G. Dolique, J.L. Oudar, I. Abram [*Phys. Rev. B (USA)* vol.41 (1990) p.3688]

[11]   S. Hunsche, K. Leo, H. Kurz, K. Kohler [*Phys. Rev. B (USA)* vol.50 (1994) p.5791]

[12]   H.T. Grahn, H. Schneider, W.W. Ruhle, K. von Klitzing, K. Ploog [*Phys. Rev. Lett. (USA)* vol.64 (1990) p.2426]

[13]   R.A. Hopfel et al [*Phys. Rev. B (USA)* vol.47 (1993) p.10943]

[14]   D.Y. Oberli, D.R. Wake, M.V. Klein, J. Klem, T. Henderson, H. Morkoc [*Phys. Rev. Lett. (USA)* vol.59 (1987) p.696]

[15]   M.C. Tatham, J.F. Ryan, C.T. Foxon [*Phys. Rev. Lett. (USA)* vol.63 (1989) p.1637]

[16]   [*Opt. Quantum Electron. (UK)* vol.26 (1994) p.S647-855]

[17]   P.W.M. Blom, P.J. van Hall, J.E.M. Haverkort, J.H. Wolter [*Phys. Rev. B (USA)* vol.47 (1993) p.2072]

[18]   B. Deveaud et al [*Opt. Quantum Electron. (UK)* vol.26 (1994) p.S679]

[19]   P.W.M. Blom, J. Claes, J.E.M. Haverkort, J.H. Wolter [*Opt. Quantum Electron. (UK)* vol.26 (1994) p.S667]

[20]   B. Deveaud, F. Clerot, A. Regreny, K. Fujiwara, K. Mitsunaga, J. Otha [*Appl. Phys. Lett. (USA)* vol.55 (1989) p.2446]

[21]   S. Morin, B. Deveaud, F. Clerot, A. Regreny, K. Fujiwara, K. Mitsunaga [*IEEE J. Quantum Electron. (USA)* vol.27 (1991) p.1669]

[22]   M.R.X. Barros, P.C. Becker, D. Morris, B. Deveaud, A. Regreny, F. Beisser [*Phys. Rev. B (USA)* vol.47 (1993) p.10951]

[23]   M. Gurioli et al [*Phys. Rev. B (USA)* vol.44 (1991) p.3115]

[24]   L.E. Oliveira, M. de Dios-Leyva [*Phys. Rev. B (USA)* vol.48 (1993) p.15092]

[25]   G. Bongiovanni, J.L. Staehli [*Phys. Rev. B (USA)* vol.46 (1992) p.9861]

[26]   B. Deveaud, F. Clerot, K. Fujiwara, K. Mitsunaga [*Appl. Phys. Lett. (USA)* vol.58 (1991) p.1485]

[27]   S. Goodnick, W.H. Knox [*Int. Conf. on Quantum Electronics Technical Digest Series (USA)* vol.9 (1992) p.246]

[28]   W. Sha, T.B. Norris, W.J. Schaff, K.E. Meyer [*Phys. Rev. Lett. (USA)* vol.67 (1991) p.2553]

[29]   W.H. Knox [*Appl. Phys. A (Germany)* vol.53 (1991) p.503]

[30]   L. Esaki, R. Tsu [*IBM J. Res. Dev. (USA)* vol.14 (1970) p.61]

[31]   J. Shah [in *Spectroscopy of Semiconductor Nanostructures* Eds G. Fasol, A. Fasolino, P. Lugli (Plenum, New York, 1989) p.535-60]

[32]   J. Shah [*Hot Carriers in Semiconductor Nanostructures* (Academic Press, Boston, 1992) ch.IV.1]

[33]   L.L. Chang, E.E. Mendez, C. Tejedor (Eds) [*Resonant Tunneling in Semiconductors* (Plenum, New York, 1991)]

[34]   T.C.L.G. Sollner, E.R. Brown, J.R. Soderstrom, T.C. McGill, C.D. Parker, W.D. Goodhue [in *Resonant Tunneling in Semiconductors* Eds L.L. Chang, E.E. Mendez, C. Tejedor (Plenum, New York, 1991) p.487]

[35]   M. Tsuchiya, T. Matsusue, H. Sakaki [*Phys. Rev. Lett. (USA)* vol.59 (1987) p.2356]

[36]   T.B. Norris, X.J. Song, W.J. Schaff, L.F. Eastman, G. Wicks, G.A. Mourou [*Appl. Phys. Lett. (USA)* vol.54 (1988) p.60]

[37]   R. Ferreira, G. Bastard [*Phys. Rev. B (USA)* vol.40 (1989) p.1074]

[38]   S.M. Goodnick, P. Lugli [in *Hot Carriers in Semiconductor Nanostructures* Ed. J. Shah (Academic Press, Boston, 1992) ch.III.1]

[39]   D.Y. Oberli et al [*Phys. Rev. B (USA)* vol.40 (1989) p.3028]

[40]   K. Leo et al [*Phys. Rev. B (USA)* vol.42 (1990) p.7065]

[41]   T.B. Norris, N. Vodjdani, B. Vinter, E. Costard, E. Bockenhoff [*Phys. Rev. B (USA)* vol.43 (1991) p.1867]

[42]   R. Ferreira, G. Bastard [*Europhys. Lett. (Switzerland)* vol.10 (1989) p.279]

[43]   K. Leo et al [*IEEE J. Quantum Electron. (USA)* vol.28 (1992) p.2498]

[44]   H.G. Roskos et al [*Phys. Rev. Lett. (USA)* vol.68 (1992) p.2216]

[45]   B. Deveaud, J. Shah, T.C. Damen, B. Lambert, A. Chomette, A. Regreny [*IEEE J. Quantum Electron. (USA)* vol.24 (1988) p.1641]

[46]   B. Deveaud et al [*Appl. Phys. Lett. (USA)* vol.59 (1991) p.2168]

[47]   F. Bloch [*Z. Phys. (Germany)* vol.52 (1928) p.555]

[48]   M. Dignam, J.E. Sipe, J. Shah [*Phys. Rev. B (USA)* vol.49 (1994) p.10502]

[49]   P. Leishing et al [*Phys. Rev. B (USA)* vol.50 (1994) p.14389]

[50]   J. Feldman et al [*Phys. Rev. B (USA)* vol.46 (1992) p.7252]

[51]   K. Leo, P. Haring Bolivar, F. Bruggemann, R. Schwedler, K. Kohler [*Solid State Commun. (USA)* vol.84 (1992) p.943]

[52]   C. Waschke, H.G. Roskos, R. Schwedler, K. Leo, H. Kurz, K. Kohler [*Phys. Rev. Lett. (USA)* vol.70 (1993) p.3319]

[53]   L.C. Andreani [*Solid State Commun. (USA)* vol.77 (1991) p.61]

[54]   D.S. Citrin [*Phys. Rev. B (USA)* vol.47 (1993) p.3832]

[55]   D.S. Citrin [in *Confined Electrons and Photons* Eds E. Burstein, C. Weisbuch (Plenum, 1995)]

[56]   B. Deveaud, F. Clerot, N. Roy, K. Satzke, B. Sermage, D.S. Katzer [*Phys. Rev. Lett. (USA)* vol.67 (1991) p.2355]

[57]   A. Vinattieri, J. Shah, T.C. Damen, D.S. Kim, L.N. Pfeiffer, L.J. Sham [*Solid State Commun. (USA)* vol.88 (1993) p.189]

[58]   A. Vinattieri [*Phys. Rev. B (USA)* vol.50 (1994) p.10868]

[59]   B. Sermage, S. Long, B. Deveaud, D.S. Katzer [*J. Phys. (France)* vol.C5 (1993) p.19]

[60]   J. Lee, E.S. Koteles, M.O. Vassell [*Phys. Rev. B (USA)* vol.33 (1985) p.5512]

[61]   T.C. Damen, K. Leo, J. Shah, J.E. Cunningham [*Appl. Phys. Lett. (USA)* vol.58 (1991) p.1902]

[62]   J. Feldman et al [*Phys. Rev. Lett. (USA)* vol.59 (1987) p.2337]

[63]   T.C. Damen, J. Shah, D.Y. Oberli, D.S. Chelma, J.E. Cunningham, J.M. Kuo [*J. Lumin. (Netherlands)* vol.45 (1990) p.1818]

[64]   W.H. Knox [*Phys. Rev. Lett. (USA)* vol.54 (1985) p.1306]

5.3 Confined and interface optical and acoustic phonons in quantum wells, superlattices and quantum wires

M.A. Stroscio, G.J. Iafrate, H.O. Everitt, K.W. Kim, Y. Sirenko, S. Yu, M.A. Littlejohn and M. Dutta

July 1995

A   INTRODUCTION

The effects of carrier confinement have been studied extensively in nanoscale and mesoscopic systems. However, to model carrier energy loss in nanoscale and mesoscopic systems, it is essential that calculations of carrier scattering by longitudinal-optical phonons and acoustic phonons take into account that confinement also changes the strength and spatial properties of longitudinal-optical phonons.

Electron interactions with longitudinal-optical (LO) phonon modes in quantum wells, superlattices and quantum wires are affected strongly by the changes in the Fröhlich and deformation potential Hamiltonians caused by phonon confinement and localisation, as well as by the changes in the electronic wave function due to the confining potential. The presence of heterointerfaces in nanoscale structures produces large changes in dielectric and elastic properties and gives rise to confined and interface modes. For the case of optical phonons, these interface modes shall be denoted as both interface (IF) modes and as surface optical (SO) modes. For the case of optical phonons, confined and interface modes arise because the abrupt changes in dielectric constant near heterointerfaces in nanostructures make it impossible to satisfy the full bulk dispersion relation which normally restricts LO phonon frequencies to those where the dielectric constant vanishes. In the case of acoustic phonons, differing elastic constants result in similar confinement effects; for the case of acoustic phonons it is necessary to maintain continuity in both the elastic displacements of the lattice and normal traction forces at each heterojunction. These same phenomena also lead to the establishment of interface and confined modes in superlattices, quantum wires and quantum dots; quasi-one-dimensional quantum wires have recently been fabricated [1-3] for a variety of purposes including quantum wire lasers [2,3]. Furthermore, wire-like regions are critical elements of proposed mesoscopic devices [4,5] and of novel structures suggested for the suppression of carrier - LO-phonon scattering [6]. The accurate treatment of the optical-phonon modes in quantised systems is essential for understanding the electronic properties of quantum wells, superlattices, quantum wires and quantum dots. Many of the basic mechanisms leading to interface and confined optical phonon modes in quantum wells, superlattices and quantum wires were elucidated in early pioneering papers describing confined and interface phonons in ionic slabs [7-10] as well as carrier interactions [11,12] with the phonon modes established in these ionic slabs. As discussed in this Datareview, acoustic phonon confinement may play a special role in mesoscopic devices.

B   QUANTISED PHONONS IN QUANTUM WELLS AND SUPERLATTICES

In recent years, both microscopic and macroscopic approaches to electron-optical-phonon interactions in quantum wells and superlattices [13-109] have been applied in theoretical treatments. Enhanced electron - IF-phonon (i.e. electron - interface-phonon) scattering in polar semiconductors with confining dimensions less than roughly 50 Å has been indicated recently on the basis of optical measurements [51-53] and measurements of phonon-assisted tunnelling currents [54]. Furthermore, recent mobility measurements [55] and Raman measurements of interface phonons involved in X-band transitions [56] are consistent with these observations. Physically, this increased scattering rate is due to the enhanced interface phonon potential which increases approximately exponentially near heterointerfaces. This exponential behaviour was first predicted within the dielectric continuum model [7,23,29] and it has been supported by agreement with recent fully microscopic calculations [41]. A microscopic theory of optical-phonon Raman scattering in quantum wells has been formulated [57] and the useful sum rule [23] for single- and double-heterojunction structures has been extended to heterostructures of arbitrary geometry [58]. Within the last few years, Bhatt et al [59] have presented a simplified microscopic model for electron - optical-phonon interactions in quantum wells and interface optical phonon potentials have been calculated for the case of structures with four heterointerfaces [60]. Bhatt et al [61] have provided additional analysis on the reduction of electron - interface-mode scattering in metal-semiconductor systems which was followed by Constantinou's analysis of similar phenomena [62] in a variety of heterostructures. Ridley [63,65] and Babiker [64] have extended the works discussed previously and direct measurements of electron - optical-phonon scattering rates in ultrathin GaAs/AlAs multiple quantum wells have been reported [66]. Other recent observations which are extremely significant include electron-energy-loss measurements of the Fuchs-Kliewer interface modes [67] and measurements of the frequency splitting between GaAs and AlAs surface phonons as a function of Al concentration in $Al_xGa_{1-x}As$ [68]. The effects of compositional disorder on phonons in layered semiconductor microstructures have been considered by Bechstedt et al [69]. Polaron binding energies and effective masses have been calculated and the relative importance of the different optical phonon modes has been investigated as a function of quantum well width by Hai et al [70]. Polar optical-phonon scattering in three- and two-dimensional electron gases has been analysed in a particularly simple form [71] which is suited for relaxation-time approximations and perhaps for the inclusion of confined phonon effects. Bhatt et al [72] calculated longitudinal-optical-phonon lifetimes in GaAs in a form which is intended for generalisation to the case of confined phonons. A number of these recent advances have been summarised in [73]. The role of confined and interface optical phonons in Γ-X transitions in short-period superlattice transitions has been discussed by Dutta [74] and others [75,81]. Localisation and diffusion effects of phonons in superlattices have been investigated experimentally [82] and treated theoretically by Fertig and Reinecke as well as by Molteni et al [83]. Ridley and Babiker [84] have given a brief critique of continuum theories of optical phonons and polaritons in superlattices and Zhu [85] commented that the Fuchs-Kliewer interface modes do possess Fröhlich potentials. Chamberlain et al [86] have presented a separate continuum model of optical phonons in GaAs/AlAs superlattices in terms of linear combinations of optical phonon modes and the apparent incompatibility between mechanical and electrostatic boundary conditions has been analysed [87].

Progress in both the theory and experimental investigation of acoustic phonon modes in quantum wells and superlattices has been rapid in recent years [88-109]. Bannov et al [88,89] and Mitin et al [90] have considered the cases of confined acoustic phonons in free-standing quantum wells as well as acoustic phonon scattering in low dimensional structures. The development of acoustic phonon packets in superlattices has been studied experimentally by Mizuno et al [91] and a generalised piezoelectric scattering rate for electrons in a two-dimensional electron gas has been given by Stroscio and Kim [92]; this last contribution points to the existence of previously uncalculated acoustic-phonon effects in both bulk and quantum well structures. Kochelap and Gülseren have modelled the localisation of acoustical modes due to the electron-phonon interaction within a two-dimensional electron gas [93,94]. Earlier theoretical works on confined and interface acoustic phonons are found in [95,96] which, however, did not provide prescriptions for quantising the phonon displacement amplitudes. Hillebrands et al [97] have provided evidence for the existence of guided longitudinal acoustic phonons in ZnSe films on GaAs and Wybourne and co-workers [98-101] have presented evidence for confined acoustic phonon effects. Acoustic phonons have been investigated in superlattices through the use of Raman scattering [102a-102d] to study acoustic phonon folding and disorder induced effects. Sapega et al [103] have observed folded acoustic modes in GaAs/Al$_x$Ga$_{1-x}$As multiple quantum wells in the presence of strong magnetic fields. Acoustic-phonon-assisted migration of localised excitons in GaAs/AlGaAs in multiple-quantum-well structures [104] point to the possible role of confined acoustic phonon effects in exciton dynamics. Tamura and co-workers [105a-105f] have considered resonant transmission of acoustic phonons in double-barrier systems as well as acoustic phonon transmission and interference in superlattices; He et al [106,107] have considered optical-absorption effects on light scattering by acoustic phonons in superlattices as well as light scattering by longitudinal-acoustic phonons in superlattices and have performed a coupled Brillouin-Raman study of folded acoustic modes in GaAs/AlAs superlattices [107]. Boudouti and Djavari-Rouhani [108] used a Green's function method to study acoustic waves of shear horizontal polarisation in finite superlattices. Finally, Tamura has calculated anomalously long lifetimes for high-energy Rayleigh surface acoustic phonons [109].

In summary, the extensive literature on interface and confined phonons in quantum wells and superlattices [7-109] serves as the foundation for modelling phonons in quantum wires.

C   QUANTISED PHONONS IN QUANTUM WIRES

Carrier-phonon interactions in quantum-wire structures are introduced in this section with a brief discussion indicating that many interesting phenomena [110-115] arise in these structures as a result of confinement effects which are independent of the transition from bulk LO phonons to interface and confined phonons. Furthermore, this section highlights early work [116-119] on interface and confined optical phonon modes in confined polar-semiconductors including structures with wire-like geometries. A survey of efforts to model interface and confined optical phonons in quantum wires [120-135] is given along with discussions highlighting the current understanding of: carrier - optical-phonon scattering rates in quantum wires with variable cross-section [134,135], hot-phonon effects and dynamical screening [136,137], and how optical phonon modes in quantum wires are modified [131] by the presence of metal-semiconductor heterointerfaces at selected boundaries of these nanoscale structures [138-140].

It is well known that substantial size effects such as singularities in the optical phonon emission rate [110] result from the one-dimensional density of electronic states in quantum wires. In addition, it has been shown [111] that subband population inversions occur in quantum wires when subband spacings are equal to the optical phonon energy. Furthermore, it has been demonstrated that resonant inter-subband optical phonon scattering (RISOPS) occurs in quasi-one-dimensional structures even in the absence of azimuthal symmetry [112]. Related, and important, size effects in quantum wires include enhanced differential mobility [113] and anomalous carrier cooling [114]. Confinement effects in polar-semiconductor quantum wires [64] lead to a wide variety of new phenomena [110-114] even when confined and interface optical phonon interactions are approximated with the bulk LO phonon interaction Hamiltonian. Future research efforts could revisit these effects for the case where phonon confinement is included.

The quantised optical phonon modes in ionic crystals of finite size have been derived for a variety of confinement geometries [116-119] on the basis of the dielectric continuum model [7-10]. The geometries considered in these studies included cylindrical quantum wires but not rectangular quantum wires for which it has been known for some time that it is not possible to derive closed-form analytical results [141] describing all of the allowed optical phonon modes. Since 1989, there has been a resurgence in interest in the optical phonon modes in quantum wires [120-137] as a result of revolutionary techniques for fabricating quantum wires [1-3] as well as novel quantum-wire device concepts [2-6,115]. The confined optical phonon modes in a rectangular polar-semiconductor quantum wire were derived [120] by imposing the condition that the bulk phonon potential vanishes at the lateral heterointerfaces of the rectangular quantum wire; a vanishing optical phonon potential at such heterointerfaces is consistent with the so-called electrostatic boundary condition of the earlier dielectric continuum models [7-12]. By imposing this so-called electrostatic boundary condition on the bulk LO phonon interaction Hamiltonian [120] it follows immediately that the well-known confined modes of a polar-semiconductor quantum well [7-12,16,23,29,31] are established in each of the confined lateral dimensions of the rectangular quantum wire. Additionally, confined and interface phonons have been modelled for cylindrical quantum wires [121,122] and the polarisation eigenvectors of surface-optical (SO) phonon modes were derived for a free-standing rectangular quantum wire in vacuum [123]. Using well-known approximation techniques for treating dielectric rectangular waveguides [141], approximate solutions for the SO phonons (alternatively designated as interface (IF) phonons) in rectangular quantum wires were derived [124,125]. Scattering rates for emission and absorption in rectangular GaAs quantum wires embedded in AlAs were computed from these approximate modes [73,74] in the approximation of the Golden rule. In these calculations, carriers are taken to be free in the axial direction and bound in the ground states of finitely deep quantum wells in the transverse directions. For structures as small as a 40 Å/40 Å GaAs quantum wire of infinite length in the axial direction, it was found that there are strong one-dimensional density-of-states effects in the rates for surface-optical (SO) phonon emission. The strong SO-phonon emission rates for the symmetric interface modes near the AlAs-like LO and GaAs-like transverse-optical (TO) phonon frequencies are more than an order of magnitude larger than the corresponding bulk LO phonon emission rates. Scattering rates as a function of the electron energy in the axial direction in a 40 Å/40 Å GaAs quantum wire are depicted in FIGURE 1. FIGURE 1 illustrates clearly the strong one-dimensional density-of-states effects in interface phonon emission which occur at the energies of the low-frequency and high-frequency symmetric SO phonon modes of 33 meV and 50 meV, respectively.

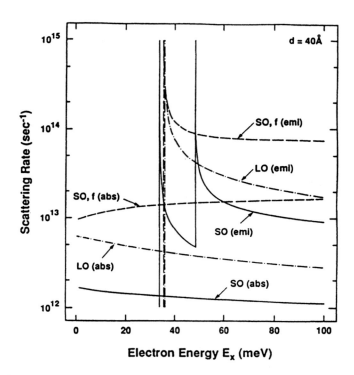

FIGURE 1. Scattering rates as a function of electron energy in a rectangular quantum wire. Confinement dimensions are 40 Å in both the two transverse directions.

This enhanced scattering from interface modes is not surprising in light of related theoretical results [28,29,39,41,42,44,47,49,50] and experimental results [51-56] for systems where interface modes arise from confinement in only one dimension. However, if the lateral dimensions of the quantum wire are varied sinusoidally in amplitude by only about 10 %, the narrow one-dimensional density-of-states emission peaks become much less pronounced and have averaged values close to those of bulk carrier - LO-phonon interaction rates [129,134,135]. In summary, it is found that carrier - LO-phonon scattering rates in quantum wires may deviate substantially from the corresponding bulk values; however, in the more realistic case of variable quantum wire dimensions, one-dimensional effects are suppressed significantly.

Determinations of the properties of confined phonons in rectangular quantum wires have also been based on extending earlier dynamical matrix treatments for quantum wells [19,21,57] to the case of quantum wires [128,132]. These treatments predict confined modes in a rectangular quantum wire similar to those previously discussed [120] except that the product of electrostatic confined phonon modes is replaced by a product of the modified electrostatic modes known as the Huang-Zhu modes [19,21]. A comparison of Huang-Zhu, dielectric continuum, and fully microscopic phonon modes [41] indicates that the Huang-Zhu modes resemble the modes of the fully microscopic model. The modes of the microscopic dynamical matrix [128,132] model are accompanied by both transverse optical (TO) bulk-like modes and interface-like modes; for quantum wires these interface-like modes are found to be hybrids of TO and LO waves. Efforts to apply microscopic models to confined and interface modes in quantum wires are leading to substantial contributions [126,133]; the microscopic quantum-wire models for interface and confined optical phonons are based on extensions of earlier models where phonon confinement occurs in only one dimension [35,38,41,44,49,50]. Especially noteworthy is the recent success [133] in describing the coupling of interface and confined modes in rectangular quantum wires where parallel and normal phonon wavevector components are interchanged for adjacent quantum-wire heterointerfaces; in this situation there is an intrinsic coupling between interface modes and the confined modes. Furthermore, dispersion relations based on this fully microscopic model for polar-semiconductor quantum wires [133] reveal the intrinsic coupling of interface modes and confined modes in rectangular quantum wires. A similar dispersion behaviour has been predicted recently [142] for triple hybridised modes of longitudinally polarised optical modes, transversely polarised optical modes and interface polariton vibrational modes. The difficulties associated with the lack of a closed-form analytic solution for the interface optical modes of a rectangular quantum wire have been circumvented by deriving analytic solutions for the modes of quantum wires with elliptical cross-sections [130]. Furthermore, in the limit of a circular cross-section appropriate solutions are found and numerical solutions for rectangles with rounded corners reveal the presence of corner modes localised in regions of high curvature. Modes similar to these corner modes had been predicted previously [143] for phonons on stepped surfaces.

The effects of dynamical screening and hot phonons on energy relaxation in rectangular quantum wires have been studied recently [136] using techniques similar to those used earlier to model the same phenomena in quantum wells [18,25,137]. For carrier densities in the vicinity of 10^5 cm^{-1} it was found that the carrier relaxation rates due to scattering from the confined optical phonon modes [120] in 50 Å/50 Å and 200 Å/200 Å quantum wires are affected much more by hot phonon effects than by dynamical screening. Essentially the same findings have been made previously [137] for the case of quantum wells.

For carrier energies in excess of the interface LO-phonon energy, the inelastic scattering caused by carrier - interface-phonon interactions dominates over scattering due to other phonon modes when quantum-well [39,42,51] or quantum-wire [125] confinement occurs on a scale of about 50 Å or less and when carriers are confined in the extreme quantum limit; when carriers are not so strongly confined, interface phonon scattering may start to dominate over confined phonon scattering for

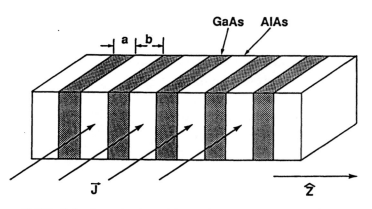

FIGURE 2. Superlattice with an electron current, J, injected parallel to the superlattice heterointerfaces. The GaAs layer is taken to have a thickness, a, and the AlAs layer has a thickness b.

dimensional scales as large as 150 Å [39]. FIGURES 2 and 3 illustrate the case where interface phonon effects begin to dominate at these larger dimensional scales; the superlattice of FIGURE 2 has a wavefunction which is strongly delocalised in the short-period case. The wavefunction may have a substantial value at the GaAs-AlAs interface where the interface phonons have their maximum values. In such a situation, the carrier - interface-phonon scattering is enhanced relative to the confined-phonon - carrier interaction as shown in FIGURE 3 where the superlattice period is taken as 2a; that is, a = b. For this case, FIGURE 3 illustrates that the ratio of carrier - interface-phonon scattering to carrier - confined-phonon scattering, Y, may be larger than unity when a times the phonon parallel wavevector, X, is less than about 3. For typical values of the phonon parallel wavevector, the value of a may be in the range of 100 to 150 Å.

As mentioned previously, establishing metal-semiconductor interfaces at the lateral boundaries of a confined polar semiconductor reduces unwanted inelastic carrier scattering caused by the interactions with interface LO-phonon modes. It has been argued that interface and confined LO phonons in mesoscopic devices may be tailored through the judicious use of metal-semiconductor interfaces in such a way as to dramatically reduce unwanted emission of interface LO phonons. Motivated by the progress in the epitaxial growth of metals in intimate contact with polar semiconductors as well as by the theory of plasmons for a quasi-two-dimensional electron gas [138-140], Stroscio et al [131] have applied the dielectric continuum model of interface phonon modes [23,29] to determine the carrier-interface-phonon interaction Hamiltonian for carriers in a confined polar-semiconductor having one or more

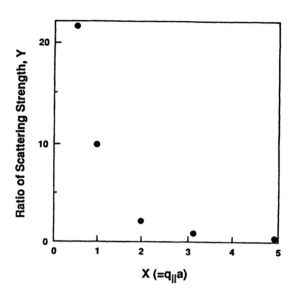

FIGURE 3. Ratio of scattering strengths, Y, as a function of the normalised parallel phonon wavevector, X.

metal-semiconductor heterointerfaces. It is demonstrated [131] that only those interface and confined modes having odd potentials about the metal-semiconductor interfaces satisfy the correct boundary conditions at the metal-semiconductor interface. Furthermore, the magnitude of the potential for the symmetric interface mode, which has the largest electron-phonon coupling constant [23], is greatly reduced. This situation is illustrated in FIGURE 4 where a slab of ionic material of thickness d is surrounded by two semi-infinite layers of polar semiconductor with a different dielectric constant. FIGURE 4 depicts a

symmetric interface mode in the 'upper' portion of the quantum well, an antisymmetric interface mode just below the symmetric interface mode, the second confined mode just below the antisymmetric interface mode, and the lowest order confined mode at the 'bottom' of the quantum well. The lower half of FIGURE 4 illustrates the modes which survive if the left half of the upper schematic is replaced with a perfectly conducting metal; as shown, only the antisymmetric interface modes and the confined modes with an even number of half wavelengths survive. These modes survive because they satisfy the required metal-semiconductor boundary conditions [131]; specifically

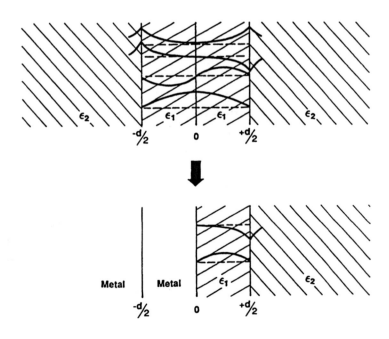

FIGURE 4. Schematic of the phonon modes in semiconductor-semiconductor and metal-semiconductor structures.

these modes vanish at the metal-semiconductor interface as they must since the optical phonon fields are screened very effectively by the free electron gas in the metal.

The assumption of an ideal metal with infinite conductivity at the metal-semiconductor interface is not essential to realise this result since the typical plasmon frequency in the metal is normally two orders of magnitude greater than the LO phonon frequency in the semiconductor. It was shown recently [144] that the practical limit to this interfacial screening of interface LO phonons is set by the Thomas-Fermi screening length in the metal; hence, the 'free' electron gas in the metal is capable of screening the LO phonon fields emanating in the semiconductor after they have penetrated only a few Å into the metal. In practice, it may be very desirable to form an epitaxial metal-semiconductor interface and to select metal-semiconductor systems such as Al/InAs [145] where Fermi level pinning does not result in the depletion of the semiconductor layer; indeed, Al/InAs is an example of a system which exhibits an inversion in the semiconductor.

As argued previously [7,12], the components of the optical phonon polarisation vectors for dimensionally-confined longitudinal-optical phonon modes have opposite parities for polarisation components parallel to the semiconductor interface and for polarisation components normal to the interface. Since the average tangential components of the optical-phonon electrical field have to vanish at the metal-semiconductor heterointerface and since the phonon polarisation vector is proportional to the phonon electric field, it is evident that only optical phonon modes with even polarisation vectors normal to the metal-semiconductor interface will survive. Furthermore, since the Laplacian of the Fröhlich potential for a given phonon mode is proportional to the divergence of the polarisation vector for that mode, it follows that only optical phonon modes with Fröhlich potentials which are odd about the metal-semiconductor heterointerface satisfy the correct boundary conditions at the metal-semiconductor interface. Indeed, of the well-known symmetric and antisymmetric interface modes [7,23,29] only the antisymmetric mode will survive. This situation is depicted schematically in FIGURE 4. In the limit of no dispersion, the antisymmetric interface mode has a Fröhlich potential which scales as one divided by the square root of cosh(qd) x sinh(qd) where q is the magnitude of the parallel phonon wavevector at the metal-semiconductor interface and d is the thickness of the semiconductor slab. For small values of qd, the Fröhlich potential of the antisymmetric interface mode then decreases exponentially with qd. Thus, it follows that the carrier - optical-phonon scattering rate decreases to zero at the metal-semiconductor interface and that the strength of the Fröhlich potential for the antisymmetric interface mode in the semiconductor slab decreases exponentially as qd increases.

Recent advances in the area of confined and interface phonons in quantum wires have been surveyed by Stroscio [146]. Jiang and Leburton [147] considered confined and interface phonon scattering in finite barrier GaAs/AlGaAs quantum wires and showed that under appropriate conditions the total contributions of LO and SO phonon scattering resemble closely the GaAs bulk LO phonon scattering rate. Enderlein [148] and Wang and Lei [149] have modelled optical-phonon modes and their interactions with carriers in circular quantum wires. Based on a generalisation of the results of [87], Comas et al [150] have modelled optical phonons and electron-phonon interactions in quantum wires. For the case of metal-encapsulated quantum wires Chiu and Stroscio [151], Stroscio et al [152,153] and Dutta et al [154] have shown that the symmetric interface modes are suppressed strongly. Hot phonon effects in GaAs quantum wires have been considered by Mickevicius et al [155] and

Gaska et al [156] have modelled hot-electron relaxation dynamics in GaAs quantum wires. Bottleneck effects due to confined phonons in reduced-dimensional systems have been examined by de la Cruz et al [157]. Finally, Sirenko et al [158] have demonstrated the ballistic propagation of interface optical phonons due to the relatively strong dispersion of the interface modes near zone centre.

For the case of acoustic phonons in quantum wires, Stroscio and Kim [159] gave an early calculation of piezoelectric scattering from confined acoustic modes in cylindrical quantum wires. This early work provided a prescription for normalising the amplitudes of acoustic mode displacements which has been used for a variety of structures [160]. In conjunction with the classical acoustic mode solutions of Morse [161] for a rod with a rectangular cross-section, this normalisation procedure facilitated the calculation of acoustic phonons in rectangular quantum wires as well as the deformation potential interactions in such wires due to the approximate compressional acoustic modes in such structures [162,163]. The same normalisation procedure was used to calculate the quantised acoustic phonon modes in quantum wires [164]. The techniques presented in this paper are directly applicable for quantising the acoustic phonon modes available from a number of studies [165,166]. The confined and interface acoustic phonon modes existing in cylindrical quantum wires have been analysed by Nishiguchi [167,168] but these studies do not consider the quantisation of the various acoustic modes. Recently Yu et al [169] have applied the quantisation procedure of [159] and [164] to calculate electron - acoustic-phonon scattering rates in cylindrical GaAs quantum wires due to the deformation potential for two cardinal boundary conditions: free-surface and clamped-surface boundary conditions. In some cases, results indicate strong resonant enhancements in electron - acoustic-phonon scattering at the thresholds for quantised acoustic phonon emission. For example, FIGURE 5 illustrates enhanced scattering for such

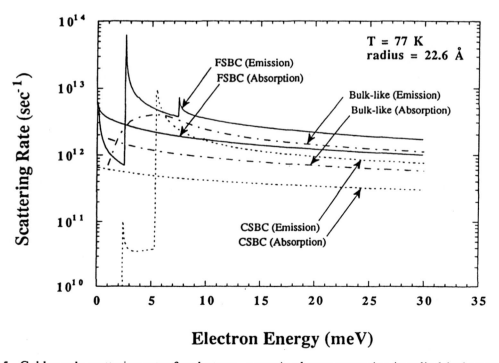

FIGURE 5. Golden rule scattering rates for electron - acoustic-phonon scattering in cylindrical quantum wires for acoustic phonon modes determined for three separate cases: free-standing boundary condition (FSBC) (displacements are unrestricted on the perimeter of the cylinder); clamped-surface boundary condition (CSBC) (displacements are set to zero on the perimeter of the cylinder); and bulk acoustic phonons.

modes in a free-standing cylindrical quantum wire; all rates in FIGURE 5 were calculated at 77 K. Finally, Sirenko et al [170] have used the elastic continuum model to calculate the acoustic modes in hollow cylinders embedded in fluid; these models have been evaluated for the biologically important case of microtubes which are long hollow cylindrical macromolecules with a diameter of the order of 25 nm and serve as a major component of cytoskeleton in eukariotic cells.

D   CONCLUSION

After being dormant for two decades, the theory of interface and confined optical phonon modes in polar-semiconductor quantum wires has advanced dramatically since 1989. This progress has been stimulated by the impressive advances in the technologies underlying the fabrication of one-dimensional nanostructures as well as by concepts for novel devices which incorporate quantum-wire components. There is still not complete agreement between the various continuum models for confined and interface optical phonon modes in quantum wells and quantum wires; however, the recent success of the fully microscopic model as well as innovative approaches to treating hybridised modes provide powerful tools for future refinements and advances in the theory of interface and confined optical phonon modes in polar-semiconductor nanostructures. The recent efforts to understand the properties of confined and interface optical phonons have been extended to the study of confined and interface acoustic phonons in quantum wells, superlattices and quantum wires. The international effort better to understand the properties of these acoustic modes is stimulated to a large extent by the potential importance of these modes in mesoscopic devices.

ACKNOWLEDGEMENTS

The authors gratefully acknowledge the extremely beneficial collaborations with Dr. H.L. Grubin, Professor V.V. Mitin, Dr. R. Mickevicius, Dr. N. Bannov, Professor S. Das Sarma, Professor S. Teitsworth, Dr. P. Turley, Dr. X.T. Zhu and Dr. Reza Haque. This effort was supported, in part, by the U.S. Army Research Office and by the Office of Naval Research. Ms. P. Lassiter's excellent preparation of the manuscript is acknowledged gratefully.

REFERENCES

[1]   M. Watt, C.M. Sotomayer-Torres, H.E.G. Arnotand, S.P. Beaumont [*Semicond. Sci. Technol. (UK)* vol.5 (1990) p.285-90]
[2]   M. Tsuchiya, J.M. Gaines, R.H. Yan, R.J. Simès, P.O. Holtz, L.A. Coldren [*Phys. Rev. Lett. (USA)* vol.62 (1989) p.466-9]
[3]   M. Tsuchiya, J.M. Gaines, R.H. Yan, R.J. Simes, P.O. Holtz, L.A. Coldren [*J. Vac. Sci. Technol. B (USA)* vol.7 (1989) p.315-8]
[4]   F. Sols, M. Macucci, U. Ravaioli, K. Hess [*Appl. Phys. Lett. (USA)* vol.54 (1989) p.350-3]
[5]   F. Sols, M. Macucci [*Phys. Rev. B (USA)* vol.41 (1990) p.11887-91]
[6]   H. Sakaki [*Jpn. J. Appl. Phys. (Japan)* vol.28 (1989) p.L314-6]
[7]   R. Fuchs, K.L. Kliewer [*Phys. Rev. (USA)* vol.140 (1965) p.A2076-88]
[8]   K.L. Kliewer, R. Fuchs [*Phys. Rev. (USA)* vol.144 (1966) p.495-503]
[9]   K.L. Kliewer, R. Fuchs [*Phys. Rev. (USA)* vol.150 (1966) p.573-88]
[10]   R. Fuchs, K.L. Kliewer, W.J. Pardee [*Phys. Rev. (USA)* vol.150 (1966) p.589-96]

[11]   A.A. Lucas, E. Kartheuser, R.G. Badro [*Phys. Rev. B (USA)* vol.7 (1970) p.2488-99]

[12]   J.J. Licari, R. Evrard [*Phys. Rev. B (USA)* vol.15 (1977) p.2254-64]

[13]   F.A. Riddoch, B.K. Ridley [*J. Phys. C, Solid State Phys. (UK)* vol.16 (1983) p.6971-82]

[14]   R. Lassing [*Phys. Rev. B (USA)* vol.30 (1984) p.7132-7]

[15]   F.A. Riddoch, B.K. Ridley [*Physica B+C (Netherlands)* vol.134B (1985) p.342-6]

[16]   L. Wendler [*Phys. Status Solidi B (Germany)* vol.129 (1985) p.513-30]

[17]   N. Sawaki [*J. Phys. C. Solid State Phys. (UK)* vol.19 (1986) p.4965-75]

[18]   L. Wendler, R. Pechstedt [*Phys. Status Solidi B (Germany)* vol.141 (1987) p.129-50]

[19]   K. Huang, B. Zhu [*Phys. Rev. B (USA)* vol.38 (1988) p.2183-6]

[20]   H. Chu, S.-F. Ren, Y.-C. Chang [*Phys. Rev. B (USA)* vol.37 (1988) p.10746-55]

[21]   K. Huang, B. Zhu [*Phys. Rev. B (USA)* vol.38 (1988) p.13377-86]

[22]   H. Akera, T. Ando [*Phys. Rev. B (USA)* vol.40 (1989) p.2914-31]

[23]   N. Mori, T. Ando [*Phys. Rev. B (USA)* vol.40 (1989) p.6175-88]

[24]   T. Tsuchiya, H. Akera, T. Ando [*Phys. Rev. B (USA)* vol.39 (1989) p.6025-33]

[25]   L. Wendler, R. Haupt, V.G. Grigoryan [*Physica B (Netherlands)* vol.167 (1990) p.91-9; *Physica B (Netherlands)* vol.167 (1990) p.101-12]

[26]   M.H. Degani, O. Hipolito [*Superlattices Microstruct. (UK)* vol.5 (1989) p.141-4]

[27]   K.W. Kim, M.A. Stroscio, J.C. Hall [*J. Appl. Phys. (USA)* vol.67 (1990) p.6179-83]; M.A. Stroscio, K.W. Kim, J.C. Hall [*Superlattices Microstruct. (UK)* vol.7 (1990) p.115-8]

[28]   Guo-qiang Hai, M.F. Peeters, J.T. Devreese [*Phys. Rev. B (USA)* vol.42 (1990) p.11063-72]

[29]   K.W. Kim, M.A. Stroscio [*Appl. Phys. Lett. (USA)* vol.68 (1990) p.6289-92]

[30]   R. Chen, D.L. Lin, T.F. George [*Phys. Rev. B (USA)* vol.41 (1990) p.1435-92]

[31]   S. Rudin, T.L. Reinecke [*Phys. Rev. B (USA)* vol.41 (1990) p.7713-7]; also see S. Rudin, T.L. Reinecke [*Phys. Rev. B (USA)* vol.43 (1991) p.9298]

[32]   B.K. Ridley, M. Babiker [*Phys. Rev. B (USA)* vol.43 (1991) p.9096-101]

[33]   R. Enderlein [*Phys. Rev. (USA)* vol.43 (1991) p.14513-31]

[34]   R. Haupt, L. Wendler [*Phys. Rev. B (USA)* vol.44 (1991) p.1850-60]

[35]   H. Rucker, E. Molinari, P. Lugli [*Phys. Rev. B (USA)* vol.44 (1991) p.3463-6]

[36]   L.F. Register, M.A. Stroscio, M.A. Littlejohn [*Phys. Rev. B (USA)* vol.44 (1991) p.3850-3]

[37]   C. Guillemot, F. Clerot [*Phys. Rev. B (USA)* vol.44 (1991) p.6249-61]

[38]   P. Lugli, E. Molinari, H. Rucker [*Superlattices Microstruct. (UK)* vol.10 (1991) p.471-8]

[39]   M.A. Stroscio, G.J. Iafrate, K.W. Kim, M.A. Littlejohn, H. Goronkin, G.N. Maracas [*Appl. Phys. Lett. (USA)* vol.59 (1991) p.1093-5]

[40]   M.A. Stroscio et al [in *Nanostructures and Mesoscopic Systems* Eds W.P. Kirk, M.A. Reed (Academic Press Inc., Harcourt Brace Jovanovich Publishers, Boston, 1992) p.379-86]

[41]   H. Rucker, E. Molinari, P. Lugli [*Phys. Rev. B (USA)* vol.45 (1992) p.6747-56]

[42]   K.W. Kim, M.A. Littlejohn, M.A. Stroscio, G.J. Iafrate [*Semicond. Sci. Technol. (UK)* vol.7 (1992) p.B49-51]

[43]   M. Babiker [*Semicond. Sci. Technol. (UK)* vol.7 (1992) p.B52-9]

[44]   E. Molinari, C. Bungaro, M. Gulia, P. Lugli, H. Rucker [*Semicond. Sci. Technol. (UK)* vol.7 (1992) p.B67-72]

[45]   T. Tsuchiya, T. Ando [*Semicond. Sci. Technol. (UK)* vol.7 (1992) p.B73-6]

[46]   H. Gerecke, F. Bechstedt [*Semicond. Sci. Technol. (UK)* vol.7 (1992) p.B80-2]

[47]   N. Mori, K. Taniguchi, C. Hamaguchi [*Semicond. Sci. Technol. (UK)* vol.7 (1992) p.B83-7]

[48]   O. Al-Dossary, M. Babiker, N.C. Constantiou [*Semicond. Sci. Technol. (UK)* vol.7 (1992) p.B91-3]

[49]   H. Rucker, P. Lugli, S.M. Goodnick, J.E. Lary [*Semicond. Sci. Technol. (UK)* vol.7 (1992) p.B98-101]

[50]   P. Lugli, P. Bordone, E. Molinari, H. Rucker, A.M. de Paula, A.C. Maciel [*Semicond. Sci. Technol. (UK)* vol.7 (1992) p.B116-9]

[51]   K.T. Tsen, K.R. Wald, T. Ruf, P.Y. Yu, H. Morkoc [*Phys. Rev. Lett. (USA)* vol.67 (1991) p.2557-60]

[52] M.C. Tatham, J.F. Ryan [*Semicond. Sci. Technol. (UK)* vol.7 (1992) p.B102-8]

[53] K.T. Tsen [*Semicond. Sci. Technol. (UK)* vol.7 (1992) p.B191-4]

[54] P.J. Turley, S.W. Teitsworth [*Phys. Rev. B (USA)* vol.44 (1991) p.8181-4]; S.W. Teitsworth, C.R. Wallis, P.J. Turley, W. Li, P.K. Bhattacharya [*Proc. Int. Semiconductor Device Research Symp.* (Engineering Outreach, University of Virginia, ISBN Number: 1-880920-00-X, 1991) p.93-6]; P.J. Turley, S.W. Teitsworth [*J. Appl. Phys. (USA)* vol.72 (1992) p.2356-67]

[55] T.X. Zhu, H. Goronkin, G.N. Maracas, R. Droopad, M.A. Stroscio [*Appl. Phys. Lett. (USA)* vol.60 (1992) p.2141-3]; H. Qiang, F.H. Pollak, C.M. Sotomayer-Torres, W. Leitch, M.A. Stroscio [*Appl. Phys. Lett. (USA)* vol.61 (1992) p.1411-3]

[56] L.P. Fu, T. Schmiedel, A. Petrou, M. Dutta, P.G. Newman, M.A. Stroscio [*Phys. Rev. B (USA)* vol.46 (1992) p.7196-9]

[57] K. Huang, B.-F. Zhu, H. Tang [*Phys. Rev. B (USA)* vol.41 (1990) p.5825-42]

[58] L.F. Register [*Phys. Rev. B (USA)* vol.45 (1992) p.8756-9]

[59] A.R. Bhatt, K.W. Kim, M.A. Stroscio, J.M. Higman [*Phys. Rev. B (USA)* vol.48 (1993) p.14671-4]

[60] K.W. Kim, A.R. Bhatt, M.A. Stroscio, P.J. Turley, S.W. Teitsworth [*J. Appl. Phys. (USA)* vol.72 (1992) p.2282-7]

[61] A.R. Bhatt [*J. Appl. Phys. (USA)* vol.73 (1993) p.2338-42]

[62] N.C. Constantinou [*Phys. Rev. B (USA)* vol.48 (1993) p.11931-5]

[63] B.K. Ridley [*Phys. Rev. B (USA)* vol.47 (1993) p.4562-602]

[64] M. Babiker, N.C. Constantinou, B.K. Ridley [*Phys. Rev. B (USA)* vol.48 (1993) p.2236-43]

[65] B.K. Ridley [*Phys. Rev. B (USA)* vol.39 (1989) p.5282-6]

[66] K.T. Tsen, R. Joshi, H. Morkoc [*Appl. Phys. Lett. (USA)* vol.62 (1993) p.2075-7]

[67] P. Senet, Ph. Lambin, A.A. Lucas [*Phys. Rev. Lett. (USA)* vol.74 (1995) p.570-3]

[68] J.-L. Guyaux, P.A. Thiry, R. Sporken, R. Caudano, Ph. Lambin [*Phys. Rev. B (USA)* vol.48 (1993) p.4380-7]

[69] F. Bechstedt, H. Gerecke, H. Grille [*Phys. Rev. B (USA)* vol.47 (1993) p.13540-52]

[70] G.Q. Hai, F.M. Peeters, H.T. Devreese [*Phys. Rev. B (USA)* vol.48 (1993) p.4666-74]

[71] B.L. Gelmont, M. Shur, M.A. Stroscio [*J. Appl. Phys. (USA)* vol.77 (1995) p.657-60]

[72] A.R. Bhatt, K.W. Kim, M.A. Stroscio [*J. Appl. Phys. (USA)* vol.76 (1994) p.3905-7]

[73] M.A. Stroscio [in *NATO Advanced Research Workshop on Phonons in Semiconductor Nanostructures* Eds J.-P. Leburton, J. Pascaul, C.S. Torres (Kluwer Academic Publishers, 1993) p.13-23]

[74] M. Dutta [in *NATO Advanced Research Workshop on Phonons in Semiconductor Nanostructures* Eds J.-P. Leburton, J. Pascaul, C.S. Torres (Kluwer Academic Publishers, 1993) p.87-96]

[75] A.M. de Paula, A.C. Maciel, G. Weber, J.F. Ryan, P. Dawson, C.T. Foxon [*Semicond. Sci. Technol. (UK)* vol.7 (1992) p.3120-3]

[76] M. Dutta, M.A. Stroscio, X.-Q. Zhang [*Proc. 1993 Int. Semiconductor Device Research Symp.* Charlottesville, VA, 1-3 Dec. 1993 (Engineering Academic Outreach, University of Virginia, Charlottesville, VA, 1993) p.877-80]

[77] M. Dutta, M.A. Stroscio [*J. Appl. Phys. (USA)* vol.73 (1993) p.1693-701]

[78] M.U. Erdogan, V. Sankaran, J.W. Kim, M.A. Stroscio, G.J. Iafrate [*Phys. Rev. B (USA)* vol.50 (1994) p.2485-91]

[79] M.A. Stroscio, M. Dutta, X.-Q. Zhang [*J. Appl. Phys. (USA)* vol.75 (1994) p.1977-81]

[80] M.U. Erdogan, V. Sankaran, K.W. Kim, M.A. Stroscio, G.J. Iafrate [*Proc. SPIE (USA)* vol.2146 (1994) p.34-41]

[81] O.E. Raichev [*Phys. Rev. B (USA)* vol.49 (1994) p.5448-62]

[82] J.M. Jacob et al [*Solid State Commun. (USA)* vol.91 (1994) p.721-4]; D.S. Kim et al [*Phys. Rev. B (USA)* vol.51 (1995) p.5449-52]; D.S. Kim, A. Bouchalkha, J.M. Jacob, J.F. Zhou, J.J. Song, J.F. Klem [*Phys. Rev. Lett. (USA)* vol.68 (1992) p.1002-5]

[83]   H.A. Fertig, T.L. Reinecke [*Phys. Rev. B (USA)* vol.49 (1994) p.11168-72]; C. Molteni, L. Colombo, L. Miglio, G. Benedek [*Phys. Rev. B (USA)* vol.50 (1994) p.11684-6]

[84]   B.K. Ridley, M. Babiker [*Phys. Rev. B (USA)* vol.43 (1991) p.9096-101]

[85]   B.-F. Zhu [*Phys. Rev. B (USA)* vol.46 (1992) p.13619-23]

[86]   M.P. Chamberlain, M. Cardona, B.K. Ridley [*Phys. Rev. B (USA)* vol.48 (1993) p.14356-64]

[87]   C. Trallero-Giner, F. Garcia-Moliner, V.R. Velasco, M. Cardona [*Phys. Rev. B (USA)* vol.48 (1992) p.11944-8]

[88]   N. Bannov, V. Mitin, M.A. Stroscio [*Phys. Status Solidi B (Germany)* vol.183 (1994) p.131-42]

[89]   N. Bannov, V. Mitin, M.A. Stroscio [*Proc. 1993 Int. Semiconductor Device Research Symp.* Charlottesville, VA, 1-3 Dec. 1993 (Engineering Academic Outreach, University of Virginia, Charlottesville, VA, 1993) p.659-62]

[90]   V. Mitin, R. Mickevicius, N. Bannov, M.A. Stroscio [*Proc. 1993 Int. Semiconductor Device Research Symp.* Charlottesville, VA, 1-3 Dec. 1993 (Engineering Academic Outreach, University of Virginia, Charlottesville, VA, 1993) p.855-62]

[91]   S. Mizuno, M. Ito, S.-I. Tamura [*Jpn. J. Appl. Phys. (Japan)* vol.33 (1994) p.2880-5]

[92]   M.A. Stroscio, K.W. Kim [*Solid-State Electron. (UK)* vol.37 (???) p.181-2]

[93]   V.A. Kochelap, O. Gulseren [*J. Phys., Condens. Matter (UK)* vol.5 (1993) p.589-98]

[94]   V.A. Kochelap, O. Gulseren [*J. Phys., Condens. Matter (UK)* vol.5 (1993) p.94-9]

[95]   V.G. Grigoryan, L. Wendler [*Sov. Phys.-Solid State (USA)* vol.33 (1991) p.1193-8]

[96]   L. Wendler, V.G. Grigoryan [*Surf. Sci. (Netherlands)* vol.206 (1988) p.203-24]

[97]   B. Hillebrands, S. Lee, G.I. Stegeman, H. Cheng, J.E. Potts, F. Nizzoli [*Phys. Rev. Lett. (USA)* vol.60 (1988) p.832-5]

[98]   J. Seyler, M.N. Wybourne [*J. Phys., Condens. Matter (UK)* vol.2 (1990) p.98]

[99]   J.C. Nabity, M.N. Wybourne [*Phys. Rev. B (USA)* vol.42 (1990) p.9714-6]

[100]   J.C. Nabity, M.N. Wybourne [*Phys. Rev. B (USA)* vol.44 (1991) p.8990-6]

[101]   J. Seyler, M.N. Wybourne [*J. Phys., Condens. Matter (UK)* vol.4 (1992) p.L231-6]

[102a]   D.J. Lockwood, R.L.S. Devine, A. Rodriguez, J. Mendialdua, B.D. Rouhani, L. Dobrzynski [*Phys. Rev. B (USA)* vol.47 (1993) p.13553-60]

[102b]   V.I. Belitsky, T.R.J. Spitzer, M. Cardona [*Phys. Rev. B (USA)* vol.49 (1994) p.8263-72]

[102c]   P.V. Santos, A.K. Sood, M. Cardona, K. Ploog, Y. Ohmori, H. Okamoto [*Phys. Rev. B (USA)* vol.37 (1988) p.6381]

[102d]   Z.V. Popovic, J. Spitzer, T. Ruf, M. Cardona, R. Notzel, K. Ploog [*Phys. Rev. B (USA)* vol.48 (1993) p.1659-64]

[103]   V.F. Sapega, V.I. Belitsky, T. Ruf, H.D. Fuchs, M. Cardona, K. Ploog [*Phys. Rev. B (USA)* vol.46 (1992) p.16005-11]

[104]   H. Wang, M. Jiang, D.G. Steel [*Phys. Rev. Lett. (USA)* vol.65 (1990) p.1255-8]

[105a]   S. Mizuno, S.-I. Tamura [*Phys. Rev. B (USA)* vol.45 (1993) p.13423-30]

[105b]   S. Mizuno, S.-I. Tamura [*Phys. Rev. B (USA)* vol.45 (1992) p.734-41]

[105c]   S.-I. Tamura [*Phys. Rev. B (USA)* vol.43 (1991) p.12646-9]

[105d]   S.-I. Tamura [*Phys. Rev. B (USA)* vol.41 (1990) p.7941-4]

[105e]   S.-I. Tamura, F. Nori [*Phys. Rev. B (USA)* vol.40 (1989) p.9790-801]

[105f]   N. Nishiguchi, S.-I. Tamura [*Phys. Rev. B (USA)* vol.48 (1993) p.2515-28]

[106a]   J. He, J. Sapriel, R. Azoulay [*Phys. Rev. B (USA)* vol.40 (1989) p.1121-9]

[106b]   J. He, B. Djafari-Rouhani, J. Sapriel [*Phys. Rev. B (USA)* vol.37 (1988) p.4086-98]

[107]   J. Sapriel, J. He, B. Djafari-Rouhani, R. Azoulay, F. Mollot [*Phys. Rev. B (USA)* vol.37 (1988) p.4099-105]

[108]   E.H. El Boudouti, B. Djafari-Rouhani [*Phys. Rev. B (USA)* vol.49 (1994) p.4586-92]

[109]   S.-I. Tamura [*Phys. Rev. B (USA)* vol.30 (1994) p.610-7]

[110]   J.-P. Leburton [*J. Appl. Phys. (USA)* vol.56 (1984) p.2850-5]

[111]   S. Briggs, D. Jovanovic, J.-P. Leburton [*Appl. Phys. Lett. (USA)* vol.54 (1989) p.2012-4]

[112]   S. Briggs, J.-P. Leburton [*Superlattices Microstruct. (UK)* vol.5 (1989) p.145-8]

[113] S. Briggs, J.-P. Leburton [*Phys. Rev. B (USA)* vol.38 (1988) p.8163-70]

[114] J.-P. Leburton [*Phys. Rev. B (USA)* vol.45 (1992) p.11022-30]

[115] J.-P. Leburton, D. Jovanovic [*Semicond. Sci. Technol. (UK)* vol.7 (1992) p.B202-9]

[116] R. Englman, R. Ruppin [*Phys. C: Proc. Phys. Soc. (UK)* vol.1 (1968) p.614-29]

[117] R. Englman, R. Ruppin [*Phys. C: Proc. Phys. Soc. (UK)* vol.1 (1968) p.630-43]

[118] R. Englman, R. Ruppin [*Phys. C: Proc. Phys. Soc. (UK)* vol.1 (1968) p.1515-31]

[119] R. Ruppin, R. Englman [*Rep. Prog. Phys. (UK)* vol.33 (1970) p.149-96]

[120] M.A. Stroscio [*Phys. Rev. B (USA)* vol.40 (1989) p.6428-31]

[121] N.C. Constantinou, B.K. Ridley [*Phys. Rev. B (USA)* vol.41 (1990) p.10622-6]

[122] N.C. Constantinou, B.K. Ridley [*Phys. Rev. B (USA)* vol.41 (1990) p.10627-31]

[123] M.A. Stroscio, K.W. Kim, M.A. Littlejohn, H. Chuang [*Phys. Rev. B (USA)* vol.42 (1990) p.1488-91]

[124] M.A. Stroscio, K.W. Kim, M.A. Littlejohn [*Proc. SPIE (USA)* vol.1362 (1990) p.566-79]

[125] K.W. Kim, M.A. Stroscio, A. Bhatt, R. Mickevicius, V.V. Mitin [*J. Appl. Phys. (USA)* vol.70 (1991) p.319-27]

[126] S.-F. Ren, Y.-C. Chang [*Phys. Rev. B (USA)* vol.43 (1991) p.11857-63]

[127] M.A. Stroscio, K.W. Kim, S. Rudin [*Superlattices Microstruct. (UK)* vol.10 (1991) p.55-8]

[128] B.-F. Zhu [*Phys. Rev. B (USA)* vol.44 (1991) p.1926-9]

[129] R. Mickevicius, V.V. Mitin, K.W. Kim, M.A. Stroscio [*Superlattices Microstruct. (UK)* vol.11 (1992) p.277-80]

[130] P.A. Knipp, T.L. Reinecke [*Phys. Rev. B (USA)* vol.45 (1992) p.9091-102]

[131] M.A. Stroscio, K.W. Kim, G.J. Iafrate, M. Dutta, H.L. Grubin [*Philos. Mag. Lett. (UK)* vol.65 (1992) p.173-6]

[132] B.-F. Zhu [*Semicond. Sci. Technol. (UK)* vol.7 (1992) p.B88-90]

[133] C. Bungaro, P. Rossi, F. Lugli, L. Rota, E. Molinari [*Proc. SPIE (USA)* vol.1677 (1992) p.55-8]

[134] R. Mickevicius, V.V. Mitin, K.W. Kim, M.A. Stroscio, G.J. Iafrate [*J. Phys., Condens. Matter (UK)* vol.4 (1992) p.4959-70]

[135] R. Mickevicius, V.V. Mitin, K.W. Kim, M.A. Stroscio [*Semicond. Sci. Technol. (UK)* vol.7 (1992) p.B299-301]

[136] S. Das Sarma, V.B. Campos, M.A. Stroscio, K.W. Kim [*Semicond. Sci. Technol. (UK)* vol.7 (1992) p.B60-6]; V.B. Campos, S. Sarma, M.A. Stroscio [*Phys. Rev. B (USA)* vol.46 (1992) p.3849-53]

[137] S. Sarma, J.K. Jain, R. Jalabert [*Phys. Rev. B (USA)* vol.41 (1990) p.3561-71]

[138] J.P. Harbison, T. Sands, N. Tabatabaie, W.K. Chan, L.T. Florez, V.G. Keramidas [*Appl. Phys. Lett. (USA)* vol.53 (1988) p.1717-9]

[139] J.P. Harbison, T. Sands, R. Ramesh, L.T. Florez, B.J. Wilkens, V.G. Keramidas [*J. Cryst. Growth (Netherlands)* vol.111 (1988) p.978-83]; N. Tabatabaie, T. Sands, J.P. Harbison, H.L. Gilchrist, V.G. Keramidas [*Appl. Phys. Lett. (USA)* vol.53 (1988) p.2528-30]

[140] W.H. Backes, F.M. Peeters, F. Brosens, J.T. Devreese [*Phys. Rev. B (USA)* vol.45 (1992) p.8437-42]

[141] E.J. Marcatili [*Bell Syst. Tech. J. (USA)* vol.48 (1969) p.2071-102]

[142] B.K. Ridley [*Proc. SPIE (USA)* vol.1675 (1992) p.492-7]

[143] P. Knipp [*Phys. Rev. B (USA)* vol.43 (1991) p.6908-23]

[144] A.R. Bhatt et al [*J. Appl. Phys. (USA)* vol.73 (1992) p.2338-43]

[145] U. Mishra [private communication (1992)]

[146] M.A. Stroscio [in *Phonons in Semiconductor Nanostructures* Eds J.-P. Leburton, J. Pascual, C.S. Torres (Kluwer Academic Press, 1993) p.13-23]

[147] W. Jiang, J.-P. Leburton [*J. Appl. Phys. (USA)* vol.74 (1993) p.1652-9]

[148] R. Enderlein [*Phys. Rev. B (USA)* vol.47 (1993) p.2162-75]

[149] X.F. Wang, X.L. Lei [*Phys. Rev. B (USA)* vol.49 (1994) p.4780-9]

[150] F. Comas, C.T. Giner [*Phys. Rev. B (USA)* vol.47 (1993) p.7602-5]

[151] C.-J. Chiu, M.A. Stroscio [*Superlattices Microstruct. (UK)* vol.13 (1993) p.401-4]

[152] M.A. Stroscio, G.J. Iafrate, K.W. Kim, A.R. Bhatt, M. Dutta, H.L. Grubin [in *Phonon Scattering in Condensed Matter VII* Eds M. Meissner, R.O. Pohl (Springer-Verlag, 1993) p.341-2]

[153] M.A. Stroscio, K.W. Kim, G.J. Iafrate, M. Dutta, H.L. Grubin [*Proc. Int. Semiconductor Device Research Symp.* Charlottesville, VA, 4-6 Dec. 1991 (Engineering Academic Outreach, University of Virginia, 1991) p.87-90]

[154] M. Dutta, H.L. Grubin, G.J. Iafrate, K.W. Kim, M.A. Stroscio [U.S. Patent 5,264,711 dated 23 Nov. 1993]

[155] R. Mickevicius, R. Gaska, V. Mitin, M.A. Stroscio, G.J. Iafrate [*Semicond. Sci. Technol. (UK)* vol.9 (1994) p.889-92]

[156] R. Gaska, R. Mickevicius, V. Mitin, M.A. Stroscio, G.J. Iafrate, H.L. Grubin [*J. Appl. Phys. (USA)* vol.76 (1994) p.1021-8]

[157] R.M. de la Cruz, S.W. Teitsworth, M.A. Stroscio [*Superlattices Microstruct. (UK)* vol.13 (1993) p.481-6]

[158] Y.M. Sirenko, M.A. Stroscio, K.W. Kim, V. Mitin [*Phys. Rev. B (USA)* vol.51 (1995) p.9863-6]

[159] M.A. Stroscio, K.W. Kim [*Phys. Rev. B (USA)* vol.48 (1993) p.1936-8]

[160] M.A. Stroscio, G.J. Iafrate, K.W. Kim, S. Yu, V. Mitin, N. Bannov [*Proc. 1993 Int. Semiconductor Device Research Symp.* Charlottesville, VA, 1-3 Dec. 1993 (Engineering Academic Outreach, University of Virginia, Charlottesville, VA, 1993) p.873-5]

[161] R.W. Morse [*J. Acoust. Soc. Am. (USA)* vol.22 (1950) p.219-23]

[162] K.W. Kim, S. Yu, M.U. Erdogan, M.A. Stroscio, G.J. Iafrate [*Proc. on Ultrafast Phenomena in Semiconductors* Los Angeles, CA, 27-28 Jan. 1994 (SPIE - The International Society for Optical Engineering, 1994) p.77-86]

[163] S. Yu, K.W. Kim, M.A. Stroscio, G.J. Iafrate, A. Ballato [*Phys. Rev. B (USA)* vol.50 (1994) p.1733-8]

[164] M.A. Stroscio, K.W. Kim, W. Yu, A. Ballato [*J. Appl. Phys. (USA)* vol.76 (1994) p.4670-5]

[165] R.A. Waldron [*IEEE Trans. Microw. Theory Tech. (USA)* vol.17 (1969) p.893-904]

[166] E.A. Ash, R.M. De La Rue, R.F. Humphryes [*IEEE Trans. Microw. Theory Tech. (USA)* vol.17 (1969) p.882-92]

[167] N. Nishiguchi [*Jpn. J. Appl. Phys. (Japan)* vol.33 (1994) p.2858-8]

[168] N. Nishiguchi [*Phys. Rev. B (USA)* vol.50 (1994) p.10970-80]

[169] S. Yu, K.W. Kim, M.A. Stroscio, G.J. Iafrate [*Phys. Rev. B (USA)* vol.51 (1995) p.4695-8]

[170] Y. Sirenko, M.A. Stroscio, K.W. Kim [*Phys. Rev. E (USA)* submitted for publication (1995)]

5.4 Electronic specific heat of GaAs/AlGaAs multilayers

G. Strasser

February 1996

A INTRODUCTION

During the last two decades there has been much interest in the physical properties of two-dimensional electron gas systems in a magnetic field B. If the magnetic field is applied perpendicular to the plane of electrical confinement it leads to full quantisation of the electron motion. The energy spectrum consists of Landau levels (LL) separated by the cyclotron energy $\hbar\omega_c$ ($\omega_c = eB/m^*$). In an unperturbed system the LLs are discrete and highly degenerate, with $1/2\pi l^2$ ($l = (\hbar/eB)^{1/2}$) possible states in each level. For an electron concentration n_s the filling factor ν is defined as $\nu = n_s h/eB$, including spin splitting.

In a real system Landau levels are broadened due to scattering by impurities, phonons and other mechanisms. Therefore, in a perpendicular B, the LLs consist of a series not of ideal δ functions, but of broadened peaks centred at the discrete LL energies. In the simplest approximation the levels are described by a level width Γ. In a high magnetic field ($\hbar\omega_c \ll \Gamma$) real gaps appear between the LLs and lead to an oscillatory structure of practically all physical quantities.

The most fundamental quantity underlying all these physical properties of 2DEG systems is the form of the density of states (DOS). The most pronounced effects are the quantum Hall effect [1] and the fractional quantum Hall effect [2]. Current theoretical understanding of the quantum Hall effect [3,4] and the fractional quantum Hall effect [2] relies on the assumptions about the DOS. The distinction between localised and extended electronic states being essential, the form of the density of states can be obtained directly by measurement of thermodynamic quantities such as specific heat or magnetisation, where localised and extended states contribute equally in equilibrium. Measurements of transport and optical properties also yield information on the DOS; however, the extraction of the pure DOS is more complicated [5].

Different methods have been applied to determine the DOS of 2DEG systems in high magnetic fields. Beside specific heat [6,7] and magnetisation [8], a capacitance measurement [9,10] and the determination of activation energies in the Hall plateau range [10] were applied. In this Datareview we concentrate on specific heat measurements only.

B EXPERIMENTAL METHODS

The most direct method to determine the DOS is the measurement of the specific heat given by $C_v = dF/dT$, where F is the free energy. An externally induced temperature change leads to a reordering of the electrons at the Fermi energy. The heat capacity of the electron system is proportional to the density of states at the Fermi energy. Two different techniques, a heat pulse technique and an AC calorimetry technique, are reported to measure the specific heat of a 2DEG system in GaAs/GaAlAs MQWs.

B1   Heat Pulse Technique

A heat pulse technique [11,12] was applied to determine the electronic specific heat by Gornik et al [6,13]. Short heat pulses were used to heat the sample adiabatically, which implies a controlled thermal connection between sample and heat bath. Thermal isolation of the sample was achieved by hanging the sample on four 5 - 10 μm thick superconducting Nb wires. The whole arrangement was mounted in a vacuum isolated tube (see FIGURE 1(a)). A 100 Å thick NiCr film served as a heater, and a 1000 to 2000 Å thick Au-Ge film (8% Au) as a temperature detector [14], shown schematically in FIGURE 1(a). The sample was heated with electric pulses which were considerably shorter than the thermal time constant of the sample-bath system. The temperature sensitivity of the detector film was about 1 MΩ/K at 4.2 K and 5 MΩ/K at 2 K resulting in a maximum resolution of 0.1 mK.

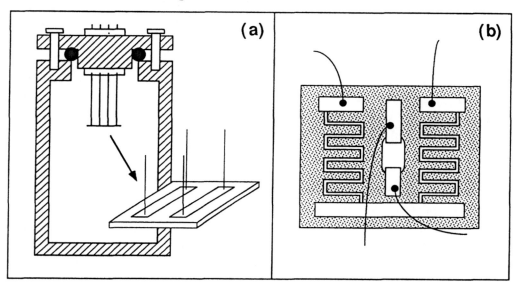

FIGURE 1.   (a) Sample arrangement for the heat pulse technique.
(b) Sample arrangement for the AC calorimetry technique.

The experiments were performed on two different multilayer materials: sample 1 consisted of 172 modulation doped quantum well layers on SI substrate (200 Å GaAs wells and 200 Å AlGaAs barriers); mobility at 4.2 K was 40000 cm^2/V s, and carrier concentration 6.3 x 10^{11}/cm^2. Sample 2 consisted of 94 quantum wells (220 Å GaAs/500 Å AlGaAs), with a mobility at 4.2 K of 80000 cm^2/V s and a carrier concentration of 7.7 x 10^{11}/cm^2. On both samples the substrate was lapped and etched down. Total sample thickness was 10 μm for sample 1 and 20 μm for sample 2.

B2   AC Calorimetry

The technique of AC calorimetry was developed for layered materials [15,16] and applied to systems with small heat capacities such as multiple quantum well structures by Wang et al [7,17]. As in the heated pulse technique the samples were thinned and a heater and a thermometer were evaporated on the sample. As a heater 250 - 400 Å gold on a 60 Å chromium (adhesive) layer was used in a meander pattern to ensure homogeneous heating. The thermometer was made of a 1000 to 2000 Å thick Au-Ge film (~10% gold) [14] (see FIGURE 1(b)). Samples were mounted by soldering ball-bonded wires to commercial eight pin headers, and pressed into a socket inside a resealable brass capsule.

In the AC calorimetry technique the sample heater is driven by a sinusoidal voltage. Under steady state conditions the constant component of heat flux will dissipate out through the wires (17 μm thick gold wires). The temperature gradient across the sample has to be negligible compared to that from sample to holder and/or bath.

To study heat capacity only the AC component of the heat flux has to be considered. Temperature oscillations dependent on thermal properties of the sample occur at twice the heater voltage frequency. The frequency has to be adjusted to be fast enough to measure the temperature response before the heat dissipates to the holder (or bath) and to be slow enough to be uniform within the sample. Correlation techniques lead to a high precision of 0.01%.

The experiments were performed on MQWS material consisting of 75 (modulation doped) layers of 175 Å GaAs wells and 460 Å AlGaAs barriers on SI substrate. Mobility was 100000 cm^2/V s, carrier concentration $9 \times 10^{11}/cm^2$ per layer, and sample thickness was 20 μm.

The same experiments were repeated on high mobility MQWS material with low carrier concentrations to take spin splitting into account [17]. Sample parameters for the latter examination were: 85 layers of 175 Å GaAs wells and 600 Å AlGaAs barriers. Mobility was 4.8×10^5 cm^2/V s, and carrier concentration was $1.7 \times 10^{11}/cm^2$ per layer. (A more detailed description of these samples is given in [18].)

C   RESULTS

C1   Heat Pulse Data

Calculations by Zawadzki and Lassnig [19-21], neglecting g-value enhancement and magnetic field dependent broadening, predicted a pronounced oscillatory character for a Gaussian density of states, consisting of intra- and inter-Landau-level contributions.

With this model the first observation of the magnetic field dependent electronic specific heat [6,13] could be described quite well. Data and calculations are shown in FIGURE 2 (taken from [6]). FIGURE 2(a) shows the oscillatory behaviour of sample 1 [6] at 4.2 K and 1.5 K, respectively. ΔR denotes the change of the temperature versus magnetic field (averaged over 10 runs); due to the heat pulse the sample temperature rose by 0.5 K at 4.2 K (0.03 K at 1.5 K). $ΔR_F$ shows the background DC resistance of the detector film on an extended scale. Sample temperature oscillations are clearly observed with 'spikelike' behaviour at even filling factors ν. A similar behaviour is observed for sample 2 [6]. Additional spikes are observed for this sample (FIGURE 2(b)), interpreted as inter-Landau-level contributions, which appear at higher temperatures, in agreement with the theoretical prediction [20]. Calculations of the expected experimental results with different forms of the DOS are performed and shown in

FIGURE 2. G stands for a Gaussian DOS, L for a Lorentzian DOS. The best fit was achieved for a Gaussian density of states sitting on a flat background of states.

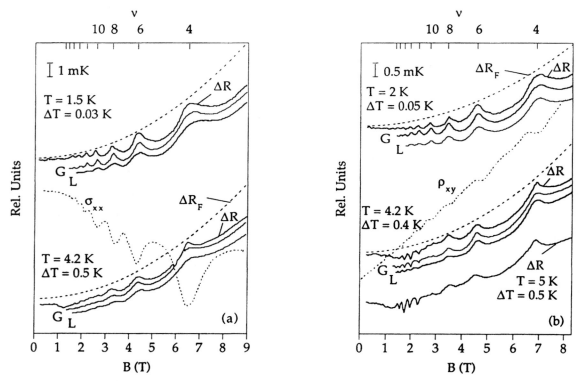

FIGURE 2. (a) Temperature oscillations of the thermometer as a function of magnetic field for sample 1; ΔR measured; ΔR_F background DC resistance (extended scale); G (L) fit with a Gaussian (Lorentzian) DOS; $\Gamma = 2.5$ meV, 25% constant background-DOS, all data for 1.5 K and 4.2 K. ν is the filling factor including spin splitting ($\nu = n_s h/eB$). (b) Temperature oscillations for sample 2; $\Gamma = 1.5$ meV, 20% constant background-DOS, all data given for 2 K and 4.2 K. ν is the filling factor. The lowest curve shows the measured ΔR for 5 K. From [6].

C2   AC Calorimetry Data

Heat capacity data achieved by the AC calorimetry technique (Wang et al [7]) differ qualitatively from the measurements described above. Their data cannot be fitted by assuming B-independent level broadening. The best fit for the data from [7] leads to Landau-level width oscillations with variation of B, in agreement with a model in which screening of the scattering potential and the energy level broadening influence each other self-consistently [22-25].

Wang et al [7] checked their measurement signals by sophisticated routines to be sure there was purely AC heating of the sample (a detailed description is given in [7]). The changes in heat capacitance of their sample are shown in FIGURE 3(b), obtained at 1.3 K. These experimental results cannot be fitted with a constant Γ and a constant DOS background x, and the self-consistent screening for long range impurity scattering has to be taken into account. A self-consistent calculation with Γ varying periodically with 1/B in the Gaussian model with no background states can describe this data very well. The oscillatory features of $\Gamma(B)$ used to fit the data are shown in FIGURE 3(a); the calculation as well as the data show a quite different shape compared to FIGURE 2. The interlevel peaks at higher T shown in FIGURE 2(a) could not be resolved by Wang et al at higher T. By using different sample geometries in studies of frequency or time dependence they conclude they are able to thoroughly elucidate the interplay between 2DEG heat capacity and coupling to phonons. This is not necessarily valid for the data reported in [6], although a direct comparison is not possible.

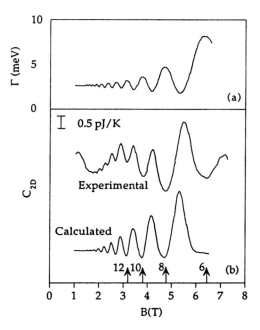

FIGURE 3. (a) Γ(B) used for the calculated heat capacitance plotted in (b). (b) Measured and calculated heat capacitance (shifted), T = 1.3 K, arrows indicate even filling factors ν. From [7].

In a more recent report by Wang et al [17] AC calorimetry was applied to MQWs with low carrier densities and high mobilities to measure specific heat in the lowest spin-split Landau-levels. Analysing the density of states derived from these data Wang et al [17] found that many-body interactions have to be considered to explain the B dependence of the spin splitting and the broadening of the Landau levels. Fits of the heat capacity are consistent with experimental observation that an enhancement of the g-factor oscillates with 1/B; therefore from the experimental observation of the specific heat, the exact B dependence of the parameters g and Γ cannot be determined. This is demonstrated in FIGURE 4, where two different calculations for the specific heat of a 2DEG (FIGURE 4(b)) are shown. Dashed lines indicate a fit with a constant g(B) factor, while solid lines are self-consistently determined. In FIGURE 5 the measured data at 0.55 K are compared to the calculations to prove the statement (both figures from [17]).

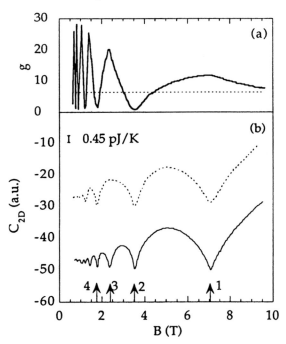

FIGURE 4. (a) g(B) used for the calculation of the specific heat (shown in FIGURE 4(b)): dashed line: g(B) = 6 (constant value); solid line: g(B) self-consistently determined. (b) Calculation of the specific heat at 0.55 K; Γ = 1 meV: dashed line: constant g-factor; solid line: g(B) self-consistently determined. Arrows indicate integer filling factors ν for the MQWS sample. From [17].

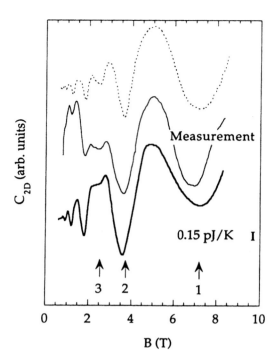

FIGURE 5. Calculated and measured $\Delta C_{2D}(B)$. Two different sets of fitting parameters are used to fit the experimental data (top solid line); parameters for the calculated lines are: dashed line: constant g-factor (g = 13); solid line: g(B) self-consistently determined. Lines are shifted for clarity; T = 0.55 K, arrows indicate integer filling factors ν. From [17].

D   CONCLUSION

Heat capacitance measurements of two-dimensional electrons in multi-quantum-well structures at high magnetic fields are used to determine the density of states. To explain the broadening of the Landau-levels many-body interactions have to be considered. The given analysis is intrinsically limited, because two parameters (Γ and g) are fitted from the measurement of the specific heat versus the magnetic field. Therefore, from heat measurements in MQW systems alone an exact density of states cannot be derived.

REFERENCES

[1]   K. von Klitzing, G. Dorda, M. Pepper [*Phys. Rev. Lett. (USA)* vol.45 (1980) p.494]

[2]   D.C. Tsui, H.L. Störmer, A.C. Gossard [*Phys. Rev. Lett. (USA)* vol.48 (1982) p.1559]

[3]   R.B. Laughlin [*Phys. Rev. B (USA)* vol.23 (1981) p.5632]

[4]   R.B. Laughlin [*Phys. Rev. Lett. (USA)* vol.50 (1983) p.1395]

[5]   K. von Klitzing, G. Ebert [*Physica B (Netherlands)* vol.117/118 (1983) p.682]

[6]   E. Gornik, R. Lassnig, G. Strasser, H.L. Störmer, A.C. Gossard, W. Wiegmann [*Phys. Rev. Lett. (USA)* vol.54 (1985) p.1820]

[7]   J.K. Wang, J.H. Campbell, D.C. Tsui, A.Y. Cho [*Phys. Rev. B (USA)* vol.38 (1988) p.6174]

[8]   J.P. Eisenstein, H.L. Störmer, V. Narayanamurti, A.Y. Cho, A.C. Gossard, C.W. Tu [*Phys. Rev. Lett. (USA)* vol.55 (1985) p.875]

[9]   T.P. Smith, B.B. Goldberg, P.J. Stiles, M. Heiblum [*Phys. Rev. B (USA)* vol.32 (1985) p.2696]

[10]   D. Weiss, E. Stahl, G. Weimann, K. Ploog, K. von Klitzing [*Surf. Sci. (Netherlands)* vol.170 (1986) p.285]

[11]   R. Backmann et al [*Rev. Sci. Instrum. (USA)* vol.43 (1972) p.205]

[12]   S. Alterovitz, G. Deutscher, M. Garschenson [*J. Appl. Phys. (USA)* vol.46 (1975) p.3637]

[13]   E. Gornik, R. Lassnig, G. Strasser, H.L. Störmer, A.C. Gossard [*Surf. Sci. (Netherlands)* vol.170 (1986) p.277]

[14]   B.W. Dodson, W.L. McMillan, J.M. Mochel, R.C. Dynes [*Phys. Rev. Lett. (USA)* vol.46 (1981) p.46]

[15]   P.F. Sullivan, G. Seidel [*Phys. Rev. (USA)* vol.173 (1968) p.679]

[16]   J.K. Wang, J.H. Campbell [*Rev. Sci. Instrum. (USA)* vol.59 (1988) p.2031]

[17]   J.K. Wang, D.C. Tsui, M. Santos, M. Shayegan [*Phys. Rev. B (USA)* vol.45 (1992) p.4384]

[18]   M. Shayegan, J.K. Wang, M. Santos, T. Sajoto [*Appl. Phys. Lett. (USA)* vol.54 (1989) p.27]

[19]   W. Zawadzki, R. Lassnig [*Surf. Sci. (Netherlands)* vol.142 (1984) p.225]

[20]   W. Zawadzki, R. Lassnig [*Solid State Commun. (USA)* vol.50 (1984) p.537]

[21]   W. Zawadzki [*Physica B (Netherlands)* vol.127 (1984) p.388]

[22]   S. Das Sarma, X.C. Xie [*Phys. Rev. Lett. (USA)* vol.61 (1988) p.738]

[23]   X.C. Xie, Q.P. Li, S. Das Sarma [*Phys. Rev. B (USA)* vol.42 (1990) p.7132]

[24]   Qiang Li, X.C. Xie, S. Das Sarma [*Phys. Rev. B (USA)* vol.40 (1989) p.1381]

[25]   V. Sa-yakanit, N. Choosiri, H.R. Glyde [*Phys. Rev. B (USA)* vol.38 (1988) p.1340]

CHAPTER 6

OPTICAL PROPERTIES

6.1 Photoluminescence, photoluminescence excitation and absorption spectroscopy of GaAs-based single, coupled and multiple quantum wells

D.C. Reynolds

June 1995

A   INTRODUCTION

Photoluminescence (PL), photoluminescence excitation (PLE), and absorption have been used extensively to analyse the intrinsic as well as the extrinsic properties of quantum wells, superlattices, and single heterostructures. Intrinsic or free exciton formation is observed in most well-formed heterostructures. The free excitons have been applied with a great deal of success in probing the intrinsic properties of these structures. Bound excitons have been applied equally successfully in probing the impurity and defect structure of many of the same structures. These investigations have compiled a huge body of data from which selective samplings will be presented in this Datareview.

B   INTRINSIC PROPERTIES OF QUANTUM WELL (QW) STRUCTURES

B1   Brief Description of QW Structures

Carriers in a QW may be confined to certain energies by potential barriers, resulting in new optical and transport properties. The allowed carrier energy levels are determined by quantisation effects, when the confined regions are sufficiently small. One very common heterostructure is GaAs/AlGaAs. By cladding the GaAs layer with AlGaAs barriers, the electrons and holes are confined within the GaAs well, resulting in a modification of their energy levels in the well. The confining potential is determined from the differences in bandgap energies of the barrier layer material and the well material. The barrier material $Al_xGa_{1-x}As$ has a direct bandgap for $x < 0.45$, with a bulk bandgap which is $1.25x$ eV greater than that of GaAs. In the QW, the difference in bandgap energies is divided between the conduction band and the valence band; the percentage contribution to each band is a measure of the confining barrier for that band. If infinite confining barriers are assumed, the allowed minimum energies for electrons are given by

$$E_n = h^2 n^2 / 8m^* L^2 \qquad (1)$$

where h is Planck's constant, n is an integer marking the number of half-wavelengths of the confined electron, m^* is the effective mass, and L is the well thickness. These energy levels for electrons in the conduction band are shown in FIGURE 1.

The energy levels in the valence band are also modified. In bulk GaAs, the light- and heavy-hole valence bands are degenerate at $K = 0$. From EQN (1) it is seen that the energy of the confined particles is different for different masses. The layered structure has reduced the cubic symmetry of bulk GaAs to uniaxial symmetry. The optical transitions thus become non-degenerate as shown in FIGURE 1. Transitions between the conduction and valence subbands, created by optical excitation, are due to excitons associated with the various subbands. The excited electrons are bound to the holes by Coulomb interaction [1]. The line shape of the PL and absorption QW transitions is excitonic; verification of these transitions as being assigned to light- and heavy-hole free-exciton transitions has been made from polarisation measurements [2].

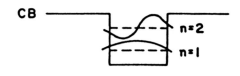

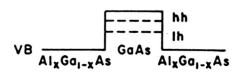

FIGURE 1. Energy band edge and electron and hole quantum state energy levels in GaAs/AlGaAs quantum wells.

Repeating the growth of the layers shown in FIGURE 1 results in multi-quantum well (MQW) structures, the number of wells being equal to the number of repeated cycles. If the barrier layer thicknesses are large enough the wells do not interact; as the barrier thicknesses are reduced coupled wells result. When the dimensions of the multilayered periodic structures approach the atomic spacings of the constituent materials, superlattices are formed.

B2   Excitons in GaAs/AlGaAs QWs

The early work of Dingle et al [3] showed that excitons played a more dominant role in the optical transitions in QWs than was the case in bulk GaAs. Absorption and emission measurements [4,5] on QWs showed that the light- and heavy-hole degeneracy at K = 0 in bulk GaAs was removed in QWs. Estimates of the exciton binding energy in a QW were first made by Dingle et al [3] and Miller et al [5]. Structure was observed in the exciton spectra, which could be identified with excited or continuum states from which the exciton binding energy was determined. A binding energy of ~9 meV for a QW $L_z \leq 100$ Å was obtained. Exciton excited states have been subsequently observed by other investigators from which the exciton binding energies can be estimated [6-10]. These binding energies have been compared with calculated exciton binding energies with good agreement [6,8,11-21]. The exciton binding energies, from a representative calculation [14], as a function of well size for two different x-values are shown in FIGURE 2. In this calculation the conduction and valence band discontinuities were taken as 85 and 15%, respectively. The values of 60 and 40% are now more generally accepted; however, the exciton binding energies are not strongly dependent on the choice of band offsets.

The intersubband transitions from a number of GaAs/Al$_{0.25}$Ga$_{0.75}$As MQW structures having well widths in the range 35 - 245 Å were measured by photocurrent response [22]. The photocurrent response is proportional to the exciton transition strength. Measured and calculated exciton transition energies as a function of well width for several MQW structures are shown in FIGURE 3. The measured values were based on the x-values and well sizes provided by the crystal grower. The agreement with calculated values is quite good.

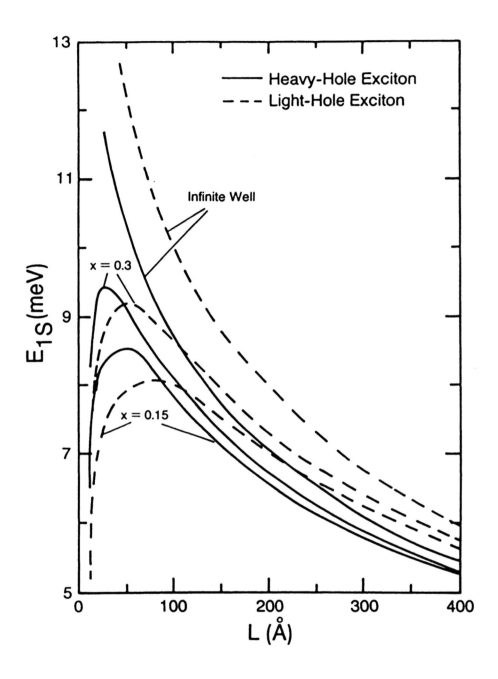

FIGURE 2. Variation of the binding energy of the ground state E_{1s}, of a heavy-hole exciton (solid lines) and light-hole exciton (dashed lines) as a function of GaAs quantum-well size (L) for Al concentrations x = 0.15 and 0.3, and for an infinite potential well.

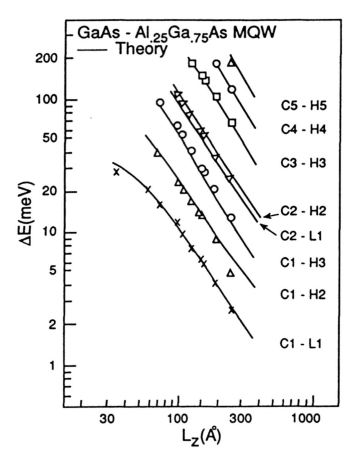

FIGURE 3. Measured and calculated exciton transition energies as a function of L_z for GaAs/Al$_{0.25}$Ga$_{0.75}$As multiple-quantum-well structures. ΔE is the energy measured from the C1-H1 exciton transition energy. The solid lines are the theoretical values calculated for symmetrical wells with ΔE_g (x = 0.25).

B3   Binding Energies of Excitons Associated with Coupled QW Structures

The interaction between closely-spaced QW structures has been investigated by optical techniques [23-25]. Detailed PL and PLE measurements were performed on narrow, pseudomorphic, nominally symmetric, coupled In$_{0.08}$Ga$_{0.92}$As/GaAs double (QW) structures [25]. From these measurements the binding energies of excitons associated with symmetric and antisymmetric wave functions were determined. In nominally symmetric coupled systems the electron and hole subband levels are split into levels associated with symmetric and antisymmetric combinations of isolated QW wave functions. Therefore, the isolated QW heavy-hole level is split into symmetric and antisymmetric heavy-hole levels associated, respectively, with symmetric and antisymmetric combinations of isolated well wave functions, while the light-hole and electron subband levels are split similarly. The binding of a 'symmetric electron' (e$_s$) and a 'symmetric heavy-hole' (hh$_s$) give rise to a symmetric heavy-hole free-exciton (S-HHFE). Similarly, an antisymmetric heavy-hole free exciton (A-HHFE) is formed from an 'antisymmetric heavy-hole' (hh$_A$) and an antisymmetric electron (e$_A$). The binding of a 'symmetric electron' and a 'symmetric light-hole' gives rise to a symmetric light-hole free exciton (S-LHFE), while an antisymmetric light-hole free-exciton (A-LHFE) is formed from an 'antisymmetric light-hole' and 'antisymmetric electron'. A list of the calculated valence subband to conduction subband transition energies and best fit x-values, along with observed excitonic and subband edge transition energies is given in TABLE 1. Also included are the exciton binding energies (ΔE), both calculated and experimental. In this experiment the well sizes were provided by the crystal grower from which the best fit x-values were determined. These x-values were previously shown to agree well with actual x-value determination obtained via X-ray diffraction measurements. The S-HHFE and S-LHFE binding energies were found to be greater than the respective A-HHFE and A-LHFE binding energies. This is believed to be due to the greater confinement energy of the symmetric states.

TABLE 1. Experimental and calculated transition energies and their differences ΔE for well width L_Z = 14, 10 and 4 ML double QW structures with 'coupling' barrier widths L_B = 120, 30, and 20 Å. The best fit x-values are also given. The exciton transitions are labelled; the associated subband transition for A-LHFE, for example, is (lh_A, e_A).

L_Z (ML)	Transition (see text)	Exciton energy expt. (eV)	Subband energy calc. (eV)	Subband energy expt. (eV)	ΔE (meV) calc.	ΔE (meV) expt.	L_B (Å)	x-value
14ML	S-HHFE	1.4784	1.4850	1.4864	6.6	8.0	120	0.075
14ML	A-HHFE	1.4898	1.4925	1.4965	2.7	6.7	120	0.075
14ML	S-LHFE	1.4983	1.5037		5.4		120	0.075
14ML	A-LHFE	1.5087	1.5125		3.8		120	0.075
14ML	S-HHFE	1.4615	1.4699		8.4		30	0.085
14ML	A-HHFE	1.4984	1.5016		3.2		30	0.085
14ML	S-LHFE	1.4882	1.4939		5.7		30	0.085
14ML	S-LHFE	1.4569	1.4665		9.6		20	0.085
10ML	S-HHFE	1.4841	1.4918	1.4920	7.7	7.9	120	0.080
10ML	A-HHFE	1.4977	1.5014		3.7		120	0.080
10ML	S-LHFE	1.5010	1.5072		6.2		120	0.080
10ML	S-HHFE	1.4726	1.4813	1.4810	8.7	8.4	30	0.085
10ML	S-LHFE	1.4952	1.5005		5.3		30	0.085
10ML	S-HHFE	1.4644	1.4735	1.4723	9.1	7.9	20	0.095
10ML	S-LHFE	1.4901	1.4967		6.6		20	0.095
4ML	S-HHFE	1.5044	1.5101		5.7		120	0.080
4ML	S-LHFE	1.5103	1.5149		4.6		120	0.080
4ML	S-HHFE	1.4988	1.5057		6.9		30	0.080
4ML	S-LHFE	1.5088	1.5147		5.9		30	0.080
4ML	S-HHFE	1.4959	1.5028	1.5028	6.9	6.9	20	0.085
4ML	S-LHFE	1.5072	1.5118		4.6		20	0.085

B4   Magneto-Optical Analysis of Conduction and Valence Band Effective Masses

In an applied magnetic field electrons and holes, after reaching thermal equilibrium through nonradiative processes, generally recombine via Landau levels with the same quantum number n (0, 1, ...), i.e. with $\Delta n = 0$, according to selection rules. It was observed [26] that a new class of recombination between electrons and holes, namely, off-diagonal transitions (i.e. $\Delta n > 0$) in strained modulation-doped QW structures was occurring. Similar transitions have also been observed in lattice-matched structures [27-29]. These transitions have been attributed to the breakdown of the $\Delta n = 0$ selection rule due to electron and hole Landau level mixing caused by impurity-carrier interactions. The strengths of these transitions have been calculated using diagrammatic techniques [30]. A modulation-doped GaAs/Al_xGa_{1-x}As single heterostructure was investigated [29] by PLE spectra, demonstrating Landau level oscillations.

The extrapolation of these oscillations back to zero magnetic field is shown in FIGURE 4. The transitions are labelled (n_e, n_h) where n_e and n_h designate the electron and hole Landau levels, respectively. Both allowed, $\Delta n = 0$, and forbidden, $\Delta n > 0$, transitions are observed. The parity forbidden transitions allow the values of the electron and hole masses to be determined independently. From the allowed Landau level transitions the reduced electron hole mass m^* was determined to be $0.07 \, m_e$. Using a heavy hole mass of $0.45 \, m_e$ an average value of the electron mass was $0.084 \, m_e$. From the parity forbidden transitions the average value of the electron mass was determined to be $0.085 \, m_e$.

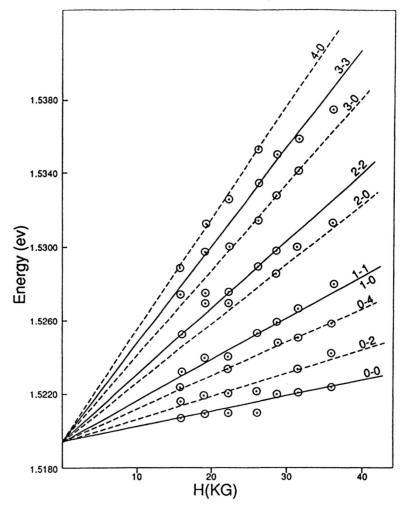

FIGURE 4. Landau fan diagram. The solid lines are allowed transitions;
the dashed lines are partly forbidden transitions.

Investigating the strained $GaAs/In_{0.15}Ga_{0.85}As/Al_{0.15}Ga_{0.85}As$ system the authors of [26] obtained a conduction band effective mass $m_e^* = 0.07 \, m_0$ and a valence band effective mass $m_v^* = 0.15 \, m_0$. These investigations demonstrate the power of parity forbidden transitions to provide independent measures of the conduction and valence band effective masses.

B5   Temperature Dependence of Light- and Heavy-Hole Free Exciton Transitions in GaAs/AlGaAs Single QWs

PL and PLE studies of single GaAs/AlGaAs QWs were carried out at various temperatures ranging from 5 K to 300 K [31]. The temperature dependence of the HHFE energies for four different well sizes, 38 Å, 64 Å, 101 Å, and 142 Å is shown in FIGURE 5. The nomenclature 1H ... 4H correlates with the narrowest to widest wells above. The HHFE energies were determined from PL measurements. The temperature dependence of the LHFE energies for the same four quantum wells is shown in FIGURE 6. The LHFE energies were determined from PLE spectra. The exciton temperature dependence follows closely that of the bandgap of bulk GaAs for all four wells. At low temperatures <40 K the exciton energies are nearly constant with temperature; at higher temperatures the exciton energy decreases with temperature. It is noted that the 2D exciton energy temperature dependence is insensitive to well width, and for a given well the HHFE and LHFE energies remain parallel to one another at all temperatures. The well sizes reported in both figures were taken from the grower.

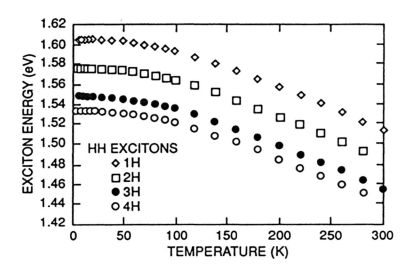

FIGURE 5. Heavy-hole exciton, E_{1H}, energies for four quantum wells as a function of temperature from 5 K to 300 K as determined from PL spectra.

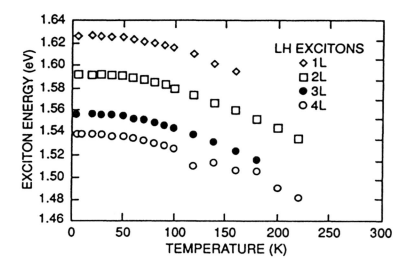

FIGURE 6. Light-hole exciton, E_{1L}, energies for four quantum wells as a function of temperature from 5 K to 220 K as determined from PLE spectra.

B6   Measured and Calculated Exciton Binding Energies in $In_xGa_{1-x}As/GaAs$ Strained QW Heterostructures

The heavy-hole exciton binding energy in narrow $In_{0.15}Ga_{0.85}As/GaAs$ QW samples was experimentally measured and compared with calculated binding energies for the same system [32]. PL and PLE measurements were used to resolve the HHFE transition from the continuum edge. The HHFE binding energies were measured for three different well sizes: 14, 25, and 48 Å. The well widths and x-values were taken from growth parameters. These measured values are shown in FIGURE 7 along with the calculated values of the binding energy as a function of well width. The calculated and observed values agree quite well.

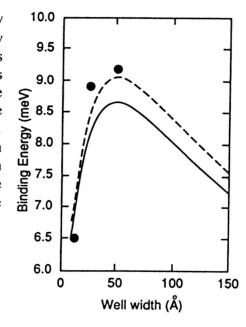

FIGURE 7. Comparison of the measured and calculated heavy-hole exciton binding energies. The solid curve was calculated using an in-plane hole mass of $0.16 \, m_0$ while the dashed curve used a value of $0.21 \, m_0$ for this parameter.

C   EXTRINSIC PROPERTIES OF QW STRUCTURES

C1   Binding Energy of Excitons to Neutral Donors at the Centre of the Well (CW), Edge of the Well (EW), and Centre of the Barrier (CB) in $Al_xGa_{1-x}As/GaAs$ QWs

Donor related complexes in QWs have been reported by several investigators [33-36]. Neutral donor bound excitons (D^o,X) as well as ionised donor bound excitons (D^+,X) were observed at the centre of the wells by Liu et al [36]. Both Nomura et al [35] and Liu et al [36] plot the variation of the binding energy of (D^o,X) transitions as a function of well size in $Al_xGa_{1-x}As/GaAs$ QWs. They found that the binding energy decreased as the well size increased, in agreement with the variational calculations of Kleinman [37].

A systematic PL study was made of the binding energy of (D^o,X) in several wells of varying sizes and as altered by the physical location of the neutral donors [38]. The $Al_xGa_{1-x}As/GaAs$ samples investigated were either nominally undoped, or Si-doped in the CW, EW, or CB. The binding energies of excitons to neutral donors at the various positions in the wells as a function of the HHFE energy and the well size are shown in FIGURE 8. The well sizes were calculated from the measured HHFE emission energies using the theoretical results of Greene et al [14]. The binding energies of (D^o,X) complexes increase as the well size is reduced for all three doping locations. However, for well widths of approximately 100 Å, they tend to reach their respective maximum values and then decrease as the well widths are further reduced. The transitions marked by filled circles in FIGURE 8 were speculated to be ionised-donor bound-exciton transitions associated with donors at the centre of the barrier and/or edge of the well.

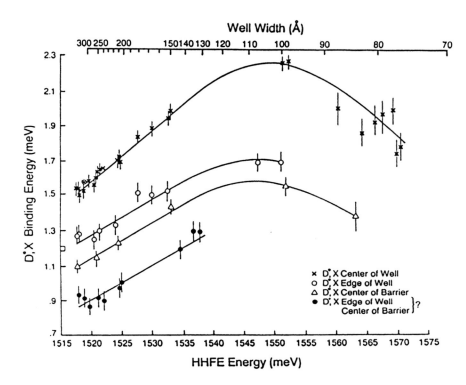

FIGURE 8. Binding energies of excitons to neutral donors located at CW (indicated by X), EW(0), and CB(Δ) as a function of HHFE energy. A scaling containing the calculated well sizes corresponding to HHFE energies is included at the top of the figure. Included are data (indicated by filled circles) speculated to be the (D^+,X) binding energy of excitons to EW and/or CB ionised donors as a function of HHFE energy. The binding energy of (D^o,X) in bulk GaAs is indicated by the empty square and the 300 Å well data of Nomura et al [35] by ⊗.

C2   Conduction Band-to-Acceptor and Donor-to-Acceptor Transitions in Coupled QWs

A number of impurity studies have been conducted on molecular beam epitaxy (MBE) grown GaAs/AlGaAs heterostructures [33,39-47]. Many of the studies used undoped MBE material which is usually p-type; and the observed acceptor related peaks resulted from conduction band-to-acceptor $(e-A^o)$ rather than donor-to-acceptor (D^o-A^o) transitions. A rather detailed study was made by Skromme et al [48] on organometallic chemical vapour deposition (OMCVD) grown GaAs/AlGaAs superlattices. The material was undoped and also n-type, and both $(e-A^o)$ and (D^o-A^o) recombinations were observed. The PL spectra showed a broad lineshape due to acceptors at various positions along the growth axis with distinct peaks which are due to acceptors in the centres of the wells (A_C) and in the centres of the barriers (A_B). Acceptor binding energies were determined to be 13.5 meV in the barriers and 31.9 meV in the wells.

C3   Binding Energy of Residual Donors in $Al_xGa_{1-x}As/GaAs$ Quantum Wells

The binding energy of donors in quantum wells has received considerable theoretical [12,49-52] and experimental attention [33,40,53,54]. The binding energy of residual donors in nominally 300 Å-wide GaAs/$Al_xGa_{1-x}As$ QWs was determined from the results of low-temperature PL, PLE, and resonant excitation measurements [54]. The well width and x-value were provided by the crystal grower. The centre of each QW was δ-doped with Be acceptors, resulting in ionisation of residual donors and allowing the observation of free-heavy-hole-to-donor (D^o,h) transitions. The n = 1 (D^o,h) transition for CW donors was observed in PL, and the n = 2 (D^o,h) transition for CW donors was observed in PLE; it was proposed that both 2s and 2p± excited states of the donor were being observed. The calculated binding energy of the 2s and 2p± excited states can be added to the measured transition energy from the heavy-hole subband to the respective excited states. This method gave a donor binding energy of 8.0 ± 0.5 meV. The calculated binding energy is 8.7 meV. The observed energy separation between the 2s and 2p± excited states of the donor was 0.7 meV, in good agreement with the calculated value of 0.5 meV.

C4   Binding Energy of Excitons to Neutral Donors in $In_{0.1}Ga_{0.9}As$/GaAs QWs

The well-width dependence of the binding energy of excitons to neutral donors in narrow, pseudomorphic single QW structures was obtained from low-temperature PL measurements [55]. It was found that the binding energy varies smoothly with well width for wells that vary from two monolayers (5.7 Å) to ten monolayers (28.5 Å) as shown in FIGURE 9. The results are understood on the basis of simple physical arguments. From FIGURE 9 a linear increase in the binding energy of the donor as the well size increases is observed. It would be expected that as the well size increases, a maximum in the binding energy will be reached for some critical well width and will then decrease tending toward that of bulk $In_{0.1}Ga_{0.9}As$/GaAs. One must be aware that as the well thickness increases a critical thickness for strain relief will be reached, and the system may no longer grow pseudomorphically. As the wells become very narrow the binding energy tends toward that observed in the GaAs barrier.

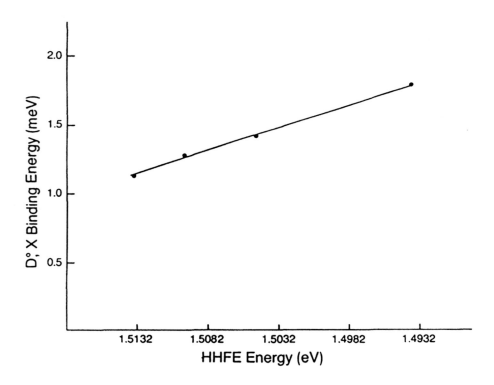

FIGURE 9. Variation of E_{BE} with HHFE transition energy for the two-, four-, six-, and ten-monolayer-wide $In_{0.1}Ga_{0.9}As$/GaAs QW structures.

C5 Energy Variation of Several Intrinsic and Extrinsic QW Transitions as a Function of Well Size

A number of early investigations of the intrinsic and extrinsic properties of GaAs/AlGaAs QWs have been reported [1,5,30,34,41,56-60]. The PL spectra from four $Al_{0.25}Ga_{0.75}As$/GaAs MQWs having well thicknesses of 100, 200, 300, and 400 Å and barrier thicknesses of 100 Å were investigated [61]. The energy variation of the QW transitions as a function of well thickness L_Z is shown in FIGURE 10. The free and bound transitions extrapolate to the analogous transitions in bulk GaAs. The light-hole free exciton is only observed in the 400 Å well; the light hole energy is calculated for the rest of the wells. The free-to-bound transitions were calculated from the following expression:

$$E_{D^0,h} = E_e - E_{D^0} + E_h + E_g \qquad (2)$$

where $E_{D^0,h}$ is the energy of the free-to-bound transition, E_e is the electron subband energy, E_h is the hole subband energy, E_g is the bandgap, and E_{D^0} is the neutral-donor binding energy. The energies E_e, E_h and E_{D^0} are all theoretical values [59,60]. The appropriate values are taken for both the light- and heavy-hole exciton and are plotted as the xs in FIGURE 10. The solid dots are the experimental values. The QW transitions were tracked as L_Z was varied.

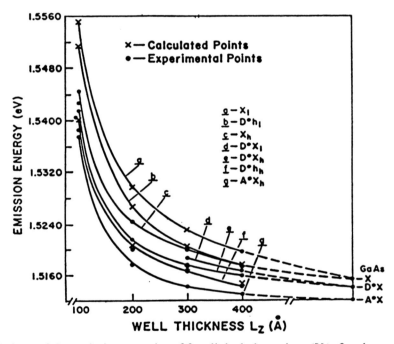

FIGURE 10. Variations of the emission energies of free light-hole exciton (X_l), free heavy-hole exciton (X_h), light-hole exciton bound to a neutral donor ($D^0.X_l$), heavy-hole exciton bound to a neutral donor ($D^0.X_h$), heavy-hole exciton bound to a neutral acceptor ($A^0.X_h$), neutral-donor-light-hole (D^0h_l), and neutral-donor-heavy-hole (D^0h_h) transitions as a function of GaAs well thickness (L_Z).

D CONCLUSION

Optical characterisation is a powerful technique for analysing the properties of heterostructure systems. In view of this, a large spectrum of properties has been investigated resulting in a large amount of data. To accommodate this situation this Datareview has focused on some of the more pertinent intrinsic and extrinsic properties of the more common heterostructure systems, although admittedly the focus is somewhat arbitrary. A partial listing of some of the properties not included is: time-resolved measurements, effects of screening, interface structure, type II heterostructures, phonon structure, band offsets, bi-excitons in heterostructure systems, effects of electric fields, and many-body singularities. All of these properties are associated also with a variety of heterostructure systems. This clearly demonstrates that there is a substantial amount of data yet to be compiled.

REFERENCES

[1]   R. Dingle [in *Festkorperprobleme Advances in Solid State Physics* Ed. H.J. Queisser, vol.XV (Pergamon, Oxford, 1975) p.21]

[2]   R.C. Miller, D.A. Kleinman, A.C. Gossard [*Inst. Phys. Conf. Ser. (UK)* no.43 (1979) ch.27]

[3]   R. Dingle, W. Wiegnann, C.H. Henry [*Phys. Rev. Lett. (USA)* vol.33 (1974) p.827]

[4]   C. Weisbuch, R.C. Miller, R. Dingle, A.C. Gossard, W. Wiegnann [*Solid State Commun. (USA)* vol.37 (1981) p.219]

[5]   R.C. Miller, D.A. Kleinman, W.A. Nordland Jr., A.C. Gossard [*Phys. Rev. B (USA)* vol.22 (1980) p.863]

[6]   R.C. Miller, D.A. Kleinman, W.T. Tsang, A.C. Gossard [*Phys. Rev. B (USA)* vol.24 (1981) p.1134]

[7]   M.H. Meynadier, C. Delalande, G. Bastard, M. Voos, F. Alexandre, J.L. Lievin [*Phys. Rev. B (USA)* vol.31 (1985) p.5539]

[8]   P. Dawson, K.J. Moore, G. Duggan, H.I. Ralph, C.T.B. Foxon [*Phys. Rev. B (USA)* vol.34 (1986) p.6007]

[9]   K.J. Moore, P. Dawson, C.T. Foxon [*Phys. Rev. B (USA)* vol.34 (1986) p.6002]

[10]   D.C. Reynolds et al [*Phys. Rev. B (USA)* vol.37 (1988) p.3117]

[11]   G. Bastard, E.E. Mender, L.L. Chang, L. Esaki [*Phys. Rev. (USA)* vol.26 (1982) p.1974]

[12]   R.L. Greene, K.K. Bajaj [*Solid State Commun. (USA)* vol.45 (1983) p.831]

[13]   Y. Shin Ozuka, M. Matsuura [*Phys. Rev. B (USA)* vol.28 (1983) p.4878]

[14]   R.L. Greene, K.K. Bajaj, D.E. Phelps [*Phys. Rev. B (USA)* vol.29 (1984) p.1807]

[15]   T.F. Jiang [*Solid State Commun. (USA)* vol.50 (1984) p.589]

[16]   C. Priester, G. Allan, M. Lannoo [*Phys. Rev. B (USA)* vol.30 (1984) p.7302]

[17]   J.A. Brum, G. Bastard [*J. Phys. B (UK)* vol.30 (1984) p.7302; *J. Phys. C, Solid State Phys. (UK)* vol.18 (1985) p.L789]

[18]   G.D. Sanders, Y.C. Chang [*Phys. Rev. B (USA)* vol.31 (1985) p.6892; vol.32 (1985) p.5517]

[19]   K.S. Chan [*J. Phys. C, Solid State Phys. (UK)* vol.19 (1986) p.L125]

[20]   D.A. Broido, L.J. Sham [*Phys. Rev. B (USA)* vol.34 (1986) p.7919]

[21]   U. Ekenberg, M. Altarelli [*Phys. Rev. B (USA)* vol.35 (1987) p.7585]

[22]   P.W. Yu et al [*Phys. Rev. B (USA)* vol.35 (1987) p.9250]

[23]   A. Torabi, K.F. Brennan, C.J. Summers [*Proc. SPIE (USA)* vol.792 (1987)]

[24]   Q.X. Zhao, B. Monemar, T. Westgaard, B.O. Fimland, K.J. Ohannessen [*Phys. Rev. B (USA)* vol.46 (1992) p.12853]

[25]   D.C. Reynolds, K.R. Evans, B. Jogai, C.E. Stutz, P.W. Yu [*Solid State Commun. (USA)* vol.86 (1993) p.339]

[26]   S.K. Lyo, E.D. Jones, J.F. Klem [*Phys. Rev. Lett. (USA)* vol.61 (1988) p.2265]

[27]   E.D. Jones, S.K. Lyo, J.F. Klem, J.E. Schirber, S.Y. Lin [in *Gallium Arsenide and Related Compounds* Ed. W.T. Lindley (Institute of Physics and Physical Society, Bristol, 1992) p.407]

[28]   Q.X. Zhao, P.O. Holtz, B. Monemar, T. Lundstrom, J. Wallin, G. Landgren [*Phys. Rev. B (USA)* vol.48 (1993) p.11890]

[29]   D.C. Reynolds, D.C. Look, B. Jogai, C.E. Stutz, R. Jones, K.K. Bajaj [*Phys. Rev. B (USA)* vol.50 (1994) p.11710]

[30]   S.K. Lyo [*Phys. Rev. B (USA)* vol.40 (1989) p.8418]

[31]   Y.J. Chen, E.S. Koteles, J. Lee, J.Y. Chi, B.S. Elman [*Proc. SPIE (USA)* vol.792 (1987) p.162]

[32]   K.J. Moore, G. Duggan, K. Woodbridge, C. Roberts [*Phys. Rev. B (USA)* vol.41 (1990) p.1090]

[33]   B.V. Shanabrook, J. Comas [*Surf. Sci. (Netherlands)* vol.142 (1984) p.504]

[34]   D.C. Reynolds et al [*Phys. Rev. B (USA)* vol.29 (1984) p.7038]

[35]   Y. Nomura, K. Shinazaki, M. Ishii [*J. Appl. Phys. (USA)* vol.58 (1985) p.1864]

[36]   X. Liu, A. Petrou, B.D. Mcombe, J. Ralston, G. Wicks [*Phys. Rev. B (USA)* vol.38 (1988) p.8522]

[37]   D.A. Kleinman [*Phys. Rev. B (USA)* vol.28 (1983) p.871]

[38]   D.C. Reynolds et al [*Phys. Rev. B (USA)* vol.40 (1989) p.6210]

[39]   R.C. Miller, A.C. Gossard, W.T. Tsang, O. Munteanu [*Phys. Rev. B (USA)* vol.25 (1982) p.387]

[40]   R.C. Miller, A.C. Gossard, W.T. Tsang, O. Munteanu [*Solid State Commun. (USA)* vol.43 (1982) p.519]

[41]   B. Lambert, B. Deveaud, A. Regreny, G. Talalaeff [*Solid State Commun. (USA)* vol.43 (1982) p.443]

[42]   R.C. Miller, W.T. Tsang, O. Munteanu [*Appl. Phys. Lett. (USA)* vol.41 (1982) p.374]

[43]   W.T. Masselink et al [*J. Vac. Sci. Technol. B (USA)* vol.2 (1984) p.117]

[44]   B. Deveaud, J.Y. Emery, A. Chomette, B. Lambert, M. Baudet [*Appl. Phys. Lett. (USA)* vol.45 (1984) p.1078]

[45]   M.H. Meynadier, J.A. Brum, C. Delalande, M. Voos, F. Alexandre, J.L. Lievin [*J. Appl. Phys. (USA)* vol.58 (1985) p.4307]

[46]   P.J. Pearah et al [*Appl. Phys. Lett. (USA)* vol.47 (1985) p.166]

[47]   Zh.I. Alferov et al [*Sov. Phys.-Semicond. (USA)* vol.19 (1985) p.439]

[48]   B.J. Skromme, R. Bhat, M.A. Koza [*Solid State Commun. (USA)* vol.66 (1988) p.543]

[49]   G. Bastard [*Phys. Rev. B (USA)* vol.24 (1981) p.4714]

[50]   G. Mailhot, Y.C. Chang, T.C. McGill [*Phys. Rev. B (USA)* vol.26 (1982) p.4449]

[51]   G. Bastard, E.E. Mendez, L.L. Chang, L. Esaki [*Solid State Commun. (USA)* vol.45 (1983) p.367]

[52]   P. Lane, R.L. Greene [*Phys. Rev. B (USA)* vol.33 (1985) p.587]

[53]   R.C. Miller, A.C. Gossard [*Phys. Rev. B (USA)* vol.28 (1983) p.3645]

[54]   D.C. Reynolds, K.R. Evans, C.E. Stutz, K.K. Bajaj, P.W. Yu [*Phys. Rev. B (USA)* vol.44 (1991) p.8869]

[55]   D.C. Reynolds, K.R. Evans, C.E. Stutz, P.W. Yu [*Phys. Rev. B (USA)* vol.44 (1991) p.1839]

[56]   L. Esaki, R. Tsu [*IBM J. Res. Rev. (USA)* vol.14 (1970) p.61]

[57]   B.A. Vojak, N. Holonyak Jr., D.W. Laidig, K. Hess, J.J. Coleman, P.D. Dapkus [*Solid State Commun. (USA)* vol.35 (1980) p.477]

[58]   W.T. Masselink et al [*Appl. Phys. Lett. (USA)* vol.44 (1984) p.435]

[59]   R.L. Greene, K.K. Bajaj [*Solid State Commun. (USA)* vol.45 (1983) p.825]

[60]   S. Chaudhuri, K.K. Bajaj [*Phys. Rev. B (USA)* vol.29 (1984) p.1803]

[61]   D.C. Reynolds et al [*Phys. Rev. B (USA)* vol.29 (1984) p.7038]

6.2 Modulation spectroscopy of III-V quantum wells and superlattices

F.H. Pollak

June 1995

A   INTRODUCTION

Modulation spectroscopy (particularly contactless modes) is a major tool for the study and characterisation of semiconductor microstructures (quantum wells, superlattices, quantum dots, etc.) in addition to bulk/thin film semiconductors, semiconductor interfaces (heterojunctions, semiconductor/vacuum, semiconductor/metal) and process-induced damage [1-4]. It also can be used to investigate the effects of external perturbations such as temperature, electric and magnetic fields, hydrostatic pressure, uniaxial stress, etc. Recent studies also have clearly demonstrated its considerable potential in the evaluation of important device parameters for semiconductor structures such as heterojunction bipolar transistors (HBTs), pseudomorphic high electron mobility transistors (PHEMTs), quantum well lasers, vertical cavity surface emitting lasers (VCSELs), multiple quantum well (MQW) infrared detectors, etc. [5,6].

B   BACKGROUND OF THE FIELD

Modulation spectroscopy is an analogue method for taking the derivative of the optical spectrum (reflectance or transmittance) of a material by modifying in some manner the measurement conditions [1-4]. Instead of measuring the optical reflectance (or transmittance) of the material, the derivative with respect to some parameter is evaluated. The spectral response of a material can be modified directly by applying a repetitive perturbation such as electric field (electromodulation), heat pulse (thermomodulation), or stress (piezomodulation). The periodic variation of the measurement conditions gives rise to sharp, differential-like spectra in the region of interband (intersubband) transitions. Therefore, modulation spectroscopy emphasises relevant spectral features and suppresses uninteresting background effects. The ability to perform a lineshape fit is one of the great advantages of modulation spectroscopy. Since for the modulated signal the features are localised in photon energy it is possible to account for the lineshapes to yield accurate values of important quantities such as the energies and broadening parameters of intersubband transitions, Fermi energies, electric fields, etc. For example, even at room temperature it is possible to determine the energies of spectral features to within a few meV.

A particularly useful form of modulation spectroscopy is electromodulation (EM) since it is sensitive to surface/interface electric fields and can be performed in contactless modes that require no special mounting of the sample. The sensitivity of EM methods to surface/interface electric fields has proven to be one of its most important properties. For sufficiently high built-in electric fields the EM spectrum can display an oscillatory behaviour above the band gap called Franz-Keldysh oscillations (FKOs). The period of these FKOs is a direct measure of the built-in electric field.

Contactless electromodulation can be performed using (a) photoreflectance (PR), (b) an approach which uses a capacitor-like arrangement (contactless electroreflectance (CER)) or (c) electron-beam electroreflectance. In PR, modulation of the electric field in the sample is caused by photo-excited electron-hole pairs created by the pump source (laser or other light source) which is chopped at frequency Ω_m.

Modulation spectroscopy is a very powerful tool for investigating many of the fundamental aspects of compositional single (SQW)/multiple (MQW) quantum wells and superlattices (SLs) as well as doping SLs including band offset, well and barrier widths, excitons, strain, coupling (and decoupling) between wells, miniband formation, two-dimensional electron gas (2DEG) effects, zone-folding in short-period SLs, built-in electric fields, etc. A large variety of III-V, II-VI and GeSi systems have been studied including both lattice-matched and strain-layer configurations. In addition, the effects of various external perturbations can be studied.

C   GaAs/GaAlAs MQW

As an example of the type of information that can be obtained from modulation spectroscopy displayed in FIGURE 1 are the PR spectra (dotted lines) at 300 K from four GaAs/Ga$_{0.82}$Al$_{0.18}$As MQWs all having well width $L_z = 71$ Å, but with different barrier widths ($L_B = 201$ Å, 150 Å, 99 Å and 71 Å) [1,3]. The arrow at the bottom of the figure is the energy of the direct (E_0) gap of the Ga$_{0.82}$Al$_{0.18}$As barrier material as determined from PR on a thick epilayer. Thus, all features above this energy are due to transitions involving unconfined states. The solid lines are fits to an appropriate lineshape function. Indicated by arrows and listed in TABLE 1 are the transition energies obtained.

In order to identify the origins of the large number of features in the PR spectra, a theoretical calculation was performed of both the transition energies and the matrix elements (intensity) using an envelope function model. Exciton binding energy corrections were taken into account. Also listed in TABLE 1 are the results of this calculation. The best overall agreement was found using a conduction band offset parameter Q_c ($= \Delta E_c/(\Delta E_c + \Delta E_v)$) of 0.65 and the indicated values of L_z and L_B. The three features below 1.52 eV are due to 11H(Γ,π), 11L(Γ,π) and 12H(Γ,π) transitions, respectively. The notation mnH/L(Γ) or mnH/L(π) indicates a transition between the mth conduction and nth valence subbands of heavy (H)- or light (L)-hole-like character at the minizone centre (Γ)

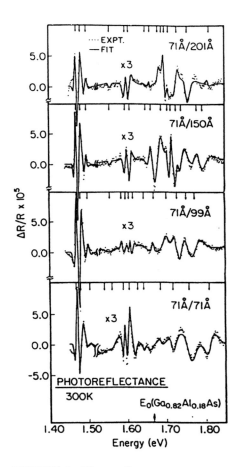

FIGURE 1. Photoreflectance spectra at 300 K (dotted lines) of a series of four GaAs/Ga$_{0.82}$Al$_{0.18}$As MQWs having the same L_z (71 Å) but different L_B. The solid lines are fits to an appropriate lineshape function yielding the energies indicated by arrows.

or edge (π). Transitions denoted by mnH/L(Γ,π) are essentially degenerate, i.e. no miniband dispersion because of the relatively large L_B.

TABLE 1. Experimental and theoretical values of the various confined and unconfined quantum transitions of several GaAs/Ga$_{0.82}$Al$_{0.18}$As multiple quantum wells with different barrier layer thicknesses. The unconfined states occur above E$_0$ (Ga$_{0.82}$Al$_{0.18}$As) = 1.673 eV as indicated by the solid line.

Transition	L_z/L_B 71 Å/201 Å		L_z/L_B 71 Å/150 Å		L_z/L_B 71 Å/99 Å		L_z/L_B 71 Å/71 Å	
	Expt.	Theory	Expt.	Theory	Expt.	Theory	Expt.	Theory
11H(Γ,π)	1.469	1.467[a]	1.469	1.467[a]	1.469	1.466[a]	1.467	1.465[a]
11L(Γ,π)	1.479	1.481[b]	1.480	1.481[b]	1.479	1.480[b]	1.479	1.478[b]
12H(Γ,π)	1.494	1.502[c]	1.495	1.502[c]	1.497	1.502[c]	1.502	1.501[c]
13H(Γ)	1.555	1.547[c]	1.555	1.546[c]	1.555	1.546[c]	1.540	1.544[c]
13H(π)	1.555	1.548[c]	1.555	1.549[c]	1.555	1.551[c]	1.560	1.556
21L(π)	1.595	1.590[d]	1.595	1.589[d]	1.587	1.586[d]	1.589	1.584[d]
21L(Γ)	1.595	1.591[d]	1.595	1.592[d]	1.599	1.597[d]	1.604	1.605[d]
22H(π)	1.604	1.609[d]	1.607	1.608[d]	1.607	1.605[d]	1.604	1.601[d]
22H(Γ)	1.604	1.6010[d]	1.607	1.611[d]	1.618	1.616[d]	1.626	1.626[d]
22L(π)	1.644	1.649[d]	1.647	1.648[d]	1.638	1.644[d]	1.642	1.639[d]
22L(Γ)	1.658	1.655[d]	1.662	1.658[d]	1.666	1.667[d]	1.685	1.689[e]
33H(Γ)	1.680	1.684[e]	1.682	1.690[e]	1.704	1.704[e]	1.718	1.722[e]
33L(Γ)	1.689	1.693[e]	1.705	1.704[e]	1.728	1.731[e]	1.759	1.766[e]
34H(π)	1.696	1.696[e]	1.713	1.713[e]	1.748	1.755[e]	1.809	1.812[e]
43H(π)	1.706	1.712[e]	1.727	1.734[e]	1.778	1.776[e]		
33L(π)	1.706	1.712[e]	1.740	1.741[e]				
45H(Γ)	1.728	1.743[e]	1.779	1.788[e]				
54H(Γ)	1.754	1.757[e]	1.793	1.799[e]				
56H(π)	1.806	1.809[e]						

(a) Exciton binding energy of 8 meV; (b) exciton binding energy of 10 meV; (c) exciton binding energy of 7 meV; (d) Exciton binding energy of 6 meV; (e) exciton binding energy of 4.5 meV.

The first structure above 1.52 eV, identified as 12H, is fairly sharp for the 71 Å/201 Å and 71 Å/150 Å material but becomes markedly broader for the 71 Å/99 Å case and splits into a doublet for the smallest L_B. As L_B decreases there is increased miniband dispersion resulting in a well defined doublet for L_z = 71 Å. Note also that 12H has become broader for the narrowest L_B, indicating the onset of miniband dispersion.

The 22L transition, which shows distinct miniband dispersion (see TABLE 1), is interesting since both Γ and π are confined for the three samples with the widest L_B. For the narrowest barrier material 22L(Γ) becomes unconfined while 22L(π) is still confined. The energy separation between 22L(Γ) and 22L(π) is a sensitive function of L_B, even when both are confined, and hence can be used to evaluate this quantity.

D   EFFECTS OF EXTERNAL PERTURBATIONS

This optical method can be used to examine the effects of external perturbations such as electric and magnetic fields, temperature (energy gaps and broadening parameters), hydrostatic pressure, uniaxial stress, etc. on the transitions of various quantum well structures [1-3].

D1   Temperature

Plotted in FIGURE 2 is the temperature dependence of the various features observed in the PR of a GaAs/Ga$_{0.7}$Al$_{0.3}$As MQW with L$_z$ = 260 Å [7].   The features labelled 1-16 originate from Γ-band transitions, i.e. electrons and holes both in the GaAs wells.   For peak 16 the transition originates in the spin-orbit split valence band.   The dashed curves (1,2,4 and 6a) correspond to 1S exciton transitions while the solid lines are for the band-to-band continuum transitions of 11H.   This observation makes it possible to determine the binding energies of the 11H, 22H, 33H and 44H resonances.   The peaks denoted 17-20 are from transitions involving a Γ-valence level in the GaAs well and the X-conduction band in the Ga$_{0.7}$Al$_{0.3}$As across the heterointerface.

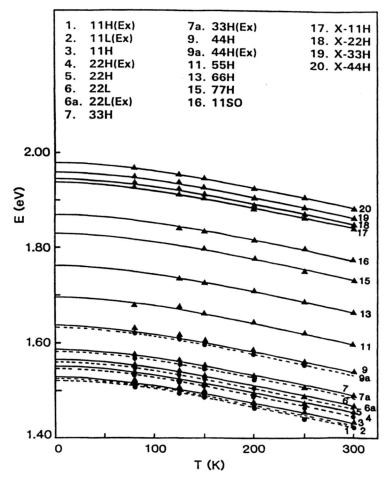

FIGURE 2.   Temperature dependence of the various quantum transitions of a GaAs/Ga$_{0.7}$Al$_{0.3}$As MQW as obtained from PR.   The symbols indicate the experimental data and the curves are fits to EQN (1). The dashed curves are for the 1S excitonic transitions and the solid curves are for the band-to-band continuum transitions.

The solid and dashed lines passing through the data in FIGURE 2 are least-squares fits to the semi-empirical Varshni equation for the temperature dependence of bandgaps in semiconductors:

$$E(T) = E(0) - \alpha T^2/(\beta+T) \tag{1}$$

where $E(0)$ is the bandgap at $T = 0$ K and α and β are Varshni coefficients. For all the Γ-band derived transitions, i.e. features 1-16, the obtained values are $\alpha = (5.40 \pm 0.2) \times 10^{-4}$ eV/K and $\beta = 204$ K. The peaks 17-20 had best fit values of $\alpha = (5.33 \pm 0.2) \times 10^{-4}$ eV/K and $\beta = 204$ K. These numbers are in very good agreement with the corresponding coefficients for the direct bandgap of bulk GaAs. Thus, for the Γ-derived states the temperature dependence of the quantum transitions tracks the band extrema of the constituent bulk well material. The behaviour also has been observed for other quantum well systems. The fact that the transitions involving the X-conduction band in the $Ga_{0.7}Al_{0.3}As$ (peaks 17-20) have the same temperature coefficients as the Γ-derived states is somewhat surprising.

D2   Hydrostatic Pressure

It has been found that the hydrostatic pressure coefficients of the energies of the subbands in SQWs and MQWs are not necessarily the same as that of the constituent bulk material, in contrast to the temperature variation [1,3]. For deep wells and subbands close to the bottom of the well (wide wells) it is found that the pressure coefficient of the energies of the intersubband transitions is indeed close to that of the bulk well material. However, for higher subbands this is no longer true. The reasons for this behaviour in terms of the pressure dependence of the effective mass, well width, exciton binding energy and barrier heights have been discussed.

Strained layer GaSb/AlSb MQWs [8] (and ZnSe/GaAs epilayers) displayed an additional influence of hydrostatic pressure. If the substrate has a different elastic constant in relation to the epitaxial quantum structure, the latter may experience a different amount of strain than the applied pressure which also can result in a uniaxial component along the growth direction. The former effect leads to a change in the pressure coefficient while the latter (uniaxial) strain will alter the separation between the HH and LH levels. Thus it is possible to pressure tune the LH-HH splitting.

D3   Uniaxial Stress

The effects of large, external uniaxial stress ($\vec{S}$) along [100] and [110] on the quantum states of a strained layer (001) $In_{0.21}Ga_{0.79}As$/GaAs SQW with $L_z = 100$ Å has been investigated using PR at 300 K [1,3]. For $\vec{S} \parallel$ [100] (~8 kbar) the stress-induced shifts of the various quantum transitions, shown in FIGURE 3, are in good agreement with a theoretical calculation based on only deformation potential theory (solid line in FIGURE 3).

The heavy-hole (out-plane) derived transitions are relatively insensitive to $\vec{S}$ because of the cancellation between the hydrostatic and shear components of the strain. At high stresses there is a small nonlinear effect due to the coupling with the spin-orbit split band. However, for the light-hole (out-plane) features the hydrostatic and shear terms add, resulting in a relatively large stress-induced shift. Thus, static external stress also can be used to discriminate between HH and LH transitions. For $\vec{S}$ ‖ [100] there were no significant changes in the intensities of the various PR features.

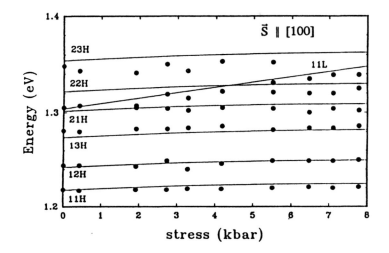

FIGURE 3. Stress-dependence of the energies of the mnH(L) features of an InGaAs/GaAs SQW for $\vec{S}$ ‖ [100]. The solid lines are a theoretical calculation of the effects of strain on the intersubband energies.

The situation for $\vec{S}$ ‖ [110] (~18 kbar) was quite different. For this lower symmetry stress direction there was a dramatic increase in the intensities of the 12H and 13H 'symmetry forbidden' features relative to the 'symmetry allowed' 11H peak. Also there was a significant red shift of 11H, 12H and 13H at high stresses. This phenomenon has been interpreted in terms of an electric field generated along the growth direction induced by the piezoelectric coupling for $\vec{S}$ ‖ [110]. Shown in FIGURE 4 are the experimental energies of the various transitions and a theoretical calculation based not only on deformation potential effects but also a quantum confined Stark effect red shift due to the piezoelectric field.

FIGURE 4. Stress-dependence of the energies of the mnH(L) features of an InGaAs/GaAs SQW for $\vec{S}$ ‖ [110]. The solid lines are a theoretical calculation of the effects of both strain and strain-generated electric field on the intersubband energies.

The effects of large, external $\vec{S}$ ‖ [100] on the interband transitions of two GaAs/GaAlAs SQWs grown on (001) Si have been observed using PR at 300 K and 77 K. The GaAs/Si structure is under an in-plane biaxial tensile stress arising during post-growth cooling from the differences in expansion coefficients. The above external stress configuration makes it possible to externally alter the LH and HH splitting. In an SQW of width 200 Å the ground state was continuously tuned from LH to HH.

E   TWO-DIMENSIONAL ELECTRON GAS EFFECTS

Photoreflectance or CER at 300 K has been used to study 2DEG effects in pseudomorphic $Ga_{1-y}Al_yAs/In_xGa_{1-x}As/GaAs$ SQWs with different materials parameters [5,6]. One such sample with nominal $y = 0.19$, $x = 0.2$ and $L_z = 100$ Å will be discussed. This structure is shown schematically in FIGURE 5. Based on the above parameters there are two confined electron and two confined heavy-hole states, with the Fermi level (E_F) occurring between the first and second electron levels. Thus, four intersubband transitions, i.e. 11H, 12H, 21H and 22H, are possible at room temperature.

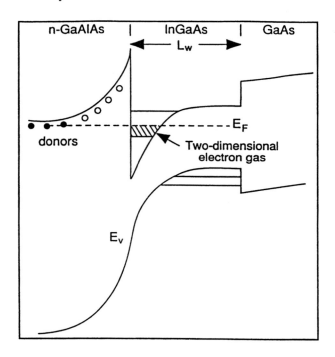

FIGURE 5. Schematic representation of the potential profile of a pseudomorphic n-GaAlAs/InGaAs/GaAs modulation-doped SQW structure.

Several features from the InGaAs portion of the sample in addition to signals from the GaAlAs and GaAs regions have been observed. The spectral features from the InGaAs channel can be explained on the basis of the derivative of a broadened step-like two-dimensional density of states, due to the screening of the excitons by the 2DEG and a Fermi level filling factor. From the details of the lineshape fit it has been possible to extract the Fermi level position and hence the 2DEG sheet density, N_s. These values for the electron sheet density are in good agreement with Hall measurements on the same samples. In addition, several intersubband transition energies in the InGaAs channel can be obtained. The GaAs and GaAlAs signals exhibit FKOs from which it is possible to obtain information about the electric fields in these regions. Also the GaAlAs signal yields the Al composition (y). Thus the entire potential profiles in such structures can be determined in a contactless manner at room temperature.

Displayed by the solid line in FIGURE 6 is the CER spectrum originating in the InGaAs section of the sample, except for the feature denoted E_0(GaAs). This resonance corresponds to the direct bandgap of GaAs and originates in the GaAs buffer/substrate. The trace of FIGURE 6 consists of four peaks, labelled mnH, positioned on a background which is due to the thermal tail of the Fermi distribution function. The dashed line in FIGURE 6 is a fit to the lineshape function mentioned above. The obtained energies of the mnH transitions are denoted by arrows. By comparing these experimental energies with a theoretical self-consistent calculation it was possible to evaluate several important materials parameters such as L_z, In composition and built-in electric field. The fit to the background yields E_F and hence N_s.

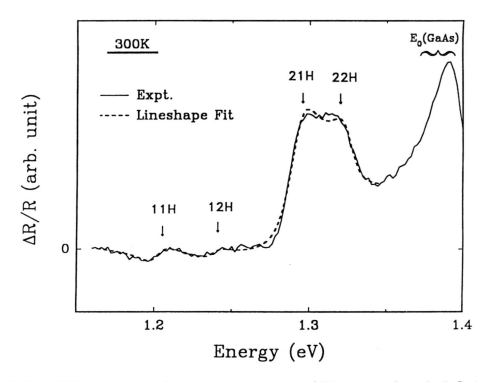

FIGURE 6. The solid line is the experimental room temperature CER spectrum from the InGaAs region of a pseudomorphic GaAlAs/InGaAs/GaAs SQW sample. The dashed line is a least-squares fit to an appropriate lineshape function. The obtained values of the intersubband energies mnH are designated by arrows.

It should be noted that at room temperature transitions to the first conduction level, (i.e. 11H and 12H), which lies below E_F, are only seen because of the significant thermal broadening of the Fermi distribution. As the temperature is lowered these resonances are no longer observed.

F   CONCLUSION

Modulation spectroscopy is a very powerful tool for investigating many of the fundamental aspects of compositional SQWs, MQWs and SLs as well as doping SLs including band offset, well and barrier widths, excitons, strain, coupling (and decoupling) between wells, miniband formation, two-dimensional electron gas (2DEG) effects, zone-folding in short-period SLs, built-in electric fields, etc. Even at room temperature a large number of sharp, derivative-like intersubband transitions can be observed. A large variety of III-V, II-VI and GeSi systems have been studied including both lattice-matched and strain-layer configurations. In addition, the effects of various external perturbations such as temperature, electric and magnetic fields, hydrostatic pressure, uniaxial stress, etc. can be evaluated.

ACKNOWLEDGEMENT

This work was supported by the National Science Foundation, Army Research Office and the New York State Foundation for Science and Technology.

REFERENCES

[1]   F.H. Pollak, H. Shen [*Mater. Sci. Eng. R Rep. (Switzerland)* vol.10 (1993) p.275-374 and references therein]

[2]   O.J. Glembocki, B.V. Shanabrook [in *Semicond. Semimet.* vol.67, Eds D.G. Seiler, C.L. Littler (Academic Press, New York, 1992) p.222-92 and references therein]

[3]   F.H. Pollak [*Handbook on Semiconductors* vol.2 Ed. M. Balkanski (North Holland, Amsterdam, 1994) p.527-635]

[4]   A.K. Ramdas [*Superlattices Microstruct.(UK)* vol.4 (1988) p.69-79]

[5]   H. Qiang, D. Yan, Y. Yin, F.H. Pollak [*Asia-Pac. Eng. J. A, Electr. Eng. (Singapore)* vol.3 (1993) p.167-98]

[6]   F.H. Pollak, H. Qiang, D. Yan, Y. Yin, W. Krystek [*J. Met. (USA)* vol.46 (1994) p.55-9]

[7]   A. Kangarula et al [*Phys. Rev. B (USA)* vol.37 (1988) p.1035-8]

[8]   B. Rockwell et al [*Mater. Res. Soc. Symp. Proc. (USA)* vol.160 (1990) p.751-6]; also [*Proc. 20th Int. Conf. Physics Semicond.* Thessaloniki, Greece, Eds E.M. Anastassakis, J.D. Joannopoulos (World Scientific, Singapore, 1990) p.929-32]

6.3 Raman spectroscopy of III-V quantum wells and superlattices

D. Gammon and R. Merlin

October 1995

A   INTRODUCTION

The purpose of this Datareview is to provide a brief guide to Raman scattering (RS) experiments in III-V superlattices (SLs) and quantum well (QW) systems. Pioneering work in this area dates back to the late 1970s. There is a large body of literature centred on the system GaAs/AlGaAs, which has been discussed in several reviews dealing with phonons [1-4], electronic excitations [5,6], impurities [7] and non-periodic structures [8], and in conference proceedings [9]. We refer the reader to these reviews for a comprehensive list of references. Because most SL and QW parameters vary widely and do not represent fundamental properties of the structures (they are controlled by the opening and closing of shutters during growth), this Datareview does not focus on numerical data.

B   PHONONS

In III-V systems, the prototypical Raman experiment is performed in the backscattering geometry on a structure grown along the [001] direction. Standard vibrational spectra show allowed folded longitudinal-acoustic (LA) modes and confined longitudinal-optical (LO) phonons. In addition, the spectra may show forbidden (disorder-induced) interface (IF) scattering near electronic resonances. Transverse acoustic (TA) and optical (TO) vibrations are not allowed in (001) backscattering, but transverse or mixed TA-LA and TO-LO modes have been observed in samples grown on, e.g. (311) substrates [10]. Allowed IF scattering has been recently reported using non-standard configurations [11].

B1   Acoustic Modes

In the continuum limit, the equation for the propagation of elastic waves in layered structures is given by [1-4]

$$\rho(z) \, \partial^2 U_s / \partial t^2 = \sum_{klm} \frac{\partial}{\partial x_k} \left[C_{sklm}(z) \frac{\partial U_s}{\partial x_m} \right] \tag{1}$$

Here, $U(r,t)$ is the displacement vector and z is the coordinate normal to the layers. The density and elastic tensor are given, respectively, by ρ and C_{sklm}. Since ρ and C_{sklm} depend only on z, the solutions to EQN (1) are of the form $U_s = U_s(z) \exp[i(k_{\parallel} r - \Omega t)]$ where $k_{\parallel}$ is a wavevector parallel to the layers. In the case where ρ and C_{sklm} are periodic functions (see [8] for Fibonacci and other non-periodic SLs), it follows from the Floquet theorem that $U_s(z) = u_s(z) \exp(ik_z z)$ where $u_s(z) = u_s(z + l)$; l is the SL period. The wavevectors of the incident (K_i) and scattered (K_s) photons are along the z-direction in the usual near-backscattering geometry. In such situations, the spectra probe modes at $k_\parallel = 0$. The frequency $\Omega(k_\parallel = 0, k_z)$ exhibits gaps at the centre and edge of the SL Brillouin zone, i.e. at $k_z = m\pi/l$ where m is an integer. From conservation of crystal momentum, one finds that the wavevectors of the acoustic modes which can participate in the scattering process are the doublets $2m\pi/l \pm |K_i - K_s|$.

For III-V SLs, the acoustic modulation is sufficiently weak that layering effects can be ignored except for a narrow range of wavevectors in the vicinity of the gaps. The concept of folded acoustic phonons, analogous to that of the reduced-zone scheme in the nearly-free electron model, refers to this limit. Let v be the average sound velocity of the relevant mode. Then, the frequencies of the Raman doublets are of the form

$$\Omega_m^\pm = m\,\frac{\pi v}{2l} \pm \frac{4\pi n(\lambda_L)}{\lambda_L} \tag{2}$$

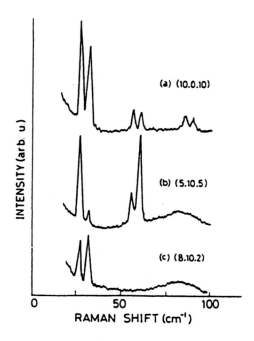

where n is the refractive index and λ_L is the laser wavelength. Typical spectra of acoustic phonons are shown in FIGURE 1 [12]. It is clear that the SL period and the refractive index can be gained from measurements of $\Omega_m^\pm$.

The selection rules in the elastic limit are determined by the elasto-optic tensor p_{sklm} defined by $\delta\chi_{sk} = p_{sklm}U_{lm}$ where $\delta\chi_{sk}$ is the change in the susceptibility tensor due to the strain. The scattering cross-section is proportional to

$$|\sum \epsilon_s^{(i)} \delta\chi_{sk} \epsilon_k^{(s)}|^2$$

where $\epsilon_s^{(i)}$ and $\epsilon_k^{(s)}$ are the components of the polarisation vector of the incident and scattered photons. Consider, for instance, (001) SLs. Here, TA phonons with, say, $U_{lm} = U_{xz}\delta_{lx}\delta_{mz}$ give $\delta\chi_{sk} = 0$ except for $\delta\chi_{zx} = \delta\chi_{xz} = p_{44}U_{xz}$. Accordingly, TA modes are forbidden in backscattering since the photons are not polarised along z. For LA modes, one has $\delta\chi_{xx} = \delta\chi_{yy} = p_{12}U_{zz}$ and $\delta\chi_{zz} = p_{11}U_{zz}$ and, otherwise, $\delta\chi_{sk} = 0$. It follows that (001) LA-backscattering is allowed in the parallel $z(x,x)\bar{z}$

FIGURE 1. Raman spectra of GaAs/AlGaAs/AlAs samples grown on (001) substrates showing scattering by folded LA phonons. Individual layer thicknesses in monolayers are indicated in the figure. The SL period (20 monolayers) is the same for the three samples [12].

configuration and that it probes p_{12} (the notation $a_i(b_i,b_s)a_s$ indicates the direction of propagation of the incident (a_i) and scattered (a_s) photons and their respective polarisations b_i and b_s). The doublets in FIGURE 1 correspond to this case.

Neglecting the weak modulation of the sound velocity, the scattering intensity for LA modes propagating along [001] is [1-4]

$$I(\Omega_m) \propto \Omega_m[n(\Omega_m) + 1]\,|P_m|^2 \tag{3}$$

where $n(\Omega_m)$ is the Bose thermal factor and

$$P_m = \int p_{12}(z) \exp(i2\pi mz/l) \, dz$$

is the Fourier transform of the elasto-optic coefficient. This expression establishes a link between RS and structural studies. In some sense, the number of doublets gives a measure of the interface quality.

For the most part, RS experiments on (001) GaAs/AlGaAs SLs show only folded LA modes (however, experiments under resonant conditions reveal anomalous continua with dips in the range of the dispersion gaps [13]; the scattering, requiring a preferential breakdown of wavevector conservation, is not well understood). Those, as well as results on other III-V compounds, are in good agreement with the standard model with regard to the positions of the Raman lines and the selection rules (see [3] for data in the vicinity of the gaps). This also applies to recent results on GaAs/AlAs samples grown on non-conventional directions (e.g. [311]) which exhibit TA or mixed TA-LA scattering [10]. Notice that the photoelastic model is not expected to account for the intensity of the doublets in very thin samples or under resonant conditions involving SL states [13].

B2   Optical Phonons

In most III-V QWs and SLs (e.g. GaAs/AlAs) the optical phonon spectrum shows confined and interface scattering [14,15]. Confined phonons are localised in individual layers. Their energies can be found easily and fairly accurately by considering wavevectors of the form [4]

$$k_m = \frac{m\pi}{(p+1)a_0}, \; 1 \leq m \leq p \tag{4}$$

where p is the number of monolayers in the layer and a_0 is the monolayer thickness. The mode energies are then given by the values of the bulk dispersion relation at k_m. This procedure gives only the SL zone centre modes; however, most confined phonons have weak dispersion for wavevectors in the plane of the QWs [4].

The confined phonon energies can be accurately measured in Raman scattering. In a backscattering experiment on a (001) GaAs/AlAs QW structure there will be a series of LO phonons close to the bulk GaAs LO phonon energy (see FIGURE 2) and another series at the AlAs energy. Notice that confined TO phonons are forbidden in this scattering geometry, and thus are much weaker in FIGURE 2.

The interface phonons arise from the dielectric mismatch between the layers (e.g. between the GaAs and AlAs), and can be described within the dielectric continuum model [4,15]. In contrast to the confined modes these phonons are not allowed in the normal backscattering geometry. They also have large wavevector dispersion. In order for interface phonon scattering to be allowed it is necessary to transfer the wavevector perpendicular to the SL axis. This can be done in a backscattering geometry off the edge of the SL [11] (FIGURE 3). Frequently, forbidden interface phonon scattering is observed (FIGURE 2) because of a breakdown of wavevector conservation through, for example, interface roughness [15].

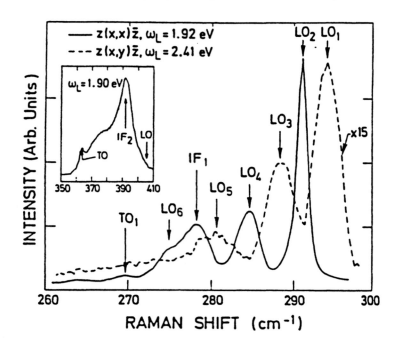

FIGURE 2. Resonant (solid line) and off-resonance (dashed line) spectra showing confined LO phonons and the forbidden interface IF_1 mode for a GaAs(20 Å)/AlAs(60 Å) SL. AlAs-like phonons are shown in the inset [14].

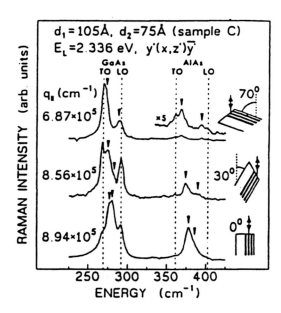

FIGURE 3. Spectra of a GaAs(105 Å)/AlAs(75 Å) sample. The component of the wavevector parallel to the layers is denoted by $q_\parallel$. Arrows indicate allowed IF phonons [11].

The nonresonant Raman scattering selection rules are such that the odd confined phonon modes are observed in $z(x',y')\bar{z}$ and the even modes are observed in $z(x',x')\bar{z}$ backscattering geometries where $z = [001]$, and x' and y' are along the [110] crystal axes [4]. Nonresonant Raman scattering experiments (i.e. such that the laser is far from an interband transition of the sample) are often done on GaAs/AlAs QWs using an argon laser (e.g. 514.5 nm). With resonant excitation, additional scattering mechanisms and higher order processes become important [4].

Raman studies of the optical phonons in semiconductor nanostructures have led to a good understanding of the confined and interface modes and how they interact with electrons [4,15]. Furthermore, time-resolved Raman scattering in pump-probe experiments has also been used to show the importance of interface phonons in cooling hot electrons [16]. It is clear that both confined and interface phonons must be included in a complete description of electron-phonon processes [4].

It is also possible to use Raman scattering of optical phonons as a local probe of the sample structure since the confined phonons are localised to a single layer. For example, shifts in the confined phonon energies provide a measure of interface roughness [17]. The energies also provide a useful measure of strain in strained layer SLs or alloy concentration in samples with alloy layers [4]. Moreover, it is possible to work with very thin layers and even with a single monolayer [18].

B3   Resonant Behaviour

Measurements of the dependence of the RS cross-section on the laser energy give information on higher-lying electronic excitations such as excitons. Resonant Raman scattering (RRS) is essentially a modulation technique bearing on changes of the optical response induced by ion displacements, i.e. the phonons are the modulation element in RRS [19]. The example in FIGURE 4 shows a large enhancement of the cross-section for LO scattering due to resonances with heavy-hole excitons [20]. RRS has been applied to the study of electric- [21] and magnetic-field [22] effects as well as the pressure behaviour [23] of SL electronic spectra. RRS distinguishes itself from other techniques in that a single excitation (e.g. HH2 in FIGURE 4) can lead to two peaks associated with resonances for the incident and scattered channel [20]. This needs to be taken into account in the interpretation of resonant data.

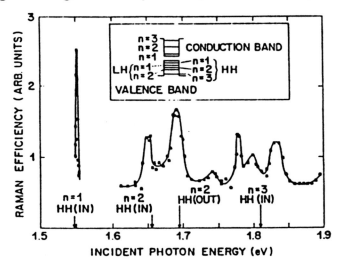

FIGURE 4. Photon energy dependence of the cross-section for scattering by LO phonons in a 102 Å QW. Ingoing and outgoing resonances with heavy-hole excitons are denoted, respectively, by IN and OUT [20].

C   ELECTRONIC EXCITATIONS

Electronic Raman scattering in QW structures and SLs is rather weak. In order to enhance the signal, it is necessary to tune the laser energy to resonate with a higher-lying transition [5,6]. Resonances associated with heavy-hole excitons and with excitations derived from the $E_0 + \Delta_0$-gap of the bulk are commonly used. A typical backscattering spectrum of a doped QW shows narrow lines due to spin-density excitations and to coupled-modes resulting from the interaction between LO phonons and the charge-density plasmon-like mode of the carriers. Away from backscattering, the spectra give intrasubband excitations such as plasmons and single-particle transitions. In addition, data on high-mobility samples show intersubband single-particle excitations. The spectrum of doped-SLs with very narrow layers reveals crossover to three-dimensional behaviour [24].

C1 Intersubband Transitions

Intersubband transitions of electrons even in a single QW can be measured under resonant excitation [25,26]. Raman scattering probes either charge-density (CDE) or spin-density excitations (SDE) of the electron gas depending on the polarisation of the light. With parallel incident and scattered light polarisations the CDE are measured while crossed polarisations probe SDE [5,6]. Electron single particle or collective excitations can be measured in Raman scattering [5,6]. In high mobility samples there are three spectral lines associated with each intersubband transition (FIGURE 5) [25,26]. In the CDE configuration a single particle excitation peak occurs at the bare intersubband energy while a collective CDE (intersubband plasmon) is shifted up from the bare transition energy by the Coulomb interaction. The intersubband plasmon couples strongly to the LO phonon forming coupled modes. In the SDE spectrum the single particle excitations are also observed along with a collective SDE which is shifted down from the bare intersubband energies.

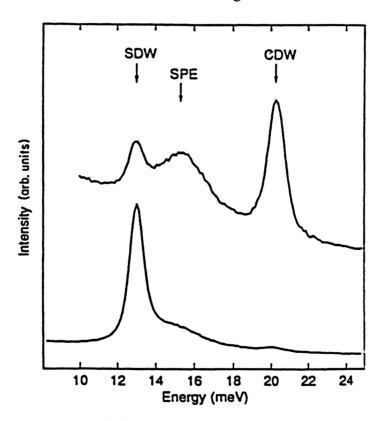

FIGURE 5. The upper trace is the $z(x'.x')\bar{z}$ spectrum showing collective charge density excitations (CDW). The bottom trace is the $z(x'.y')\bar{z}$ spectrum showing spin density excitations (SDW) [27].

Studies of these energies as a function of electron density and wavevector along the quantum well have allowed the strength of the Coulomb interactions to be determined for quasi-2D electron systems [25,26]. The splitting of the collective modes gives a good measure of the electron density while the energy of the single particle intersubband transition provides a measure of the QW width. The magnetic field dependence is discussed in [28]. The role of nonparabolicity and strain-induced electric fields in (111) oriented QWs is discussed in [29,30]. Intersubband transitions of holes have also been studied [5,6].

C2 Plasmons and Intrasubband Transitions

Plasmons and single-particle transitions represent the most common forms of intrasubband scattering of quasi two-dimensional systems [5,6]. To observe intrasubband excitations, it is necessary to use configurations other than backscattering for intrasubband frequencies vanish at $k_\parallel = 0$.

The plasma frequency of a stack of layers of period l and areal charge-density σ is given by

$$\Omega_p^2(k_\parallel, k_z) = \left(\frac{2\pi\sigma e^2}{\varepsilon_0 m'}\right)\left(\frac{k_\parallel \sinh(k_\parallel l)}{\cosh(k_\parallel l) - \cos(k_z l)}\right) \qquad (5)$$

Here, ε_0 is the dielectric constant and m' is the effective mass of the carriers. For $k_\parallel l \gg 1$, one obtains the single-layer expression $\Omega_p^2 \propto k_\parallel$ while for $k_\parallel l \ll 1$ and at $k_z = 0$, $\Omega_p^2 = 4\pi\sigma e^2/(\varepsilon_0 m' l)$ as in three dimensions. An interesting feature of the plasmon dispersion is that, at a fixed $k_z \neq 0$, the behaviour is acoustic-like, i.e. $\Omega_p \propto k_\parallel$ for $k_\parallel l \to 0$. Scattering by plasmons is shown in the spectra of FIGURE 6 [31]. Finite size effects and plasmons in non-periodic systems are considered, respectively, in [32] and [33]. The magnetic-field behaviour is discussed in [34].

The spectrum of intrasubband single-particle excitations is a continuum defined by transitions for which the wavevector transfer is $k_\parallel$. The continuum exhibits a relatively narrow maximum at $k_\parallel v_F$ and a cutoff at $k_\parallel v_F + \hbar k_\parallel^2/2m'$; v_F is the Fermi velocity of the carriers. Intrasubband spectra of a structure with $\sigma \sim 7 \times 10^{11}$ cm^{-2} are shown in FIGURE 7 [35].

C3 Shallow Impurities

Quantum confinement leads to shifts and splittings of the impurity spectrum which become important when the QW width L is of the order of the impurity Bohr radius. The impurity binding energy generally increases with decreasing L. In addition,

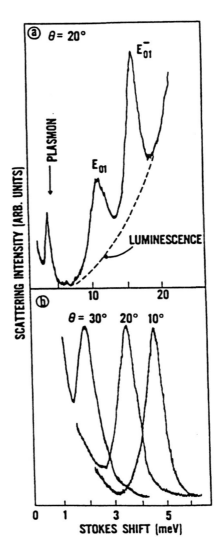

FIGURE 6. (a) Spectra showing plasmon and intersubband scattering for a modulation-doped GaAs/AlGaAs structure. θ is the angle of incidence. (b) Dependence of the plasmon energy on θ [31].

confinement gives rise to the appearance of a series of resonant impurity states reflecting the subband ladder. The magnitude and nature of these effects depend on the parameters of the structure and on the position of the impurity in the well. The latter manifests itself as inhomogeneous broadening of spectral features. Typical results on acceptors are shown in FIGURE 8 [36]. Raman work prior to 1990 is reviewed in [7]. The latter includes a discussion of effects due to magnetic field and uniaxial stress.

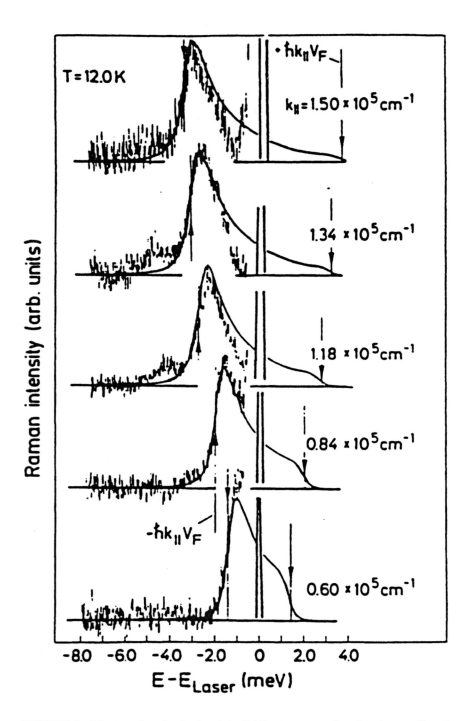

FIGURE 7. Measured and calculated (solid line) spectra showing intrasubband single-particle excitations for a modulation-doped GaAs structure; $k_\parallel$ is the component of the scattering wavevector parallel to the layers [35].

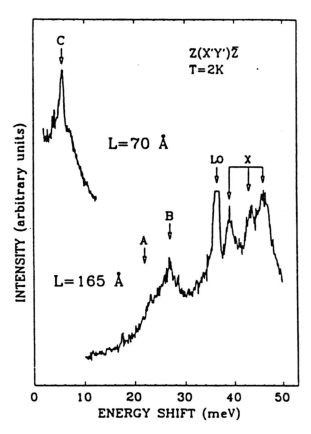

FIGURE 8. Raman spectra of Be-doped QW structures doped at the centre of the wells. Peaks A and B are acceptor transitions from the $1s_{3/2}(\Gamma_6)$ ground state to the excited states $2s_{3/2}(\Gamma_6)$ and $2s_{3/2}(\Gamma_7)$. respectively. C corresponds to $1s_{3/2}(\Gamma_6) \rightarrow 1s_{3/2}(\Gamma_7)$. Features labelled X are due to resonant acceptor states [36].

D   CONCLUSION

The technique of RS is a powerful tool for fundamental studies and characterisation of III-V SLs and QW structures. RS results have contributed substantially to the understanding of low-energy excitations of artificial structures, and they continue to do so in areas of current interest, such as the fractional quantum Hall effect [37]. On the applied side, Raman data give direct information on a wide range of intrinsic and extrinsic parameters providing also a measure of sample quality.

ACKNOWLEDGEMENT

This work was supported in part by the Office of Naval Research and by the U.S. Army Research Office under Contract No. DAAL$03-92-0233$.

REFERENCES

[1]   M.V. Klein [*IEEE J. Quantum Electron. (USA)* vol.22 (1986) p.1760-70]
[2]   J. Sapriel, B. Djafari-Rouhani [*Surf. Sci. Rep. (Netherlands)* vol.10 (1989) p.189-275]
[3]   J. Menendez [*J. Lumin. (Netherlands)* vol.44 (1989) p.285-314]

[4]   B. Jusserand, M. Cardona [in *Light Scattering in Solids V, Topics in Appl. Phys.* vol.66, Eds M. Cardona, G. Guntherodt (Springer, Berlin, 1988) ch.3 p.49-152]

[5]   G. Abstreiter, R. Merlin, A. Pinczuk [*IEEE J. Quantum Electron. (USA)* vol.22 (1986) p.1771-84]

[6]   G. Abstreiter, A. Pinczuk [in *Light Scattering in Solids V, Topics in Appl. Phys.* vol.66, Eds M. Cardona, G. Guntherodt (Springer, Berlin, 1988) ch.4 p.153-213]

[7]   R. Merlin [*Proc. SPIE (USA)* vol.1336 (1990) p.53-63]

[8]   R. Merlin [in *Light Scattering in Solids V, Topics in Appl. Phys.* vol.66, Eds M. Cardona, G. Guntherodt (Springer, Berlin, 1988) ch.5 p.214-32]

[9]   D.J. Lockwood, J.F. Young (Eds) [*NATO ASI Ser. B, Phys. (USA)* vol.273 (1990)]

[10]   Z.V. Popovic et al [*Phys. Rev. B (USA)* vol.49 (1994) p.7577-83]

[11]   R. Hessmer et al [*Phys. Rev. B (USA)* vol.46 (1992) p.4071-6]

[12]   M. Nakayama, K. Kubota, S. Chika, H. Kato, N. Sano [*Solid State Commun. (USA)* vol.58 (1986) p.475-7]

[13]   T. Ruf, V.I. Belitsky, J. Spitzer, V.F. Sapega, M. Cardona, K. Ploog [*Phys. Rev. Lett. (USA)* vol.71 (1993) p.3035-8]; R. Merlin [*Philos. Mag. (UK)* vol.70 (1994) p.761-6]

[14]   A.K. Sood, J. Menendez, M. Cardona, K. Ploog [*Phys. Rev. B (USA)* vol.32 (1985) p.1412-4]

[15]   A.K. Sood, J. Menendez, M. Cardona, K. Ploog [*Phys. Rev. Lett. (USA)* vol.54 (1985) p.2115-8]; R. Merlin, C. Colvard, M.V. Klein, H. Morkoc, A.Y. Cho, A.C. Gossard [*Appl. Phys. Lett. (USA)* vol.36 (1980) p.43-5]

[16]   K.T. Tsen, K.R. Wald, T. Ruf, P.Y. Yu, H. Morkoc [*Phys. Rev. Lett. (USA)* vol.67 (1991) p.2557-60]

[17]   D. Gammon, B.V. Shanabrook, D.S. Katzer [*Phys. Rev. Lett. (USA)* vol.67 (1991) p.1547-50]

[18]   D.S. Katzer, B.V. Shanabrook, D. Gammon [*J. Vac. Sci. Technol. (USA)* vol.12 (1994) p.1056-8]

[19]   M. Cardona [in *Light Scattering in Solids II. Topics in Appl. Phys.* vol.50 Eds M. Cardona, G. Guntherodt (Springer, Berlin, 1982) ch.2 p.19-178]

[20]   J.E. Zucker, A. Pinczuk, D.S. Chemla, A. Gossard, W. Wiegmann [*Phys. Rev. Lett. (USA)* vol.51 (1983) p.1293-6]

[21]   A.J. Shields et al [*Phys. Rev. B (USA)* vol.46 (1992) p.6990-7001]

[22]   A. Cros, T. Ruf, J. Spitzer, M. Cardona, A. Cantarero [*Phys. Rev. B (USA)* vol.50 (1994) p.2325-32]

[23]   G.A. Kourouklis et al [*J. Appl. Phys. (USA)* vol.67 (1990) p.6438-44]

[24]   D. Kirillov, C. Webb, J. Eckstein [*J. Cryst. Growth (Netherlands)* vol.81 (1987) p.91-6]

[25]   A. Pinczuk, S. Schmitt-Rink, G. Danan, J.P. Valladares, L.N. Pfeiffer, K.W. West [*Phys. Rev. Lett. (USA)* vol.63 (1989) p.1633-6]

[26]   D. Gammon, B.V. Shanabrook, J.C. Ryan, D.S. Katzer, M.J. Yang [*Phys. Rev. Lett. (USA)* vol.68 (1992) p.1884-7]

[27]   D. Gammon, B.V. Shanabrook, J.C. Ryan, D.S. Katzer [*Phys. Rev. B (USA)* vol.41 (1990) p.12311-4]

[28]   G. Brozac, B.V. Shanabrook, D. Gammon, D.S. Katzer [*Phys. Rev. B (USA)* vol.47 (1993) p.9981-4]

[29]   B.V. Shanabrook et al [*Superlattices Microstruct. (UK)* vol.7 (1990) p.363-7]

[30]   G. Brozac et al [*Phys. Rev. B (USA)* vol.45 (1992) p.11399-402]

[31]   D. Olego, A. Pinczuk, A.C. Gossard, W. Wiegmann [*Phys. Rev. B (USA)* vol.25 (1982) p.7867-70]

[32]   A. Pinczuk, M.G. Lamont, A.C. Gossard [*Phys. Rev. Lett. (USA)* vol.56 (1985) p.2092-5]

[33]   R. Merlin, J.P. Valladares, A. Pinczuk, A.C. Gossard, J.H. English [*Solid State Commun. (USA)* vol.84 (1992) p.87-9]

[34]   A. Pinczuk, D. Heiman, A.C. Gossard, J.H. English [in *Proc. 18th Int. Conf. Physics of Semiconductors* Ed. O. Engstrom (World Scientific, Singapore, 1987) p.557-60]

[35]   G. Fasol, N. Mestres, M. Dobers, A. Fischer, K. Ploog [*Phys. Rev. B (USA)* vol.36 (1987) p.1565-9]

[36]   D. Gammon, R. Merlin, D. Huang, H. Morkoc [*J. Cryst. Growth (Netherlands)* vol.81 (1987) p.149-52]

[37]   A. Pinczuk et al [*Phys. Rev. Lett. (USA)* vol.61 (1988) p.2701-4]

6.4 Electroabsorption in III-V multiple quantum well modulators and superlattices

J.E. Cunningham

April 1995

A INTRODUCTION

Electroabsorption in 850 nm semiconductor modulators is reviewed with emphasis on state of the art developments as achieved through advanced quantum well design. This ideal modulator system further serves as a comparison to semiconductor modulators that operate at 1.55 and 1.06 μm, where recently potential applications for fibre access or with solid state lasers have emerged.

B ELECTROABSORPTION AT 850 nm

Ten years ago multiple quantum well, MQW, modulators were used to explore the nonlinear optical properties of two-dimensional excitons [1,2]. More recently, MQW modulators have become an enabling technology for practical applications in photonic switching, information processing and communications systems [3-6]. The key development in the above was the capacity of the MQW to bind an exciton so strongly that dissociation at room temperature was avoided [3]. The exciton thus becomes sufficiently robust to withstand intense electric fields applied during bias so that MQWs can operate as efficient electroabsorptive devices. In fact, in the extreme 2D limit an exciton's Bohr radius is half of that in bulk. This dimensionality improvement leads to a pronounced impact on excitonic oscillator strength and results in an absorption coefficient at threshold four times greater in pure 2D systems compared with bulk [2]. These enhancement factors are very important in applications requiring arrays of devices, e.g. with $>10^3$ MQW modulator elements. For instance, in the application described in [4] the total performance of this system is ultimately dictated by the absolute absorption strength of an MQW device through its on/off contrast ratio. (Here, receiver and transmitter elements (MQW) have been integrated with on-chip electronics (smart pixels) to communicate information on and off chip for photonic switching applications.) Lower absorptivity in each MQW element multiplies to cause overall system loss (because of the use of many identical modulator elements). This loss could be offset at the expense of increased laser power, provided its output limit has not been surpassed. Design of the MQW (i.e. at best a quasi-2D exciton) is therefore important and has been greatly aided by the increased precision of modern epitaxial growth techniques such as MBE where interfacial definition is abrupt on a monolayer scale [7].

C MULTIPLE QUANTUM WELL DESIGN

FIGURE 1 shows a comparison of room temperature photocurrent in a matrix of MQW structures that contain variable AlGaAs barrier heights and widths while maintaining the QW thickness fixed near 90 Å [8]. The MQW region is clad with optically transparent AlGaAs (x = 0.3) layers that have been doped n- and p-type respectively to make an n-i-p device. One of the remarkable observations depicted by the matrix is that 2D excitons are robust with respect to large changes in barrier design as evidenced by strong excitonic resonances in all the barrier designs. As voltage is applied these excitonic resonances red shift because of the quantum confined Stark effect, QCSE [9]. The QCSE shift in GaAs is caused mainly by the effect of field on the hole confinement because of the larger effective mass of holes than that of electrons.

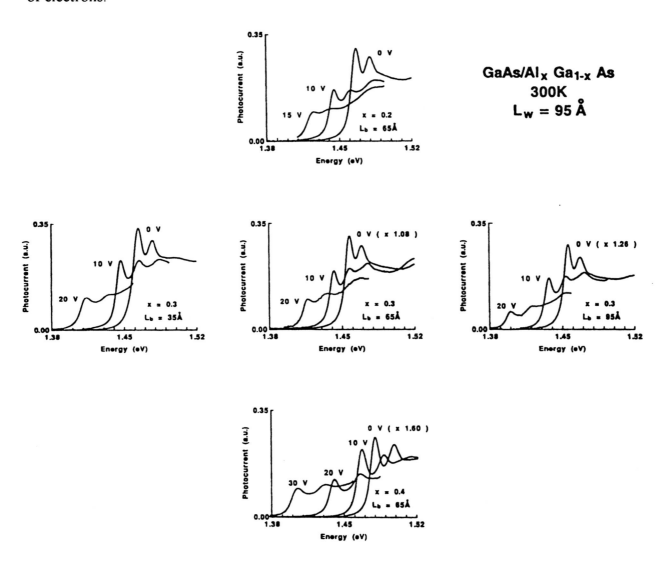

FIGURE 1. A matrix showing field dependent photocurrent versus barrier height and width for MQWs in the GaAs/AlGaAs system. The barrier height and width are identified in the figure.

Other features of the matrix are noteworthy. First, the exciton increasingly field ionises for progressively lower barrier height and width samples. We take as a reference the specimen having a prominent barrier (x = 0.4 and thickness = 65 Å). Here, the exciton is well preserved for bias voltage as high as 30 V. By contrast, going vertically up or to the left in the diagram the exciton is observed to degrade with field. This weakening of the exciton in specimens with lower barrier height and width is caused by thermionic emission and tunnelling of carriers out of the well. The rates of carrier escape by the latter processes increase with field and shorten the exciton lifetime in the QW [10]. Secondly, going vertically up in the matrix we enter the so called 'shallow quantum well' regime. Shallow QWs exhibit rapid carrier escape while retaining good excitonic absorption [11]. They further provide physical conditions in which excitonic transformations from 2D to 3D can be explored [12,13]. Thirdly, by going laterally across the diagram (from right to left) excitonic absorption crosses from 'MQW like' to 'coupled quantum well' behaviour, i.e. the wavefunction spills into adjacent QWs. When the barriers become even thinner we enter the Wannier-Stark regime of superlattices. To the extreme left of superlattices lies the regime where Bloch oscillations in a semiconductor superlattice were first observed [14,15].

D   SUPERLATTICES

D1   Wannier-Stark Superlattices

Although Bloch oscillations and solid state superlattices were proposed years ago their experimental demonstration has only recently progressed [17]. In fact, solid state superlattice systems evolved from work originally proposed by Wannier in connection with Bloch oscillations in periodic solids [18,19]. These so called Wannier-Stark superlattices have energy eigenvalues that have been exactly quantised by electric field F to be $E_0 + neF\lambda$, where λ is the superlattice period and n is a superlattice index. The index n represents the number of intervening superlattice periods between where a hole and an electron are confined. E_0 is an energy eigenvalue when $F = 0$. In superlattices the energy eigenvalues shift linearly with applied field and consequently differ from the quadratic shifts (QCSE) characteristic of MQW structures.

FIGURE 2(a) shows electroabsorption spectra at 10 K from an $Al_{0.30}Ga_{0.70}As/GaAs$ superlattice with a period length of 110 Å and barrier width of 15 Å [16,20]. Absorption in this system can be organised into three distinct types of interwell coupling regimes. At low fields, i.e. <(+1.2 V), optical transitions have superlattice characteristics since only broad light and heavy hole features are observed. Because this bias is close to flat band conditions (+1.4 V), the energy levels of individual quantum wells are all in resonant alignment causing electron delocalisation over many superlattice periods. At intermediate fields, the absorption spectra show multiple peaks corresponding to Stark ladder

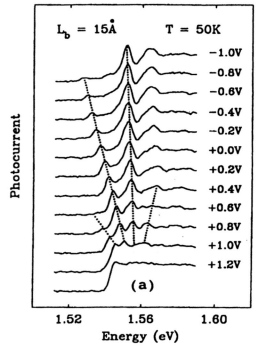

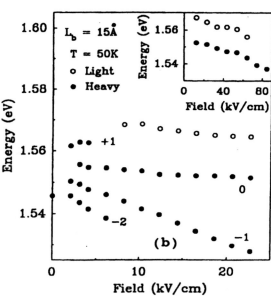

FIGURE 2. (a) Field dependent photocurrent from a GaAs/AlGaAs superlattice. Dotted lines are optical transitions from different Stark ladder transitions as they detune during bias. (b) Fan out of the Stark ladder transitions in field.

formation. Application of the field has partly broken the resonant alignment of states within the superlattice miniband because individual energy levels have been misaligned by $e\lambda F$. The destruction of resonant alignment causes electrons to become partially localised in individual quantum wells. In this regime, peaks of specific index are measured to shift linearly with applied field as shown in FIGURE 2(b). For different indices, n, the energy eigenstates increasingly separate with applied field to produce the 'fan out' states that demonstrate the Stark ladder. Here also, as applied field increases the '0th' order line of the superlattice blue shifts with increasing field by about half the miniband width of the superlattice. At high electric field (-1.0 V), the spectrum becomes characteristic of excitons isolated within individual QWs of the superlattice.

The coherence length of carriers in a superlattice is defined to be $\Delta/(eE\lambda)$ times the superlattice period, with Δ being the miniband width of electrons. In superlattices within the GaAs/AlGaAs system the band width of holes is several times smaller than that of electrons and consequently holes are confined within a single quantum well during an optical excitation. At low temperature many Stark ladder states can be observed (as high as n = 8) and imply a coherence length for electrons to be 750 Å [17]. Superlattice formation has also been reported in a variety of materials: InGaAs/InP [28], InGaAs/AlInAs on InP [29], InGaAs/GaAs [30], and HgTe/CdTe [31].

D2 Destruction of Wannier-Stark Superlattices by Exciton Formation

Excitonic effects can modify the optical spectra of semiconductor superlattices even within the Wannier-Stark regime [20,24]. Consider a superlattice that consists of a 95 Å GaAs quantum well and a 35 Å AlGaAs barrier. The electron miniband width here is 3 meV which is smaller than the excitonic binding energy of 7 meV. Photocurrent spectra at 20 K from this specimen are shown in FIGURE 3(a) versus applied bias. It exhibits coupled well behaviour as evidenced by the antilevel crossing of two distinct optical transitions versus field shown in FIGURE 3(b). This anticrossing involves two transitions between a heavy hole and either an electron state in the same QW (1) or an electron in the nearest neighbour QW (2). Fox et al [20] have shown that

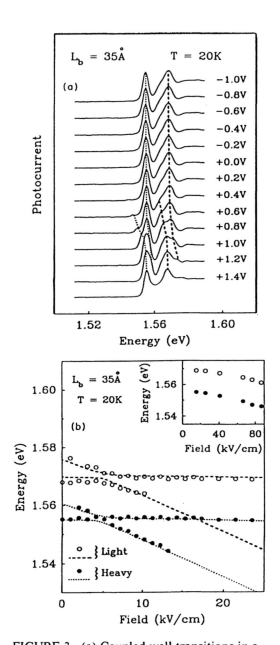

FIGURE 3. (a) Coupled well transitions in a GaAs/AlGaAs superlattice with the dashed line being the spatially direct optical transition and the dotted line being spatially indirect transitions. (b) Detuning of the direct/indirect transitions in field revealing anticrossing effects because of excitonic effects in superlattices.

in samples where the exciton binding energies are comparable to or larger than the miniband widths the Stark ladder is no longer observed even though evidence of substantial electron-hole overlap from adjacent wells can be detected. The degree to which excitonic effects optically alter the Wannier-Stark effect is determined by the relative magnitude of the electron-hole Coulomb interaction and the miniband width of the superlattice. Whichever of the two energies is greatest determines whether the system adopts CW or superlattice behaviour. This behaviour is further supported by various types of theoretical calculations [25].

D3   Shallow Quantum Well Superlattices

Even more rapid carrier escape can occur in shallow quantum well superlattices. In shallow well structures the barriers are only kT in height (room temperature) which is just sufficient to confine holes within the GaAs well [11-13]. The Coulomb field then binds electrons to spatially localised holes so that excitons form. The exciton is unstable in the field and rapidly ionises producing moderately good on/off contrast in absorption at the exciton wavelength.

To compare different modulator designs at 850 nm we note that the absolute absorption strength, when normalised to the entire intrinsic thickness, is $2.2 \, \mu m^{-1}$ for deep QWs (x = 1.0), $1.6 \, \mu m^{-1}$ for conventional modulators like those in FIGURE 1, $1.4 \, \mu m^{-1}$ for shallow QW structures and $1.0 \, \mu m^{-1}$ for superlattices.

E   LONG WAVELENGTH MQW MODULATORS

E1   1.55 μm Modulators

Long wavelengths, such as 1.55 μm, are now of special interest particularly for fibre access applications [24-30]. Several applications exist here, namely integrated laser modulator devices for long/short haul communications [31] as well as reflection modulators for fibre-to-the-home systems [32]. We focus on the latter application whereby large arrays of devices can hold down unit cost. FIGURE 4 shows the reflectivity from a device structure consisting of 1.5 μm n-type InP followed by a 200 period MQW region composed of 100 Å $In_{0.53}Ga_{0.47}As$ QWs with 80 Å InP barriers, followed by a 1 μm thick p-type InP grown on InP. Reflection was accomplished using a gold electrode deposited on the p-type InP surface with light being incident through the transparent substrate. The HWHM of the zero applied bias excitonic peak is 6.2 meV and the whole spectrum shows a good quantum confined Stark effect (QCSE), with the peak shifting by more than 50 nm without significant broadening. Unfortunately the measured absorption strength at the exciton is only 40% of that for an 850 nm MQW ($2.2 \, \mu m^{-1}$ at x = 1.0). Although an increased MQW thickness here has enabled this device to have an 8:1 contrast ratio, it does require high bias voltage (40 - 60 V) operation which is difficult to drive at high frequencies.

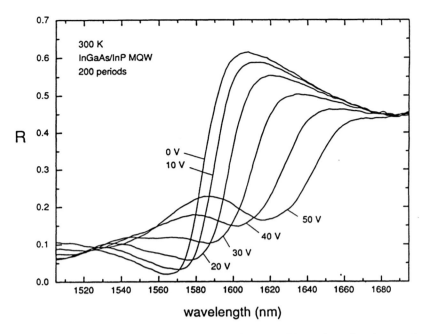

FIGURE 4. Field dependent MQW absorption at 1.55 μm in reflection mode.

E2   1.06 μm Modulators

GaAs substrate based modulators operating at 980 nm and 1.06 μm are particularly important for photonic switching since large optical powers then become available to drive arrays of MQW devices [33-35]. For comparison, a Nd:YAG laser (1.06 μm) has 30 times the output optical power of a semiconductor diode laser. These MQW devices add system flexibility because the substrate is transparent making both the front and back surfaces optically accessible. When strain balanced MQWs are used their morphologies are defect free so that continued device integration with transistors becomes an option. Furthermore, their low defect densities are advantageous over non-pseudomorphic MQW materials that contain dislocation arrays which scatter light and induce cross talk in optical channels.

1.06 μm strain balanced MQW modulators can be made on GaAs by growing compressively strained InGaAs QWs 90 Å thick (x = 0.28) with a tensile strained GaAsP barrier 65 Å thick (y = 0.53) to balance the strain. Field dependent absorption from this structure is shown in FIGURE 5. Absolute absorption strength at 1.06 μm is 60% of that in an 850 nm MQW (2.2 μm^{-1} for x = 1.0) which has been attributed to inhomogeneous broadening of the exciton line in these materials. The HWHM here is 11 meV compared to 4.5 meV for 850 nm MQWs. Nevertheless, integrated absorption from 850 nm to 1060 nm is constant as FIGURE 6 shows. In strain balanced MQWs the in-plane electron mass is enhanced by strain (30%) which preserves both excitonic binding energy (8.3 meV) and integrated excitonic oscillator strength. Furthermore, strain balanced materials show an improved capacity to optically modulate light signals through the quantum confined Stark effect compared to 850 nm MQWs. This is attributed to a light-heavy hole splitting of 140 meV that is caused by a large component of shear strain in these materials.

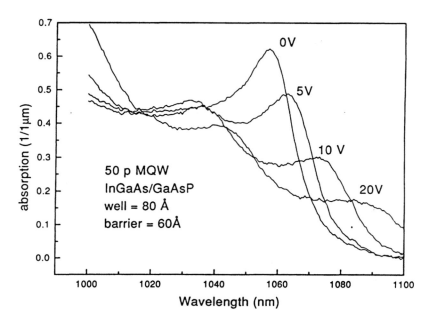

FIGURE 5. Field dependent absorption at 1.06 μm in a strain balanced MQW that consists of 50 p of an InGaAs (x = 0.29) QW and a GaAsP (y = 0.67) barrier.

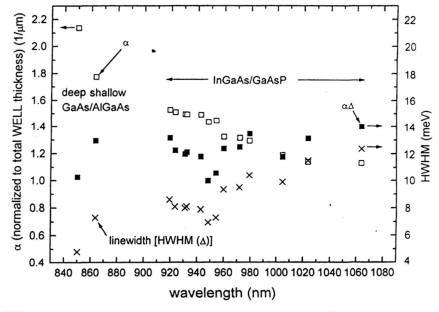

FIGURE 6. A comparison of linewidth (cross), absorption coefficient at exciton position (open square) and integrated oscillator strength (solid square) versus wavelength.

F   CONCLUSION

We have reviewed the current state of the art in electroabsorption in MQWs in a variety of materials for special applications. At 850 nm MQW designs can be successfully engineered quantum mechanically. Such designs also yield a physical understanding of solid state superlattices, Bloch oscillations and 2D-3D exciton transitions. Modulators also can be made at alternative wavelengths but still need more research to yield performance levels comparable to those of the 850 nm system.

REFERENCES

[1] H. Haug, S.W. Koch [*Quantum Theory of the Optical Properties of Semiconductors* (World Scientific Publishing Co., 1990)]

[2] D.A.B. Miller, D.S. Chelma, S. Schmitt Rink [*Optical Instabilities in Semiconductors* Ed. H. Haug (Academic Press, 1988) ch.13 p.325-59]

[3] D.S. Chelma, D.A.B. Miler, S. Schmitt Rink [*Optical Instabilities in Semiconductors* Ed. H. Haug (Academic Press, 1988) ch.4 p.83-120]

[4] H.S. Hinton, D.A.B. Miller [*AT&T Tech. J. (USA)* vol.71 (1992) p.82-93]

[5] D.A.B. Miller [*Opt. Lett. (USA)* vol.14 (1989) p.146]

[6] N. Streibl et al [*Proc. IEEE (USA)* vol.77 (1989) p.1954]

[7] A. Ourmazd, D.W. Taylor, J. Cunningham, C.W. Tu [*Phys. Rev. Lett. (USA)* vol.62 (1989) p.933-7]

[8] A.M. Fox, D.A.B. Miller, G. Livescu, J.E. Cunningham, W.Y. Jan [*IEEE J. Quantum Electron. (USA)* vol.27 (1991) p.2281]

[9] G. Bastard, E.E. Mendez, L.L. Chang, L. Esaki [*Phys. Rev. B (USA)* vol.28 (1982) p.3241-5]

[10] A.M. Fox, R.G. Ispasoiu, C.T. Foxon, J.E. Cunningham, W.Y. Jan [*Appl. Phys. Lett. (USA)* vol.63 (1993) p.2917-30]

[11] K.W. Goossen, J.E. Cunningham, W. Jan [*Appl. Phys. Lett. (USA)* vol.59 (1991) p.3622-3]

[12] K.W. Goossen, L.M.F. Chirovsky, R.A. Morgan, J.E. Cunningham, W. Jan [*Photonics Technol. Lett. (USA)* vol.3 (1991) p.448-51]

[13] K.W. Goossen, J.E. Cunningham, M.D. Williams, F.G. Storz, W.Y. Jan [*Phys. Rev. B (USA)* vol.45 (1992) p.13773-7]

[14] J. Feldmann et al [*Phys. Rev. B (USA)* vol.46 (1992) p.7252-6]

[15] H.G. Roskos et al [*Phys. Rev. Lett. (USA)* vol.68 (1992) p.2216-20]

[16] E.E. Mendez, G. Bastard [in *Phys. Today (USA)* (June 1994) p.34]

[17] J. Bleuse, G. Bastard, P. Voisin [*Phys. Rev. Lett. (USA)* vol.60 (1988) p.220-4]

[18] G.H. Wannier [*Elements of Solid State Theory* (Cambridge University Press, Cambridge, England, 1959)]

[19] G. Bastard [*Acta Electron. (France)* vol.25 (1983) p.147]

[20] A.M. Fox, D.A.B. Miller, J.E. Cunningham, W.Y. Jan, C.Y.P. Chao, S.L. Chuang [*Phys. Rev. B (USA)* vol.46 (1991) p.15365]

[21] J. Bleuse, P. Voisin, M. Allovon, M. Quillec [*Appl. Phys. Lett. (USA)* vol.53 (1988) p.2632-5]

[22] P.E. Bigan, M. Allovon, M. Carre, P. Voisin [*Appl. Phys. Lett. (USA)* vol.57 (1990) p.327-30]

[23] E. Riberiro, F. Cereira, ?. Roth [*Phys. Rev. B (USA)* vol.46 (1992) p.12542]

[24] I. Lo, L. Mitch, K.A. Harris [*Appl. Phys. Lett. (USA)* vol.62 (1993) p.1533-6]

[25] A.M. Fox, D.A.B. Miller, G. Livescu, J.E. Cunningham, W.Y. Jan [*Phys. Rev. B (USA)* vol.44 (1991) p.6231]

[26] M.M. Dignam, J.E. Snipe [*Phys. Rev. B (USA)* vol.43 (1991) p.4097]; A. Chomette, B. Lambert, B. Deveaud, F. Clerot, A. Regreny, G. Bastard [*Europhys. Lett. (Switzerland)* vol.4 (1987) p.461]

[27] D.R.P. Guy, D.D. Besgrove, L.L. Taylor, N. Apsley, S.J. Bass [*Proc. IEEE (USA)* vol.136 (1989) p.46]

[28] S. Goswami, P.K. Bhattacharya, J. Cowles [*IEEE Trans. Electron Devices (USA)* vol.40 no.11 (1993) p.1970-6]

[29] R.N. Pathek, K.W. Goossen, J.E. Cunningham, W.Y. Jan [*Photonics Technol. Lett. (USA)* vol.6 no.12 (1994) p.1439-41]

[30] T.L. Koch, U. Koren [*AT&T Tech. J. (USA)* vol.71 no.1 (1992) p.63-74]

[31] T. Wood, R.A. Linke, B.L. Casper, E.C. Carr [*J. Lightwave Technol. (USA)* vol.6 (1988) p.346]

[32]   N.J. Frigo et al [OFC post deadline session (USA) (1994) p.43]

[33]   J.E. Cunningham, K.W. Goossen, M. Williams, W.Y. Jan [*Appl. Phys. Lett. (USA)* vol.60 (1992) p.727-30]

[34]   K.W. Goossen, J.E. Cunningham, W.Y. Jan [*Electron. Lett. (UK)* vol.19 (1992) p.1833-4]

[35]   K.W. Goossen, J.E. Cunningham, W.Y. Jan [*Photonics Technol. Lett. (USA)* vol.5 (1993) p.1392-5]

6.5 Electro-optic effects in III-V quantum wells and superlattices

P.K. Bhattacharya

April 1996

A   INTRODUCTION

For electric fields perpendicular to the quantum well layers, the optical absorption near the bandgap energy can be shifted to lower photon energies without destroying the strong excitonic features in the absorption spectrum. The effect, known as the quantum-confined Stark effect (QCSE) [1], is of great practical interest for absorption modulators and switches. A large refractive index variation, based on QCSE, was predicted in quantum well materials [2], which makes them extremely attractive for electro-optic devices such as directional couplers and Mach-Zehnder interferometers. It is this possibility that has spurred theoretical and experimental work in this field.

B   THEORETICAL CALCULATIONS

The electro-optic properties of quantum wells are very different from those of single-layered materials. In a quantum well the valence band degeneracy is removed and this causes changes in the optical absorption. With the application of a transverse electric field, excitonic transitions do not disappear as in bulk materials but persist up to high fields. There are also changes in the energy eigenvalues of the quantum well and the coupling of electrons and holes in the subband with the same quantum number is reduced. Theoretically, it has been shown, by considering only band-to-band and excitonic transitions, that there is a large refractive index change for the TE mode in a multiple quantum well (MQW) structure, resulting in a large electro-optic coefficient [1,3-5].

Many device applications at the present time necessitate the use of pseudomorphic strained quantum wells. Electro-optic devices have also been demonstrated with such material. Biaxial strain present in the quantum wells changes the electronic and optical properties and has to be properly accounted for. To model the optical properties of semiconductor quantum wells it is important to include the strong coupling between the heavy hole

$$\left(\text{HH}, \ |\tfrac{3}{2} \pm \tfrac{3}{2} > \right)$$

and light hole

$$\left(\text{LH}, \ |\tfrac{3}{2} \pm \tfrac{1}{2} > \right)$$

states; the conduction band may be treated in the parabolic approximation. The valence band coupling is accounted for by using the 4×4 k.p Hamiltonian. The effect of strain is incorporated by including a splitting δ between the light and heavy hole diagonal elements. For example, for $In_xGa_{1-x}As$ grown on GaAs and $In_{0.53+x}Ga_{0.47-x}As$ grown on InP it is given (in eV) by $\delta = -5.966\varepsilon$, where the lattice mismatch ε is related to the excess In composition x by $\varepsilon = -0.07x$. This splitting reduces the off-diagonal mixing between the HH and LH states and changes the effective masses of the holes. In addition, the strain-induced splitting dramatically changes the difference between the light and heavy hole subband energies. The Hamiltonian is solved in the presence of strain and electric fields. We also solve for the electron states in the presence of fields and compute the transition energies. The electric field shifts both the light and heavy hole transition energies quadratically. In addition, the strain strongly affects the zero-field transition energies for heavy and light holes. In the lattice matched case the light hole bandgap is slightly larger than that of the heavy hole because of the larger quantisation energy associated with its lighter mass. The application of compressive strain, though, greatly increases this separation; applying tensile strain reduces the light hole energy below that of the heavy hole. The splitting strongly influences the optical properties because of the polarisation dependence of the selection rules applying to optical transitions. TE mode ($\hat{x}$ polarised) light couples three times more strongly to HH states than to LH states and TM mode ($\hat{z}$ polarised) light couples only to the LH state; growth is in the z direction. Hence, the separation between HH and LH states will strongly affect the absorption spectra and the associated birefringence. After the band structure has been obtained, the excitonic binding energies and oscillator strengths are calculated.

To determine the effect of strain and electric fields on the refractive index, the Kramers-Kronig transform is applied to the absorption spectra, yielding the result

$$n(\omega_0) - 1 = \frac{c}{\pi}P \int_0^\infty \frac{\alpha(\omega)d\omega}{\omega^2 - \omega_0^2} \tag{1}$$

where ω_0 is the frequency of the incident light. The excitonic effects dominate the electro-optic effect when the input light energy is close to the bandgap.

C   EXPERIMENTAL RESULTS

As with bulk materials, phase modulation measurements are made with single mode optical waveguides. In this case, the quantum wells are incorporated in the guiding layer. The waveguide is excited with light which has a polarisation oriented at 45° to the direction of the applied electric field to equally excite both TE and TM modes in the guide. The phase shift between these modes, as the wave propagates through the guide of length l, is given by

$$\Delta\phi = \frac{\pi l}{\lambda} n_0^3 \left[r_{41}^l E_z + (s_{12}^q - s_{11}^q)E_z^2 \right] \tag{2}$$

Here s_{11}^q and s_{12}^q are the quadratic electro-optic coefficients and r_{41}^l is the linear electro-optic coefficient describing the index ellipsoid for the quadratic effect and the linear effect, respectively.

There is another important difference between bulk and quantum well materials. In bulk semiconductors there is no anisotropy and therefore there is no built-in birefringence. The layering of multiple materials with alternating indices of refraction will induce a birefringence with an extraordinary index of refraction, which is perpendicular to the layer plane. The mean refractive indices

$$n_{\parallel}^2 = \frac{d_1 n_1^2 + d_2 n_2^2}{d_1 + d_2} \tag{3}$$

and

$$n_{\perp}^2 = \frac{n_1^2 n_2^2 (d_1 + d_2)}{d_1 n_2^2 + d_2 n_1^2} \tag{4}$$

can be obtained by examining the boundary value problem for plane waves, which relates to the polarisation of the optical electric field vector parallel or perpendicular to the quantum well. In EQNS (3) and (4), n_1 and d_1 refer to the refractive index of the well material and the thickness of the well, respectively. Likewise, n_2 and d_2 refer to the index of refraction of the barrier material and the barrier thickness, respectively. For a multiple quantum well, d_1 is replaced by $N_z d_1$, where N_z is the number of wells. Similarly d_2 is replaced by $N_b d_2$, where N_b is the number of barriers. The built-in birefringence in the MQW can be expressed as

$$\Delta \phi_0 = (\beta_{TE} - \beta_{TM}) \, l = \Delta n l \left(\frac{2\pi}{\lambda} \right) \tag{5}$$

where

$$\Delta n = n_{\parallel} - n_{\perp} \tag{6}$$

The total phase shift in an MQW can therefore be expressed as

$$\Delta \phi = \Delta n l \left(\frac{2\pi}{\lambda} \right) + \frac{\pi l}{\lambda} n_0^3 \left[\mp r_{41}^l E_z + (s_{12}^q - s_{11}^q) E_z^2 \right] \tag{7}$$

The expected strong quadratic electro-optic effect, for photon energies close to the HH excitonic resonance, has been absorbed for lattice-matched and strained GaAs/AlGaAs, InGaAs/GaAs, and InGaAs/InGaAsP/InP MQW systems [5-11]. Typical values of the linear and quadratic electro-optic coefficients in compressively strained $In_xGa_{1-x}As/Al_{0.2}Ga_{0.8}As$ MQWs and compressively and tensilely strained $In_{0.53+x}Ga_{0.47-x}As/InGaAsP$ MQWs are shown in TABLES 1 and 2, respectively.

TABLE 1. Measured linear and quadratic electro-optic coefficients in $In_xGa_{1-x}As/Al_{0.2}Ga_{0.8}As$ strained MQWs.

x	λ (μm)	ΔE (meV)	r_{41}^l (10^{-12} m/V)	$(s_{12}^q - s_{11}^q)$ (10^{-20} m^2/V^2)
0	0.8965	92	1.08	3.48
0.03	0.9215	100	1.09	9.09
0.06	0.9345	89	1.10	15.54
0.09	0.9460	77	1.29	34.40

TABLE 2. Measured linear and quadratic electro-optic coefficients in $In_{0.53+x}Ga_{0.47-x}As/InGaAsP$ strained MQWs.

x	λ (μm)	r^l_{41} (10^{-12} m/V)	$(s^q_{12} - s^q_{11})$ (10^{-20} m^2/V^2)
-0.08	1.584	2.83	146
0	1.65	3.74	282
0.07	1.67	4.10	389

Based on the enhanced electro-optic effect in MQWs, efficient low-voltage phase modulators, directional coupler switches and Mach-Zehnder interferometers have been demonstrated for operation with 0.8, 1.3 and 1.55 μm photoexcitation [12-15].

D ELECTRO-OPTIC EFFECTS IN SUPERLATTICES AND ASYMMETRIC QUANTUM WELLS

There is one report on the electro-optic properties of monolayer superlattices [16]. The linear electro-optic coefficient can be large in asymmetric (either coupled or triangular wells) quantum wells due to the expected large linear change of the excitonic transitions with electric field. Theoretical calculations and experimental results on devices based on asymmetric wells have been reported by several authors [17-20].

E CONCLUSION

The electro-optic properties of quantum wells and superlattices are very different from those in bulk semiconductors, due to the QCSE. This is a relatively new area of development in the field of multilayered materials. The enhanced electro-optic effect in these materials can lead to novel and efficient integrated-optical devices.

REFERENCES

[1]   H. Yamamoto, M. Asada, Y. Suematsu [*Electron. Lett. (UK)* vol.21 (1985) p.579]
[2]   D.A.B. Miller et al [*Phys. Rev. Lett. (USA)* vol.52 (1984) p.2173]
[3]   J.S. Weiner, D.A.B. Miller, D.S. Chelma [*Appl. Phys. Lett. (USA)* vol.50 (1987) p.842]
[4]   T. Hiroshima [*Appl. Phys. Lett. (USA)* vol.50 (1987) p.968]
[5]   J. Pamulapati, J.P. Loehr, J. Singh, P.K. Bhattacharya, M.J. Ludowise [*J. Appl. Phys. (USA)* vol.69 (1991) p.4071]
[6]   M. Glick, F. Reinhart, G. Weimann, W. Schlapp [*Appl. Phys. Lett. (USA)* vol.48 (1986) p.989]
[7]   M. Glick, D. Pavuna, F.K. Reinhart [*Electron. Lett. (UK)* vol.23 (1987) p.1235]
[8]   H. Nagai, M. Yamanishi, Y. Kan, I. Suemune [*Electron. Lett. (UK)* vol.22 (1986) p.888]
[9]   J.E. Zucker, T.L. Hendrickson, C.A. Burrus [*Appl. Phys. Lett. (USA)* vol.52 (1988) p.945]
[10]   U. Das, Y. Chen, P.K. Bhattacharya, P.R. Berger [*Appl. Phys. Lett. (USA)* vol.53 (1988) p.2129]
[11]   J.E. Zucker, I. Bar-Joseph, B.I. Miller, U. Koren, D.S. Chelma [*Appl. Phys. Lett. (USA)* vol.54 (1989) p.10]
[12]   J.E. Zucker, T.L. Hendrickson, C.A. Burrus [*Electron. Lett. (UK)* vol.24 (1988) p.112]

[13]   U. Koren, T.L. Koch, H. Presting, B.I. Miller [*Appl. Phys. Lett. (USA)* vol.50 (1978) p.368]

[14]   J.E. Zucker, K.L. Jones, M.G. Young, B.I. Miller, U. Koren [*Appl. Phys. Lett. (USA)* vol.55 (1989) p.2280]

[15]   J.E. Zucker, K.L. Jones, B.I. Miller, U. Koren [*IEEE Photonics Technol. Lett. (USA)* vol.2 (1990) p.32]

[16]   J.P. van der Ziel, A.C. Gossard [*J. Appl. Phys. (USA)* vol.48 (1977) p.3018]

[17]   L. Friedman, R.A. Soref [*Electron. Lett. (UK)* vol.22 (1986) p.819]

[18]   K.W. Steijn, R.P. Leavitt, J.W. Little [*Appl. Phys. Lett. (USA)* vol.55 (1989) p.383]

[19]   N. Susa [*IEEE J. Quantum Electron. (USA)* vol.31 (1995) p.92]

[20]   C. Thirstrup [*IEEE J. Quantum Electron. (USA)* vol.31 (1995) p.988]

6.6 Resonant nonlinear optical interactions in undoped GaAs quantum wells

A.C. Schaefer and D.G. Steel

January 1996

A   INTRODUCTION

The interest in direct bandgap semiconductors such as GaAs has experienced steady growth for over two decades due to the strong optically allowed transition near the band edge and because of the potential of these materials for use in the development of new devices and their direct integration with electronics.   The enhanced capability made possible by bandgap engineering through advanced material growth and processing, specifically the design of layered structures, has made these systems extremely useful in optoelectronics.

As in bulk material, the optical properties of heterostructures near the band edge are dominated by excitonic effects, which lead to distinct optical resonances and a modification of the band absorption.   However, the importance of excitonic effects is greatly enhanced in heterostructures, where quantum confinement increases the excitonic binding energy and oscillator strength.   This results in the observation of strong excitonic resonances even at room temperature in GaAs quantum wells, unlike in bulk material.

The presence of enhanced resonant features in quantum wells leads to the potential for observing strong optical nonlinearities.   Specifically, nonresonant nonlinear optical behaviour requires optical fields comparable to the Coulomb field experienced by the orbiting electron. However, as with any high Q resonator, the effective field can be enhanced by orders of magnitude when resonantly coupled.   Highly nonlinear optical behaviour can be observed in resonant systems with peak intensities as low as a few mW/cm^2.   For resonant behaviour, however, the response time is limited by the decay time of the excitation since the nonlinearities fundamentally arise from excitation-induced changes in the absorption and refraction of the material.

Interest in these systems continues, not only because of the potential for applications, but also because of the increasing effort devoted to fabricating systems with even greater confinement, such as quantum wires and quantum dots/boxes.   In addition, GaAs heterostructures remain important for fundamental physics research, as model systems for any direct gap semiconductor.   This is attested to by the relatively large foundation of understanding which has already been laid in the study of these structures.

Studies of the basic physics of the nonlinear optical response in these systems showed that the problem was considerably more complex than originally conceived, and it remains a subject of current and highly active study by many groups.   The complexity arises because the excitonic nonlinear optical response is intrinsically a many-body problem, and hence the simple pictures that were developed to describe the coherent nonlinear optical behaviour in isolated atomic systems were inadequate to describe the problem where there are strong interactions between the different resonant excitations.   The presence of interface disorder also complicates the picture. Hence, considerable effort by numerous groups was needed to develop the experimental and theoretical basis for describing nonlinear optical behaviour in semiconductors (see for example [1-5]). It should be stressed, however, that the formalism developed based on the optical Bloch equations, used so successfully in atomic and nuclear magnetic resonance (NMR) studies, is useful in providing physical insight, with proper modifications to account for the interactions (see for example [6,7]). More generally, because of the understanding of the optical Bloch equations, the development of semiconductor Bloch equations, based on the many-body interactions of the system, has greatly facilitated progress in this area [5,8,9].

In this Datareview, the current understanding of the resonant nonlinear optical response in GaAs heterostructures is summarised. Given the vast amount of research, it is impossible to mention all of the work done in this field, and therefore only the physical interactions with the most dominant effects on the optical nonlinearities will be discussed. This Datareview is divided into sections describing room temperature and low temperature studies of the resonant nonlinear response, as well as a brief section, for completeness, discussing the nonlinear response for near-resonant excitation. There will be no discussion of nonlinear magneto-optical effects. In the final section, some of the future directions of the field will be reviewed, most notably the growing interest in studying the physics of interface disorder in heterostructures on the nanometer length scale.

B   RESONANT NONLINEAR OPTICAL EFFECTS IN GaAs QUANTUM WELLS AT ROOM TEMPERATURE

In GaAs, interactions with LO phonons ionise the exciton within a picosecond [10]. This means that in bulk material the homogeneous linewidth of the exciton is greater than the binding energy, and thus the excitonic resonance is not well resolved from the band edge. In quantum wells, however, the confinement enhances the binding energy of the exciton, while for layer thicknesses greater than 100 Å, the electron-phonon interaction remains approximately the same as in bulk [10]. Therefore, sharp excitonic resonances can be resolved in quantum wells, even at room temperature, as shown in FIGURE 1.

At room temperature, nonlinear optical interactions in quantum wells at the excitonic resonance are characterised by the ionisation time of the exciton, which is approximately 300 fs [11]: before ionisation, excitonic interactions are found to dominate the nonlinear response in quantum wells, while on longer time scales the response is determined by free carrier effects. The strong nonlinearities are based on dramatic excitation-induced changes in the excitonic resonance, as shown in FIGURE 1. In this figure, it is seen that at carrier densities of approximately 10^{12} cm^{-2} (intensities of about 800 W/cm^2 with CW excitation), the excitonic absorption line can be completely saturated. Through such dramatic changes in the exciton line, changes in absorption coefficient and refractive index per e-h pair per cm^3 of 7×10^{-14} cm^2 and 3.7×10^{-19} cm^3, respectively, can be realised [12].

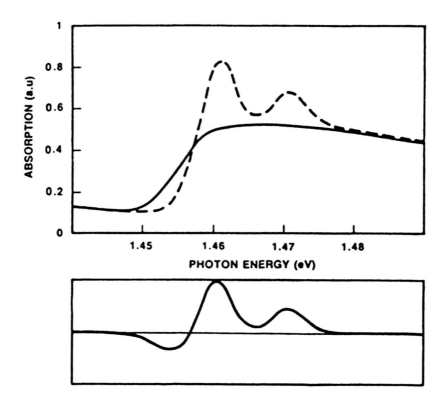

FIGURE 1. Band edge absorption spectrum of a GaAs multiple quantum well structure with (solid line) and without (dashed line) continuous excitation 32 meV above the bandgap. Note the saturation of the exciton resonances at the excitation density of 10^{11} cm^{-2}. Bottom figure: differential transmission spectrum. (From [12], used with permission.)

B1   Nonlinear Optical Response before Ionisation: Excitonic Interactions

Optical generation of excitons at room temperature in GaAs quantum wells leads to strong nonlinearities through a saturation of the excitonic absorption [11-17], shifts in the resonant energy [14,17], and broadening of the excitonic line [11,14,15]. The saturation of the excitonic absorption, which may be seen as a decrease in effective oscillator strength, is caused by a combination of screening of the Coulomb attraction between the bound e-h pairs, phase-space filling (band filling, or blocking of already filled states), and exchange effects. (The last two are the results of the Pauli exclusion principle.) In contrast to bulk GaAs, the effects of screening are greatly reduced in 2D structures, and excitonic effects such as phase-space filling and exchange are dominant sources of the saturation [10,15]. This was most clearly shown in the experiments of Knox et al [11,15], which demonstrated that the creation of excitons leads to twice as efficient bleaching of the excitonic absorption as free carriers, even though free carriers screen the Coulomb field more effectively than excitons [18]. When the excitons ionise within 300 fs, creating free carriers, the strength of the nonlinear response decreases dramatically to the response generated by direct excitation of free carriers [11].

An excitation-induced increase in the exciton resonant energy of a few meV also contributes to the nonlinear response [17,19,20]. Shifts of up to 6 meV (at an estimated density of 10^{11} cm^{-2}) [20] have been observed and were attributed in part to the excitonic exchange interaction, since no shifts were observed when free carriers were created. The magnitude of the shift was found to increase with decreasing well width [20]. A theory including exciton exchange interactions and the reduced Coulomb screening in two dimensions predicts an energy shift which is linear in excitation density and given by $\Delta E \approx 12na_{2D}^2E_{1S}$, where n is the exciton density, $a_{2D} = a_0/4$ is the 2D exciton radius and E_{1S} is the 2D binding energy [1]. In addition, the broadening of the excitonic line was attributed to exciton-exciton interactions and screening [21].

Four-wave-mixing (FWM) experiments at room temperature also verify the importance of these different excitonic effects as the source of the nonlinear response. Early work by Miller et al at low densities (≈ 30 W/cm^2 average intensity) showed that in FWM a diffraction efficiency of 10^{-4} could be achieved and arose from nonlinear absorption and refraction due to saturation, line broadening, and an energy shift [14]. Early attempts to model the nonlinear response, however, were based on saturation of the excitonic absorption (which again is due to phase space filling and screening, the latter being a lesser effect in quantum wells) as the dominant source of nonlinearity [16]. Further studies of the temporal evolution of FWM signals [22,23] showed behaviour which could not be accounted for by simple level saturation and indicated the importance of including excitonic interactions beyond phase space filling in the calculation of the nonlinear polarisation. The instantaneous frequency dynamics of transient FWM signals at low excitation densities also showed behaviour not consistent with phase space filling (the spectral distribution of the FWM signal reflected the excitonic absorption spectra and not that of the exciting laser pulse), though at higher densities ($\approx 10^{11}$ cm^{-2}) the response did become consistent with that expected for phase space filling [24]. This was interpreted as indicating the importance of Coulomb interactions among excitons in determining the nonlinear response at lower densities. The effects of excitonic interactions on the FWM signal have been studied extensively at low temperature, and will be discussed in more detail in Section C.

B2   Nonlinear Optical Response after Ionisation: Free Carrier Effects

Absorption saturation of the exciton is also seen on long time scales (>1 ps); its source, however, is band filling and screening of the Coulomb attraction due to free carriers. As mentioned previously, the nonlinear response from free carriers is not as strong as that generated from direct excitonic excitation. The nonlinear response determined by free carriers is independent of how the free carriers were generated (either directly or by excitonic ionisation) [15]. Excitonic line broadening is also observed due to free carrier screening [21,25].

Direct excitation in the band will lead to absorption saturation or hole burning in this spectral regime. The hole burning has a lifetime of ≈ 200 fs, determined by the carrier thermalisation time [15].

B3   Large Optical Nonlinearities and Fast Time Scales: Device Applications

The nonlinearities induced by optical excitation have had a major impact on the fabrication of optoelectronic devices. A typical resonant nonlinear susceptibility measured via four-wave-mixing efficiency with an intensity of only 13 W/cm^2 (diode laser source) has a value of $\chi^{(3)} = 6 \times 10^{-2}$ esu, one of the largest ever measured for a semiconductor [26]. The timescales of these nonlinearities are fast, based on the carrier thermalisation time (200 fs) [15], the exciton ionisation time (300 fs) [11] and the recombination time (5 ns) [27]. These nonlinearities have been used to mode lock semiconductor diodes (with production of pulses in the ps regime) [28]. More recently, quantum well nonlinearities have been used to mode lock solid state lasers, such as a Ti:sapphire laser with 2 ps pulses [29,30] and an NaCl colour centre laser with 275 fs pulses; this fast response was based on the excitonic ionisation time [31]. Also, utilising the induced absorption or gain created in certain spectral regimes due to the nonlinear interactions, optically bistable devices have been demonstrated (for a review, see [17]).

The rise time of the response in these materials for device applications has been shown to be limited only by the rise time of the pulse; however, unlike nonresonant excitation (which will be discussed in Section D), the ultimate relaxation time for resonant excitation is determined by the recombination time. To reduce this time well below the nanosecond time scale observed in most experiments it is necessary to reduce the recombination time. One obvious approach is to introduce a nonradiative recombination decay channel. While this lowers the quantum efficiency and reduces the long-time nonlinear optical susceptibility (in the CW limit, the nonlinear susceptibility varies linearly with the response time), the transient response time is reduced considerably. This approach was demonstrated by Smith et al [28], where recombination centres were introduced in the sample by proton bombardment, and the recombination time was reduced from 30 ns to 150 ps for a proton dose of 10^{13} cm^{-2}. Finally, even though the fundamental lifetime of carriers is limited, the recent demonstration of coherent optical control of carrier population (where a laser pulse both creates and destroys carriers) promises to lead to effective population lifetimes limited only by the laser pulsewidth [32].

The large resonant nonlinear susceptibility of GaAs quantum wells makes the observation of the coherent nonlinear response easily observable even at the greatly reduced intensities characteristic of CW frequency-domain FWM spectroscopy. However, these studies have shown that the room temperature CW nonlinear response has several contributions with different time scales, including a contribution with a slow decay corresponding to a decay time of 10 μs [27]. This component is shifted in energy, leading to interference patterns in the FWM lineshape. The origin of the slow response remains unclear but appears to be associated with effects of interface disorder. It is surprising that interface disorder would have an effect even at room temperature, when the exciton line is homogeneously broadened due to collisions with LO phonons. Nevertheless, the narrow (~MHz) nearly degenerate CW four-wave-mixing response is of obvious interest for narrowband, tunable (~720 GHz), electro-optic filters [33]. The effects of disorder become clear at low temperatures, as will be discussed in Section C.

C RESONANT NONLINEAR OPTICAL EFFECTS IN GaAs QUANTUM WELLS AT LOW TEMPERATURE

At low temperatures (below about 70 K), the phonon interaction is significantly reduced and the exciton is stable against ionisation in both quantum wells and bulk GaAs. Thus, experiments at low temperature reveal distinct excitonic physics, which is not seen at room temperature due to the fast ionisation rate, and shed light on the excitonic interactions at room temperature before the ionisation time (300 fs). Also, because of the enhancement of the exciton absorption at lower temperatures, the nonlinear susceptibility is also expected to be enhanced over that observed at room temperature.

Since the exciton is stable at low temperature, excitonic phenomena dominate the nonlinear response on all time scales for resonant excitonic excitation (and after ≈ 200 fs for free carrier excitation). As at room temperature, saturation of the exciton absorption [34,35], line broadening [36] and energy shifts (both blue and red) [19,20,36] lead to the observation of strong nonlinearities at low temperature.

Recent studies of the temporal and polarisation properties of four-wave-mixing signals reveal the importance of excitonic interactions and dynamics in determining the nonlinear response. Also, at low temperature the effects of interface disorder, introduced in even the most modern fabrication technology, become important, as evidenced by the presence of inhomogeneous broadening of the exciton line. Despite extensive investigation, the role of all of these effects on the resonant nonlinear optical signal has not been completely resolved and remains the subject of current study.

C1 Temporal and Polarisation Properties of the Resonant Nonlinear Signal at Low Temperature

Originally, the nonlinear response was modelled by the optical Bloch equations for a noninteracting two-level system, where the ground state corresponds to the ground state of the crystal (no excitons), and the upper level is the excitation of the exciton. This interpretation provided the first basis of understanding for CW and transient four-wave-mixing experiments (see for example [27,37]).

In the standard self-diffracted transient FWM geometry, two laser pulses (propagating in the directions k_1 and k_2) interfere in the sample, producing a grating which diffracts photons in the direction $2k_2 - k_1$. For noninteracting two-level systems (with local fields neglected), a perturbative solution of the optical Bloch equations gives very straightforward behaviour for the diffracted signal. In the case of the time-integrated response measured as a function of delay between the two beams ($t_2 - t_1$), the signal exists only for positive delay and decays exponentially with decay constant $T_2/2$ for a homogeneously broadened system, and $T_2/4$ for an inhomogeneously broadened system (where T_2 is the dephasing time). The time-resolved FWM signal will be a free polarisation decay for a homogeneously broadened system, emitted immediately after excitation by the second pulse; for a strongly inhomogeneously broadened system, however, the signal will be a classical photon echo, emitted after a delay given by $\tau = t_2 - t_1$ [38]. This interpretation was used successfully in early studies of relaxation and dephasing in GaAs quantum wells [25,37,39].

Further studies of the temporal evolution of resonant FWM in GaAs quantum wells, however, revealed complicated behaviour which could not be described by the noninteracting two-level model, as discussed in Section B. Observation of the so-called 'negative time delay signal' in time-integrated FWM, emitted for negative $t_2 - t_1$, was interpreted as arising from Coulomb local field effects [8,40,41] (biexcitonic effects may also give rise to this signal, as will be discussed). Time-resolved FWM in GaAs quantum wells with significant inhomogeneous broadening showed not only a delayed signal corresponding to the photon echo expected for an inhomogeneously broadened resonance, but also a prompt signal corresponding to a free polarisation decay and suggesting the contribution of a homogeneously broadened resonance [42]. Even in samples with little inhomogeneous broadening, several temporal maxima were observed [40,43], and were interpreted as arising from exciton-exciton many body interactions [8,44]. The response in some time-resolved measurements was found to be delayed by an amount determined by the dephasing rate (not a photon echo), with delays up to 1.5 ps observed [43]. Excitation-induced phase shifts of hh-lh quantum beats were observed and attributed to density-dependent dephasing and local field effects [45]. The presence of Fano resonances (discrete states coupled to a continuum) can also lead to a pronounced effect on the nonlinear signal, most notably a very fast decay of the time-integrated FWM signal, which has been interpreted as arising from quantum interference [46].

The dependence of the properties of the resonant nonlinear signal on the polarisation of the exciting beams was also found to be very complex. With reference to the non-interacting two-level system, the properties of the nonlinear signal are not expected to depend on the polarisation of the exciting beams. It was found, however, that by changing the relative polarisations of the exciting beams, completely different characteristics could be observed, including different temporal behaviour [43,45,47-50], signal strengths [47], emission spectra [51,52] and dephasing rates [47,49] (the polarisation dependence of the dephasing rates, however, was found to be sample dependent [53]). For example, as shown in FIGURE 2, for time-resolved FWM with cross-linearly polarised incident beams, a free polarisation decay indicative of a homogeneous system was observed, while for co-linearly polarised beams a photon echo, indicating an inhomogeneously broadened resonance, was seen [47]. It was determined that saturation of the photon echo in co-linearly polarised FWM allowed also for

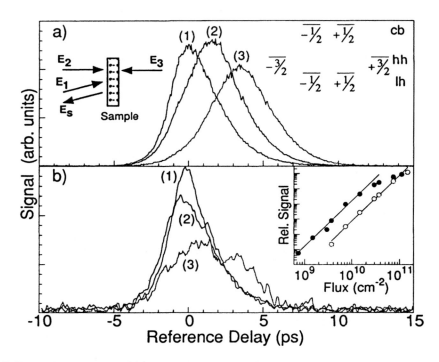

FIGURE 2. Time-resolved FWM in a GaAs multiple quantum well structure for (a) all fields co-linearly polarised and (b) E1 and E2 cross-linearly polarised. In (a), a classical photon echo, with the response delayed by the delay time between pulse 1 and 2: (1) 0 ps, (2) 2 ps and (3) 4 ps. In (b), a free polarisation decay, independent of pulse delay. Left inset is FWM geometry and right inset is magnetic substrate structure of GaAs. (From [47], used with permission.)

the observation of the free polarisation decay (seen at low intensity only in the cross-polarised geometry), giving rise to two temporal maxima in the transient FWM signal, such as that discussed above [43]. The dephasing rate measured in the cross-linearly polarised geometry was found to be up to an order of magnitude larger than that measured with co-linearly polarised beams [47,49].

There have been several proposed interpretations of the unusual polarisation dependence. Excitonic Coulomb interactions can account for the difference in transient behaviour [7,43,45,50,54,55], signal strengths [6,7,56] and measured dephasing rates [55]. Coulomb interactions, leading to an excitation-induced increase in the exciton line width, have been shown to be the dominant source of nonlinearity in bulk GaAs [6,56], and recent studies, most notably a measurement of the Stokes parameters of time-integrated FWM [57], indicate that this effect may be important in quantum wells also. A model considering disorder-induced band mixing [53,58] accounts well for the transient behaviour and differences in the dephasing rate for quantum wells with large inhomogeneous broadening, but does not account for the differences in signal strength. The presence of biexcitons (the bound state of two excitons, with measured binding energies of 1 - 2 meV in GaAs quantum wells) has also been suggested by certain observations, including temporal oscillations in differential transmission for oppositely-handed circularly-polarised pump and probe beams [48], negative time-delay signals [49], quantum beats [59,60], differences in dephasing rates measured for co- and cross-polarised FWM [49] and shifted spectral components in FWM [49,52,61-63].

The previous discussion has focused on sources of the nonlinear signal in GaAs heterostructures deriving from excitation-induced changes in the excitonic properties. It should be noted that since the nonlinear signal itself arises from excitonic features, excitonic dynamics leading to relaxation and dephasing will also result in distinct effects on the nonlinear optical signal. Some examples of this are spectral diffusion due to exciton hopping between localisation sites on the disordered interface leading to the observation of complicated lineshapes in CW FWM spectroscopy [64] and spin relaxation leading to the observation of non-resonantly excited hole-burning [65].

C2   Effects of Interface Roughness

It was recognised in early studies of quantum wells that because of the extreme confinement, excitons and free carriers are sensitive to atomic-scale roughness in the interfaces which arises through the growth process. The presence of islands of monolayer height and a lateral width of about 100 Å was confirmed through chemical lattice mapping [66] and other techniques. This disorder leads to increased scattering of carriers [67] and inhomogeneous broadening of the exciton lines [68], due to spatial potential fluctuations. In very narrow quantum wells, monolayer fluctuations cause a splitting in the excitonic resonance, as observed in luminescence [69]. The effect of disorder on the scattering rates, mobility, energy and lifetime of excitons becomes clear in nonlinear optical studies, especially at low temperature when the exciton is more easily localised.

The seminal work on the nonlinear optical properties of excitons localised by interface disorder was presented by Hegarty and co-workers [68,70,71]. Through the linear response using resonant Rayleigh scattering they showed that the homogeneous linewidth of the excitons increases dramatically as the probing laser is tuned through the inhomogeneously broadened line [68], an effect which they confirmed through hole burning experiments [70].

These measurements and the study of exciton diffusion [71] suggested the presence of an energetic mobility edge, where excitons at energies below this edge were localised in the plane of the layer and those above it were de-localised. Schultheis et al [39] were also the first to observe photon echoes due to inhomogeneously broadened excitons in GaAs quantum wells.

In frequency-domain nonlinear spectroscopy it was found that spectral diffusion, characterised by phonon-assisted migration between localisation sites, gave rise to extra contributions to the four-wave-mixing signal [64,72,73]. Measurements of differential transmission revealed optically induced excitonic absorption, characterised by a slow time scale (≈ 10 μs) and related to the localised excitons [74].

In time-domain nonlinear studies, the presence of interface disorder leads to the observation of photon echoes and other profound effects. Monolayer fluctuations, responsible for the splitting of the excitonic resonance, result in quantum beats in transient four-wave mixing [75,76]. Spectrally resolved four-wave mixing showed specifically the contributions to the transient signal from inhomogeneously broadened resonances [77]. The scale length of the interface roughness (compared to an exciton diameter, a_B) was shown to influence the nonlinear signal considerably, leading to a free polarisation decay in FWM and a corresponding homogeneously broadened spectral line when this scale length is larger than a_B, and a photon echo and inhomogeneous broadening when the scale length is less than a_B [78] (several length scales occur simultaneously in a quantum well structure [66,79,80]). Disorder may also have other impacts on the temporal and polarisation properties of the nonlinear signal, as discussed above.

D   EXCITONIC NONLINEARITIES DUE TO NEAR-RESONANT OPTICAL FIELDS

The previous discussion has focused on resonant excitonic nonlinearities because of the large values of $\chi^{(3)}$ possible through these interactions. A great amount of interest, however, has been focused on nonlinearities at the excitonic resonance induced by near-resonant pumping (for example, see [81]), mostly because of the very fast time scales of these nonlinearities (determined by the pulse width of the optical pump). The enhancement of the response rate for near-resonant excitation was first noted by Grishkowsky et al [82,83] in studies of atomic systems in the adiabatic following limit, in which the electronic excitation follows the temporal evolution of the exciting pulse.

Near-resonant excitation in a semiconductor is fundamentally different from that in atomic systems, due to Coulomb interactions [84,85]. However, like atomic systems, the excitation will also follow the electric field envelope in the adiabatic limit [86]. Near-resonant excitation leads to an optical Stark effect, giving shifts in the excitonic resonance [87], and possible changes in the excitonic oscillator strength [84,85]. In differential transmission experiments, blue shifts of 0.2 meV were realised at the hh exciton for pump intensities of ≈ 8 mW/cm^2 and a pump detuning of 30 meV [88]. This led to the observation of increases or decreases in absorption of a probe beam (depending on where the beam was tuned) of between 10 and 30%. The line shift, and thus the nonlinear response, is inversely proportional to the pump detuning from resonance, and linearly proportional to the pump intensity [89]. Again, the very fast time scale of this nonlinearity has great application to optoelectronic devices, as was shown by the operation of a GaAs/AlGaAs quantum well optical gate with a response time of less than a picosecond [90].

It should be noted that a potential problem for near-resonant excitation is that there is also some residual real excitation due to the extension of the band tail to lower energies and phonon-assisted absorption [84,88].

E   FUTURE DIRECTIONS

As evidenced by the previous discussion, confinement of excitons in GaAs quantum wells leads to strong nonlinearities. The progression to further confinement, either through the application of magnetic fields [91,92] or the engineering of 1D and 0D structures such as quantum wires, dots and microcrystallites provides new opportunities in nonlinear optical studies. It has been predicted that because of the increased confinement, the nonlinear susceptibility will be enhanced [93]. For quantum wires, values of $\chi^{(3)}$ on the order of 1 esu may be realised [94,95].

Nanoscopic studies of structures also open a new, potentially fruitful area for the study of quantum well optical nonlinearities. In situ studies by scanning tunnelling microscopy of MBE grown surfaces [96] are providing an excellent opportunity to use these systems as model materials to understand the fundamental effects of disorder and correlate micro/nanoscopic structural details with optical interactions. Microscopic imaging (with μm resolution) has allowed the observation of sharp-line photoluminescence of excitons localised on interface disorder [97-101]. The work of Hess and Betzig based on near-field microscopy has demonstrated the potential for new information to be obtained by increased spatial resolution [102]. Also, scanning tunnelling microscopy with the ultrafast response time of 2 ps (generated by optical nonlinearity) and 50 Å resolution has been demonstrated [103]. The incorporation of such methodologies with nonlinear spectroscopy provides the opportunity to understand the microscopic nature of relaxation in the absence of spatial averaging over interface disorder that occurs in typical far-field measurements.

F   CONCLUSION

This has been a brief review of the study of resonant optical nonlinearities in GaAs heterostructures. In 2D structures, excitonic interactions dominate the nonlinear response, even over the effects due to free carriers. Recent studies reveal that the temporal and polarisation properties of the response are complicated, and may have origins in Coulomb interactions, formation of biexcitons, and interface disorder. A comprehensive theoretical description of excitonic contributions to the nonlinear response in GaAs quantum wells is still a subject of intense study. The optical Bloch equations modified to include the effects of Coulomb interactions and dynamic effects of decay and dephasing still provide good guidance in interpreting the nonlinear optical signal. Resonant nonlinear optical interactions in GaAs quantum wells have shown diverse and potentially useful effects. Future study of these systems will continue to show their importance to applications and the basic understanding of III-V materials.

REFERENCES

[1]   S. Schmitt-Rink, D.S. Chemla, D.A.B. Miller [*Adv. Phys. (UK)* vol.38 (1989) p.89-188]

[2]   S. Schmitt-Rink [*Phys. Rev. B (USA)* vol.32 (1985) p.6601-9]

[3]   H. Haug, S. Schmitt-Rink [*J. Opt. Soc. Am. B (USA)* vol.2 (1985) p.1135-42]

[4]   M. Lindberg, S.W. Koch [*J. Opt. Soc. Am. B (USA)* vol.5 (1988) p.139-46]

[5]   M. Lindberg, S.W. Koch [*Phys. Rev. B (USA)* vol.38 (1988) p.3342-50]

[6]   H. Wang, K. Ferrio, D.G. Steel, P. Berman, S.W. Koch [*Phys. Rev. A (USA)* vol.49 (1994) p.1551-4]

[7]   K. Bott et al [*Phys. Rev. B (USA)* vol.48 (1993) p.17418-26]

[8]   M. Wegener, D.S. Chemla, S. Schmitt-Rink, W. Schäfer [*Phys. Rev. A (USA)* vol.42 (1990) p.5675-83]

[9]   C. Bowden, G.P. Agrawal [*Phys. Rev. A (USA)* vol.51 (1995) p.4132-9]

[10]   D.S. Chemla [*Helv. Phys. Acta (Switzerland)* vol.56 (1983) p.607-37]

[11]   W.H. Knox et al [*Phys. Rev. Lett. (USA)* vol.54 (1985) p.1306-9]

[12]   D.S. Chemla, D.A.B. Miller [*J. Opt. Soc. Am. B (USA)* vol.2 (1985) p.1155-73]

[13]   D.A.B. Miller, D.S. Chemla, D.J. Eilenberger, P.W. Smith, A.C. Gossard, W.T. Tsang [*Appl. Phys. Lett. (USA)* vol.41 (1982) p.679-81]

[14]   D.A.B. Miller, D.S. Chemla, D.J. Eilenberger, P.W. Smith, A.C. Gossard, W. Wiegmann [*Appl. Phys. Lett. (USA)* vol.42 (1983) p.925-7]

[15]   W.H. Knox, C. Hirlimann, D.A.B. Miller, J. Shah, D.S. Chemla, C.V. Shank [*Phys. Rev. Lett. (USA)* vol.56 (1986) p.1191-3]

[16]   D.S. Chemla, D.A.B. Miller, P.W. Smith, A.C. Gossard, W. Wiegmann [*IEEE J. Quantum Electron. (USA)* vol.20 (1984) p.265-75]

[17]   N. Peyghambarian, H.M. Gibbs [*J. Opt. Soc. Am. B (USA)* vol.2 (1985) p.1215-27]

[18]   G.W. Fehrenbach, W. Schäfer, R.G. Ulbrich [*J. Lumin. (Netherlands)* vol.30 (1985) p.154-61]

[19]   N. Peyghambarian et al [*Phys. Rev. Lett. (USA)* vol.53 (1984) p.2433-6]

[20]   D. Hulin et al [*Phys. Rev. B. (USA)* vol.33 (1986) p.4389-91]

[21]   L. Schultheis, J. Kuhl, A. Honold, C.W. Tu [*Phys. Rev. Lett. (USA)* vol.57 (1986) p.1635-8]

[22]   S. Weiss, M.-A. Mycek, J.-Y. Bigot, S. Schmitt-Rink, D.S. Chemla [*Phys. Rev. Lett. (USA)* vol.69 (1992) p.2685-8]

[23]   M.-A. Mycek, S. Weiss, J.-Y. Bigot, S. Schmitt-Rink, D.S. Chemla, W. Schäfer [*Appl. Phys. Lett. (USA)* vol.60 (1992) p.2666-8]

[24]   J.-Y. Bigot, M.-A. Mycek, S. Weiss, R.G. Ulbrich, D.S. Chemla [*Phys. Rev. Lett. (USA)* vol.70 (1993) p.3307-10]

[25]   A. Honold, L. Schultheis, J. Kuhl, C.W. Tu [*Phys. Rev. B (USA)* vol.40 (1989) p.6442-5]

[26]   D.A.B. Miller, D.S. Chemla, P.W. Smith, A.C. Gossard, W. Wiegmann [*Opt. Lett. (USA)* vol.8 (1983) p.477-9]

[27]   J.T. Remillard, H. Wang, D.G. Steel, J. Oh, J. Pamulapati, P.K. Bhattacharya [*Phys. Rev. Lett. (USA)* vol.62 (1989) p.2861-4]

[28]   P.W. Smith, Y. Silberberg, D.A.B. Miller [*J. Opt. Soc. Am. B (USA)* vol.2 (1985) p.1228-35]

[29]   U. Keller, W.H. Knox, H. Roskos [*Opt. Lett. (USA)* vol.15 (1990) p.1377-9]

[30]   U. Keller, W.H. Knox, G.W. t'Hooft [*IEEE J. Quantum Electron. (USA)* vol.28 (1992) p.2123-33]

[31]   M.N. Islam, E.R. Sunderman, I. Bar-Joseph, N. Sauer, T.Y. Chang [*Appl. Phys. Lett. (USA)* vol.54 (1989) p.1203-5]

[32]   A.P. Heberle, J.J. Baumberg, K. Kohler [*Phys. Rev. Lett. (USA)* vol.75 (1995) p.2598-2601]

[33]   J. Nilsen, N. Gluck, A. Yariv [*Opt. Lett. (USA)* vol.6 (1981) p.380-2]

[34]   C.V. Shank et al [*Solid State Commun. (USA)* vol.47 (1983) p.981-3]

[35]   Y. Masumoto, S. Tarucha, H. Okamoto [*J. Phys. Soc. Jpn. (Japan)* vol.55 (1986) p.57-60]

[36]   D.R. Wake, H.W. Yoon, J.P. Wolfe, H. Morkoç [*Phys. Rev. B (USA)* vol.46 (1992) p.13452-60]

[37]   L. Schultheis, A. Honold, J. Kuhl, K. Kohler, C.W. Tu [*Phys. Rev. B (USA)* vol.34 (1986) p.9027-30]

[38]   A.M. Weiner, S. De Silvestri, E.P. Ippen [*J. Opt. Soc. Am. B (USA)* vol.2 (1985) p.654-61]

[39]   L. Schultheis, M.D. Sturge, J. Hegarty [*Appl. Phys. Lett. (USA)* vol.47 (1985) p.995-7]

[40]   K. Leo et al [*Phys. Rev. Lett. (USA)* vol.65 (1990) p.1340-3]

[41]   K. Leo et al [*Phys. Rev. B (USA)* vol.44 (1991) p.5726-37]

[42]   M.D. Webb, S.T. Cundiff, D.G. Steel [*Phys. Rev. Lett. (USA)* vol.66 (1991) p.934-7]

[43]   D.-S. Kim et al [*Phys. Rev. Lett. (USA)* vol.69 (1992) p.2725-8]

[44]   W. Schäfer, F. Jahnke, S. Schmitt-Rink [*Phys. Rev. B (USA)* vol.47 (1992) p.1217-20]

[45]   A.E. Paul, W. Sha, S. Patkar, A.L. Smirl [*Phys. Rev. B (USA)* vol.51 (1995) p.4242-6]

[46]   U. Siegner, M.A. Mycek, S. Glutsch, D.S. Chemla [*Phys. Rev. Lett. (USA)* vol.74 (1995) p.470-3]

[47]   S.T. Cundiff, H. Wang, D.G. Steel [*Phys. Rev. B (USA)* vol.46 (1992) p.7248-51]

[48]   S. Bar-Ad, I. Bar-Joseph [*Phys. Rev. Lett. (USA)* vol.68 (1992) p.349-52]

[49]   H. Yaffe, Y. Prior, J.P. Harbison, L.T. Florez [*J. Opt. Soc. Am. B (USA)* vol.10 (1993) p.578-83]

[50]   G.O. Smith et al [*Solid State Commun. (USA)* vol.94 (1995) p.373-7]

[51]   K.-H. Pantke, D. Oberhauser, V.G. Lyssenko, J.M. Hvam, G. Weimann [*Phys. Rev. B (USA)* vol.47 (1993) p.2413-6]

[52]   H. Wang, J. Shah, T.C. Damen, L.N. Pfeiffer [*Solid State Commun. (USA)* vol.91 (1994) p.869-74]

[53]   D. Bennhardt, P. Thomas, E. Eccleston, E.J. Mayer, J. Kuhl [*Phys. Rev. B (USA)* vol.47 (1993) p.13485-90]

[54]   D.S. Kim, J. Shah, T.C. Damen, L.N. Pfeiffer, W. Schäfer [*Phys. Rev. B (USA)* vol.50 (1994) p.5775-8]

[55]   T. Rappen, P. Ulf-Gereon, M. Wegener, W. Schäfer [*Phys. Rev. B (USA)* vol.49 (1994) p.10774-7]

[56]   H. Wang, K. Ferrio, D.G. Steel, Y. Hu, R. Binder, S. Koch [*Phys. Rev. Lett. (USA)* vol.71 (1993) p.1261-4]

[57]   S. Patkar, A.E. Paul, W. Sha, J.A. Bolger, A.L. Smirl [*Phys. Rev. B (USA)* vol.51 (1995) p.10789-94]

[58]   H. Schneider, K. Ploog [*Phys. Rev. B (USA)* vol.49 (1994) p.17050-4]

[59]   E.J. Mayer et al [*Phys. Rev. B. (USA)* vol.50 (1994) p.14730-3]

[60]   E.J. Mayer et al [*Phys. Rev. B (USA)* vol.51 (1995) p.10909-14]

[61]   D.J. Lovering, R.T. Phillips, G.J. Denton, G.W. Smith [*Phys. Rev. Lett. (USA)* vol.68 (1992) p.1880-3]

[62]   B.F. Feuerbacher, J. Kuhl, K. Ploog [*Phys. Rev. B (USA)* vol.43 (1991) p.2439-41]

[63]   K.-H. Pantke, V.G. Lyssenko, B.S. Razbirin, H. Schwab, J. Erland, J.M. Hvam [*Phys. Status Solidi B (Germany)* vol.173 (1992) p.69-76]

[64]   H. Wang, D.G. Steel [*Phys. Rev. A (USA)* vol.43 (1991) p.3823-31]

[65]   H. Wang, M. Jiang, R. Merlin, D.G. Steel [*Phys. Rev. Lett. (USA)* vol.69 (1992) p.804-7]

[66]   A. Ourmazd, D.W. Taylor, J. Cunningham, C.W. Tu [*Phys. Rev. Lett. (USA)* vol.62 (1989) p.933-6]

[67]   C. Weisbuch, R. Dingle, A.C. Gossard, W. Wiegmann [*Solid State Commun. (USA)* vol.38 (1981) p.709-12]

[68]   J. Hegarty, M.D. Sturge, C. Weisbuch, A.C. Gossard, W. Wiegmann [*Phys. Rev. Lett. (USA)* vol.49 (1982) p.930-2]

[69]   T. Tanaka, H. Sakaki [*J. Cryst. Growth (Netherlands)* vol.81 (1987) p.153-8]

[70]   J. Hegarty, M.D. Sturge [*J. Opt. Soc. Am. B (USA)* vol.2 (1985) p.1143-54]

[71]   J. Hegarty, L. Goldner, M.D. Sturge [*Phys. Rev. B (USA)* vol.30 (1984) p.7346-8]

[72]   H. Wang, M. Jiang, D.G. Steel [*Phys. Rev. Lett. (USA)* vol.65 (1990) p.1255-8]
[73]   H. Wang, D.G. Steel [*Appl. Phys. A (Germany)* vol.33 (1991) p.514-22]
[74]   M. Jiang, H. Wang, D.G. Steel [*Appl. Phys. Lett. (USA)* vol.61 (1992) p.1301-3]
[75]   E.O. Gobel et al [*Phys. Rev. Lett. (USA)* vol.64 (1990) p.1801-4]
[76]   K. Leo, T.C. Damen, J. Shah, K. Kohler [*Phys. Rev. B (USA)* vol.42 (1990) p.11359-61]
[77]   J. Erland, K.H. Pantke, V. Mizeikis, V.G. Lyssenko, J.M. Hvam [*Phys. Rev. B (USA)* vol.50 (1994) p.15047-55]
[78]   D. Birkedal, V.G. Lyssenko, K.H. Pantke, J. Erland, J.M. Hvam [*Phys. Rev. B (USA)* vol.51 (1995) p.7977-80]
[79]   C.A. Warwick, W.Y. Jan, A. Ourmazd, T.D. Harris [*Appl. Phys. Lett. (USA)* vol.56 (1990) p.2666-8]
[80]   D. Gammon, B.V. Shanabrook, D.S. Katzer [*Phys. Rev. Lett. (USA)* vol.67 (1991) p.1547-9]
[81]   D.S. Chemla [*Rep. Prog. Phys. (UK)* vol.43 (1980) p.1192-1262]
[82]   D. Grischkowsky [*Phys. Rev. Lett. (USA)* vol.24 (1970) p.866-9]
[83]   D. Grischkowsky, N.S. Shinen, R.J. Bennett [*Appl. Phys. Lett. (USA)* vol.33 (1978) p.805-7]
[84]   C. Ell, J.F. Muller, K. El Sayed, H. Haug [*Phys. Rev. Lett. (USA)* vol.62 (1989) p.304-7]
[85]   R. Binder, S.W. Koch, M. Lindberg, W. Schäfer, F. Jahnke [*Phys. Rev. B (USA)* vol.43 (1991) p.6520-9]
[86]   R. Binder, S.W. Koch, M. Lindberg, N. Peyghamabarian, W. Schäfer [*Phys. Rev. Lett. (USA)* vol.65 (1990) p.899-902]
[87]   D.H. Mysyrowicz, A. Antonetti, A. Migus, W.T. Masselink, H. Morkoç [*Phys. Rev. Lett. (USA)* vol.56 (1986) p.2748-51]
[88]   A. Von Lehmen, D.S Chemla, J.E. Zucker, J.P. Heritage [*Opt. Lett. (USA)* vol.11 (1986) p.609-11]
[89]   S. Schmitt-Rink, D.S. Chemla [*Phys. Rev. Lett. (USA)* vol.57 (1986) p.2752-5]
[90]   D. Hulin et al [*Appl. Phys. Lett. (USA)* vol.49 (1986) p.749-51]
[91]   J.B. Stark, W.H. Knox, D.S. Chemla, W. Schäfer, S. Schmitt-Rink, C. Stafford [*Phys. Rev. Lett. (USA)* vol.65 (1990) p.3033-6]
[92]   C. Stafford, S. Schmitt-Rink, W. Schäfer [*Phys. Rev. B (USA)* vol.41 (1990) p.10000-11]
[93]   S. Schmitt-Rink, D.A.B. Miller, D.S. Chemla [*Phys. Rev. B (USA)* vol.35 (1987) p.8113-22]
[94]   F.L. Madarasz, F. Szmulowicz, F.K. Hopkins, D.L. Dorsey [*Phys. Rev. B (USA)* vol.49 (1994) p.13528-41]
[95]   F.L. Madarasz, F. Szmulowicz, F.K. Hopkins [*Phys. Rev. B (USA)* vol.52 (1995) p.8964-73]
[96]   J. Sudijono, M.D. Johnson, C.W. Snyder, M.B. Elowitz, B.G. Orr [*Phys. Rev. Lett. (USA)* vol.69 (1992) p.2811-4]
[97]   A. Zrenner, L.V. Butov, M. Hagn, G. Abstreiter, G. Bohm, G. Weimann [*Phys. Rev. Lett. (USA)* vol.72 (1994) p.3382-5]
[98]   K. Brunner, G. Abstreiter, G. Böhn, G. Tränkle, G. Weimann [*Appl. Phys. Lett. (USA)* vol.64 (1994) p.3320-2]
[99]   J.-Y. Marzin, J.-M. Gerard, A. Izrael, D. Barrier, G. Bastard [*Phys. Rev. Lett. (USA)* vol.73 (1994) p.716-9]
[100] K. Brunner, G. Abstreiter, G. Bohm, G. Trankle, G. Weimann [*Phys. Rev. Lett. (USA)* vol.73 (1994) p.1138-41]
[101] D. Gammon, E.S. Snow, D.S. Katzer [*Appl. Phys. Lett. (USA)* vol.67 (1995) p.2391-3]
[102] H.F. Hess, E. Betzig, T.D. Harris, L.N. Pfeiffer, K.W. West [*Science (USA)* vol.264 (1994) p.1740-5]
[103] S. Weiss, D.F. Ogletree, D. Botkin, M. Salmeron, D.S. Chemla [*Appl. Phys. Lett. (USA)* vol.63 (1993) p.2567-9]

6.7 Dielectric functions in III-V quantum wells

P. Snyder

June 1995

A INTRODUCTION

The primary differences between the dielectric function (ε) of a quantum well (QW) and that of a thicker layer or bulk material are near the critical point energies. At the fundamental gap (E_0) of the QW material, absorption features due to transitions between subbands in the valence and conduction bands at the Brillouin zone centre are seen [1]. Because of their close relation to well and barrier thicknesses, band offsets, and electric field effects on the confined electron and hole states, these optical transitions have been studied in great detail. Quantum confinement effects have also been seen at the higher E_1, $E_1 + \Delta_1$ critical points [2,3]. Here, instead of discrete transitions a general blue shift of the features is seen. Strain effects in pseudomorphic layers are also observed at these critical points [4]. The dielectric function of a barrier layer is generally presumed to be the same as bulk, for thicker layers (≥ 10 nm). In thinner barriers broadening of the E_1, $E_1 + \Delta_1$ structure has been observed [5], and in very thin layers (≤ 3 nm) the structure virtually disappears [6].

Much of the experimental work on optical properties has relied on transmission and photocurrent measurements, from which absorption spectra are derived. They show directly the absorption features near E_0 due to transitions between subbands in the valence and conduction bands, and the effects of an applied electric field. Many of these studies have been coupled with theoretical calculations [7], including electric field effects [8-11]. Less has been published about the full dielectric function, particularly near higher critical points. Here we review some results obtained primarily from reflection spectroscopic ellipsometry (SE), which determines the real and imaginary parts of the dielectric function [12] without the need for a Kramers-Kronig transformation. Bulk ε have been measured by SE for all III-V semiconductors of interest [13]. $Al_xGa_{1-x}As$ has been measured [14] and modelled [15] as a function of x, as have strained and unstrained $In_xGa_{1-x}As$ [4].

B GaAs/AlGaAs QUANTUM WELLS

Erman et al [2] first measured the dielectric function of single $GaAs/Al_xGa_{(1-x)}As$ ($x \cong 0.5$) quantum wells, in the E_1, $E_1 + \Delta_1$ region. They fitted the SE data with a multilayer model to determine both the thicknesses of the layers and ε of the quantum wells. A shift of the critical point energies to higher energy was observed with decreasing well thickness (FIGURE 1). They modelled the energy shifts with a quantum mechanical analysis of the square well potential, assuming a conduction band discontinuity of 85% at the Γ point. Vasquez et al [3] observed the same effect in SE measurements of $GaAs/Al_xGa_{(1-x)}As$ ($x \cong 0.3$) single QWs from 1.41 nm to 5.94 nm thick. They used a conduction band discontinuity of 65% and other modifications in the quantum mechanical square well analysis. In each of these cases reasonable agreement between theory and experiment was obtained.

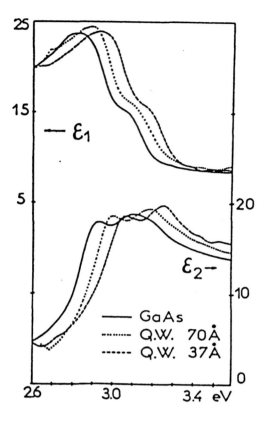

FIGURE 1. Dielectric functions of two GaAs quantum
well layers, compared with bulk GaAs. From [2].

Near E_0 it is known from planar waveguide measurements that the dielectric function is anisotropic. For light polarised parallel to the layer both the 1e-1hh and 1e-1lh transitions are allowed, while for light polarised perpendicular to the layer the 1e-1hh transition is forbidden [1]. In reflection, transmission, ellipsometry, and modulation spectroscopies, the polarisation is in or nearly in the plane of the layer, so the dielectric function may be assumed to be that for the parallel polarisation. Dielectric function measurements near E_0 of GaAs/Al$_x$Ga$_{(1-x)}$As ($x \cong 0.5$) superlattices were first made with SE by Snyder et al [16]. They showed the excitonic absorption peaks due to the 1e-1hh, 1e-1lh, and 2e-2hh transitions, and corresponding structure in the real part. Pickering et al [17] analysed SE data over a wider spectral range, including the E_0, E_1 and $E_1 + \Delta_1$, and E_2 regions, for single and multiple GaAs/Al$_x$Ga$_{(1-x)}$As ($x \cong 0.35$) quantum wells. They fitted the SE data with a multilayer model to determine layer thicknesses and ε for a single QW in the E_0 region, showing the excitonic transitions between quantised levels (FIGURE 2). They compared these energies with quantum mechanical square well calculations. At higher photon energies, above E_0 of the barrier, they found that bulk dielectric functions in the multilayer model worked well to fit the SE data. Fits provided well, barrier, cap, and native oxide thicknesses, and barrier composition [15]. The measured QW thickness and zone centre transition energies were consistent with the quantum mechanical analysis, assuming a 65% conduction band discontinuity. In a report on GaAs/AlAs superlattices, Lukes and Ploog state that the AlAs E_1 peak is broadened with decreasing barrier thickness [5]. This effect has also been observed in very thin single AlAs barrier layers, where the E_1 feature is so broadened that it virtually disappears for layers less than 3 nm thick [6].

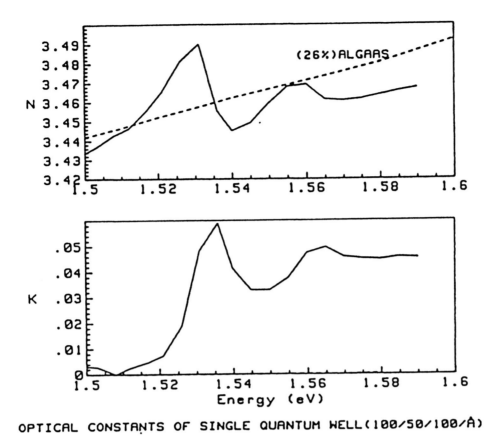

OPTICAL CONSTANTS OF SINGLE QUANTUM WELL(100/50/100/A)

FIGURE 2. Calculated optical functions for single quantum well compared with values of the average alloy composition. K for $Al_{0.26}Ga_{0.74}As$ is close to zero in this energy range. The complex dielectric function is the square of the complex refractive index: $\varepsilon_1 + i\varepsilon_2 = (N + iK)^2$. From [17].

C   OTHER III-V QW STRUCTURES

Strained single InAs/InGaAs QWs have been measured by SE [6]. The strain caused a blue shift of the $E_1 + \Delta_1$ peak, in qualitative agreement with theory. Current-voltage measurements showed resonant tunnelling due to the discrete subbands at the zone centre; however, no quantum confinement effects were seen at E_1 (i.e. no blue shift). This may be due to the fact that the InGaAs E_1 gap is only slightly larger than that for InAs, and hence only a very shallow QW exists at the E_1 gap.

D   CONCLUSION

The real and imaginary parts of the dielectric function of a QW exhibit features due to transitions between discrete subbands at the valence band maximum and conduction band minimum. The features are affected by an applied electric field, and strain. Strain also affects the E_1 and $E_1 + \Delta_1$ energies. When the E_1 gap of the QW is substantially less than that of the barrier material (as in GaAs/AlGaAs), a quantum confinement induced blue shift of that feature is seen. Conversely, in thin barrier layers the E_1, $E_1 + \Delta_1$ structure is broadened but not shifted. Away from the critical points, dielectric functions are similar to those of thicker layers or bulk.

REFERENCES

[1]   D.S. Chelma, D.A.B. Miller [*J. Opt. Soc. Am. B (USA)* vol.2 (1985) p.1155-73]

[2]   M. Erman, J.B. Theeten, P. Frijlink, S. Gaillard, J.H. Fan, C. Alibert [*J. Appl. Phys. (USA)* vol.56 (1984) p.3241-9]

[3]   R.P. Vasquez, R.T. Kuroda, A. Madhukar [*J. Appl. Phys. (USA)* vol.61 (1987) p.2973-8]

[4]   C. Pickering, R.T. Carline, M.T. Emeny, N.S. Garawal, L.K. Howard [*Appl. Phys. Lett. (USA)* vol.60 (1992) p.2412-4]

[5]   F. Lukes, K. Ploog [*Thin Solid Films (Switzerland)* vol.233 (1993) p.162-5]

[6]   C.M. Herzinger et al [*Proc. 21st Int. Symp. on Compound Semiconductors*, San Diego, California, 18-22 Sept. 1994, Ed. H. Goronkin (IOP Publishing, Bristol, 1994) p.363]

[7]   K.B. Kahen, J.P. Leburton [*Phys. Rev. B (USA)* vol.33 (1986) p.5465-72]

[8]   P.J. Stevens, M. Whitehead, G. Parry, K. Woodbridge [*IEEE J. Quantum Electron. (USA)* vol.24 (1988) p.2007-15]

[9]   K. Satzke, G. Weiser, W. Stolz, K. Ploog [*Phys. Rev. B (USA)* vol.43 (1991) p.2263-71]

[10]   J.-Th. Zettler, H. Mikkelsen, K. Leo, H. Kurz, R. Carius, A. Forster [*Phys. Rev. B (USA)* vol.46 (1992) p.15955-62]

[11]   J. Wang, J.P. Leburton, C.M. Herzinger, T.A. DeTemple, J.J. Coleman [*Phys. Rev. B (USA)* vol.47 (1993) p.4783-5]

[12]   D.E. Aspnes [in *Handbook of Optical Constants of Solids* vol.1, Ed. E.D. Palik (Academic Press, New York, 1985) p.89]

[13]   D.E. Aspnes, A.A. Studna [*Phys. Rev. B (USA)* vol.27 (1983) p.985-1009]

[14]   D.E. Aspnes, S.M. Kelso, R.A. Logan, R. Bhat [*Phys. Rev. B (USA)* vol.60 (1986) p.754-67]

[15]   P.G. Snyder, J.A. Woollam, S.A. Alterovitz, B. Johs [*J. Appl. Phys. (USA)* vol.68 (1990) p.5925-6]

[16]   P.G. Snyder et al [*Superlattices Microstruct. (UK)* vol.4 (1988) p.97-9]

[17]   C. Pickering, B.A. Shand, G.W. Smith [*J. Electron. Mater. (USA)* vol.19 (1990) p.51-8]

6.8 Refractive index in III-V quantum wells and superlattices

I. Suemune

May 1995

A INTRODUCTION

Refractive index is a key parameter for designing optoelectronic devices such as waveguides, lasers, phase and amplitude modulators, and optical switches. We discuss here the linear refractive index in quantum wells (QWs) and superlattices (SLs) that will have direct importance for these applications.

B REFRACTIVE INDEX OF ALLOY SEMICONDUCTORS

The frequency-dependent refractive index of semiconductors is given by the square root of the real part of the complex dielectric constant, $\varepsilon_1(\omega) + i\varepsilon_2(\omega)$. The real and imaginary parts are related via the well-known Kramers-Kronig relations. The imaginary part of the dielectric constant in QWs and SLs which reflects the absorption coefficient is structure dependent via the change of the absorption edge.

One simple method to determine the refractive index of QWs and SLs for the energy below the absorption edge is to employ the data for bulk semiconductors and to calculate the effective refractive index following the formula given in the next section. The refractive index in compound semiconductors below the band edge is best represented by the modified single-effective oscillator (MSEO) model [1]. The original model was given by Wemple and DiDomenico [2] where $\varepsilon_2(\omega)$ was approximated by the delta function. The accuracy of this model deteriorated at energies approaching the band edge. The MSEO model employs a better approximation to the $\varepsilon_2(\omega)$ spectrum and is given as follows:

$$n^2(E) = 1 + \frac{E_d}{E_0} + \frac{E_d}{E_0^3} E^2 + \frac{\eta}{\pi} E^4 \ln\left(\frac{2E_0^2 - E_g^2 - E^2}{E_g^2 - E^2}\right) \tag{1a}$$

$$\eta = \frac{\pi E_d}{2E_0^3\left(E_0^2 - E_g^2\right)} \tag{1b}$$

where $E = \hbar\omega$ is the photon energy. The parameters for several alloy semiconductors are given in the following:

$Al_xGa_{1-x}As$ [1]

$$E_0 = 3.65 + 0.871x + 0.179x^2 \tag{2a}$$
$$E_d = 36.1 - 2.45x \tag{2b}$$
$$E_g = 1.424 + 1.266x + 0.26x^2 \tag{2c}$$

$Ga_{1-x}In_xAs$ (x < 0.5) [3]

$$E_0 = 3.65 - 2.15x \qquad (3a)$$
$$E_d = 36.1 - 19.9x \qquad (3b)$$
$$E_g = 1.425 - 1.337x + 0.270x^2 \qquad (3c)$$

$Ga_xIn_{1-x}As_yP_{1-y}$ (x = 0.46y for lattice-matching to InP) [4,5]

$$E_0 = 3.391 + 0.524x - 1.891y + 1.626xy + 0.595x^2 (1 - y) \qquad (4a)$$
$$E_d = 28.91 + 7.54x + (-12.71 + 12.36x)y \qquad (4b)$$
$$E_g = 1.35 + 0.668x + 0.758x - (1.17 + 0.069x + 0.322x^2)y + (0.18 + 0.03x)y^2 \qquad (4c)$$

$GaAs_{1-x}P_x$ [1]

$$E_0 = 3.65 + 0.721x + 0.139x^2 \qquad (5a)$$
$$E_d = 36.1 + 0.35x \qquad (5b)$$
$$E_g = 1.441 + 1.091x + 0.21x^2 \qquad (5c)$$

$Ga_xIn_{1-x}P$ [1]

$$E_0 = 3.391 + 0.524x + 0.595x^2 \qquad (6a)$$
$$E_d = 28.91 + 7.54x \qquad (6b)$$
$$E_g = 1.34 + 0.668x + 0.758x^2 \qquad (6c)$$

The above parameters give the refractive index values of the respective semiconductors in good agreement with measurements up to near to the band edge.

C   REFRACTIVE INDEX IN MULTIPLE QUANTUM WELLS

The refractive index in multiple quantum well (MQW) layers is dependent on the polarisation of the optical field. It is convenient to replace the MQW layers by a single homogeneous layer with an effective refractive index averaged in the region.

For the case of transverse electric (TE) modes in MQW waveguides, the optimal choice for the equivalent refractive index is given by [6]

$$(n_{TE})^2 = \frac{\sum\limits_{j} n_j^2 d_j}{\sum\limits_{j} d_j} \qquad (7)$$

where d_j is the thickness of the jth layer in the MQW and n_j is the refractive index of the jth layer. The thickness of each layer is assumed to be much smaller than the optical wavelength. Another approximation method is the arithmetic average of the refractive index [7], but EQN (7) was shown to be the more accurate approximation for TE modes [8].

For transverse magnetic (TM) modes, the best approximation for the equivalent refractive index is given by [6]

$$\frac{1}{(n_{TM})^2} = \frac{\sum\limits_{j} \frac{1}{n_j^2} d_j}{\sum\limits_{j} d_j} \tag{8}$$

D   REFRACTIVE INDEX CHANGE WITH ELECTRIC FIELD

For the application to modulators and switches, the refractive index change near the absorption edge is important. The quantum confined Stark effect (QCSE) with the electric field applied perpendicular to QW planes causes the red-shift of exciton absorption and a reduction of the oscillator strength. The change in the refractive index Δn is related to the change in the absorption coefficient by the following equation [9]:

$$\Delta n(E) = \frac{c\hbar}{\pi} \int_{-\infty}^{\infty} \frac{\Delta\alpha(E',F)}{E'^2 - E^2} \, dE' \tag{9}$$

where $E = \hbar\omega$ is the photon energy. The change in the absorption coefficient induced with an electric field F, $\Delta\alpha(E',F) = \alpha(E',F) - \alpha(E',0)$, is either theoretically calculated [10] or directly measured.

The change in the refractive index as a function of the applied field is given by

$$\Delta n = -1/2 n_0^3 \left[r_{63}F + s_{13}F^2 \right] \tag{10}$$

Measurements on two types of InGaAsP/InP QW at wavelengths of 1.3 μm and 1.5 μm showed that Δn is predominantly a quadratic function of the applied field and $n_0 \approx 3.5$ and $s_{13} \approx 3 \times 10^{-14}$ cm^{2}V^{-2} were estimated at both wavelengths [11]. The detuning from the exciton resonance is important for phase modulators to minimise amplitude modulation. Δn was inversely proportional to the detuning energy ΔE and was given by $\Delta n = \eta F^2/\Delta E$. $\eta \sim 3 \times 10^{-11}$ meV cm^{2}V^{-2} was estimated in the above InGaAsP/InP QW and was approximately the same value in GaAs/AlGaAs and InGaAs/InP QWs [11]. The absolute refractive index in QWs for the high field approaches the effective bulk value given in the previous section [12].

E   REFRACTIVE INDEX CHANGE WITH INTERFACE DIFFUSION

The refractive index of an SL is modified by disordering of the structures. Disordering of the SL occurs with impurity diffusion or cap annealing. The latter is reported to be due to impurity-free vacancy diffusion [13] and is advantageous for low-loss optical waveguides. The refractive index change in a GaInAs (3 nm thick and bandgap corresponding to 1.3 μm)/InP (5 nm) SL after cap annealing at 800°C for 60 s was -1.1 $\times$ 10^{-2} for TE mode and +6.7 $\times$ 10^{-3} for TM mode, which was measured at 1.5 μm [14]. The direct relation between the interface diffusion and the change in the refractive index was calculated in [15]. In a $Ga_{0.8}In_{0.2}As$ (10 nm)/GaAs (100 nm) QW, the calculated refractive index change was approximately independent of the wavelength in the 1.2 - 1.6 μm range. It was $\sim$ -6 $\times$ 10^{-3} and $\sim$ -1.8 $\times$ 10^{-2} for the diffusion lengths of 2 nm and 4 nm respectively for the TE mode.

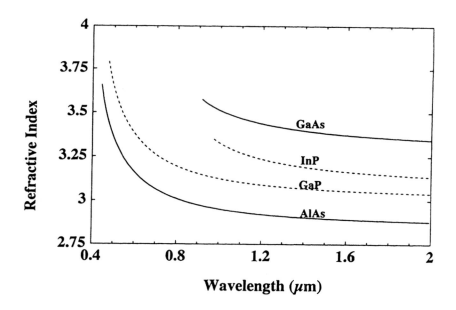

FIGURE 1. Refractive indices for binary compound semiconductors. which give the extremes for AlGaAs. GaAsP. and GaInP ternary alloys.

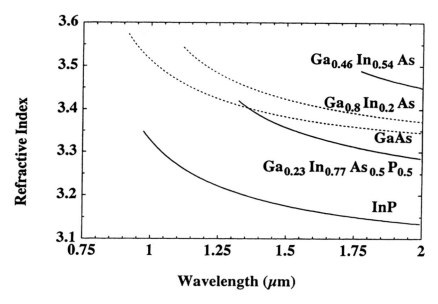

FIGURE 2. Refractive indices for $Ga_xIn_{1-x}As_yP_{1-y}$ (solid lines) with y = 0. 0.5. and 1.0 (x = 46y) and for $Ga_{1-x}In_xAs$ with x = 0. 0.2 (dashed lines). and 0.54.

F   CONCLUSION

In addition to the above features in QW and SL structures, there remain several attractive features related to the refractive index change in the quantum confined structures. Values for $\Delta n/n$ of ~ 4% and 7% were measured in GaInAs/InP quantum wires [16] and boxes [17], respectively.   A change of 17% is expected in well-defined quantum wires [18]. Carrier-induced refractive index change is another major feature of QWs.   A numerical simulator for the refractive index change caused by current injection into QW structures was developed which gives good agreement with measurements [19].

REFERENCES

[1]   M.A. Afromowitz [*Solid State Commun. (USA)* vol.15 (1974) p.59-63]

[2]   S.H. Wemple, M. DiDomenico [*Phys. Rev. B (USA)* vol.3 (1971) p.1338-51]

[3]   T. Takagi [*Jpn. J. Appl. Phys. (Japan)* vol.17 (1978) p.1813-7]

[4]   K. Utaka, Y. Suematsu, K. Kobayashi, H. Kawanishi [*Jpn. J. Appl. Phys. (Japan)* vol.19 (1980) p.L137-40]

[5]   P. Chandra, L.A. Coldren, K.E. Strege [*Electron. Lett. (UK)* vol.17 (1981) p.6-7]

[6]   G.M. Alman, L.A. Molter, H. Shen, M. Dutta [*IEEE J. Quantum Electron. (USA)* vol.28 (1992) p.650]

[7]   W. Steifer, D.S. Scifres, R.D. Burnham [*Appl. Opt. (USA)* vol.18 (1979) p.3547-8]

[8]   R.E. Smith, L.A. Molter, M. Dutta [*IEEE J. Quantum Electron. (USA)* vol.27 (1991) p.1119-22]

[9]   J.S. Weiner, D.A.B. Miller, D.S. Chelma [*Appl. Phys. Lett. (USA)* vol.50 (1987) p.842-4]

[10]   A. Bandyopadhyay, P.K. Basu [*IEEE J. Quantum Electron. (USA)* vol.29 (1993) p.2724-30]

[11]   J.E. Zucker, I. Bar-Joseph, B.I. Miller, U. Koren, D.S. Chelma [*Appl. Phys. Lett. (USA)* vol.54 (1989) p.10-2]

[12]   C.H. Lin, J.M. Meese, M.L. Wroge, C.J. Weng [*IEEE Photonics Technol. Lett. (USA)* vol.6 (1994) p.623]

[13]   D.G. Deppe et al [*Appl. Phys. Lett. (USA)* vol.49 (1986) p.510-2]

[14]   A. Wakatsuki, H. Iwamura, Y. Suzuki, T. Miyazawa, O. Mikami [*IEEE Photonics Technol. Lett. (USA)* vol.3 (1991) p.905-7]

[15]   J. Micallef, E.H. Li, B.L. Weiss [*Appl. Phys. Lett. (USA)* vol.62 (1993) p.3164-6]

[16]   K.G. Ravikumar, T. Kikugawa, T. Aizawa, S. Arai, Y. Suematsu [*Appl. Phys. Lett. (USA)* vol.58 (1991) p.1015]

[17]   T. Aizawa, K. Shinomura, S. Arai, Y. Suematsu [*IEEE Photonics Technol. Lett. (USA)* vol.3 (1991) p.907-9]

[18]   I. Suemune, L.A. Coldren [*IEEE J. Quantum Electron. (USA)* vol.24 (1988) p.1778-90]

[19]   J. Wang, J.P. Leburton, J.E. Zucker [*IEEE J. Quantum Electron. (USA)* vol.30 (1994) p.989-96]

6.9　Quantum well infrared photodetectors

B.F. Levine

June 1995

A　INTRODUCTION

In contrast to conventional long wavelength ($\lambda = 3$ - $20\,\mu m$) interband detectors, which photoexcite carriers across the bandgap, quantum well infrared photodetectors [1] (QWIPs) use intersubband transitions [2-5]. The advantages relative to the usual low bandgap materials (e.g. HgCdTe) include mature III-V (e.g. GaAs) growth and processing technologies, high uniformity, large area highly sensitive two-dimensional imaging arrays and lower cost.

B　INTERSUBBAND ENERGY LEVELS AND ABSORPTION

Intersubband absorption results from transitions between energy levels within the same conduction or valence band. These levels arise from the spatial localisation introduced in quantum wells consisting of a low-bandgap material (e.g. GaAs) surrounded by a higher-bandgap semiconductor (e.g. $Al_xGa_{1-x}As$). For infinitely high barriers and parabolic bands, the energy levels in the well are simply given by [1,2]

$$E_j = \left(\frac{\hbar^2 \pi^2}{2m^* L_w^2} \right) j^2 \tag{1}$$

where L_w is the width of the quantum well, m^* is the effective mass in the well, and j is an integer. The intersubband transition energy between the lowest and first excited state is thus

$$(E_2 - E_1) = (3\hbar^2\pi^2/2m^* L_w^2) \tag{2}$$

Furthermore, this transition has a large dipole matrix element $<z> = 16L/9\pi^2 \cong 0.18\,L_w$ and a near unity oscillator strength. More accurate calculations including finite barrier wells are shown in FIGURE 1. Thus by changing the quantum well width L_w and barrier height, i.e. Al concentration x (as given in TABLE 1), this intersubband energy can be readily varied over a wide range from $\lambda \cong 3$ - $20\,\mu m$. As shown in FIGURE 2 the absorption can be quite large with $\alpha \sim 500\,cm^{-1}$. Notice that in addition to controlling the peak position of the intersubband absorption, the spectral width can be independently selected by adjusting the energy position of the excited state relative to the top of the quantum well barriers.

It should be noted that since the oscillator strength only has a component along z, the optical electric field must also have a component parallel to z in order to induce an intersubband absorption. This can be efficiently accomplished using gratings and random scattering reflectors [6-8].

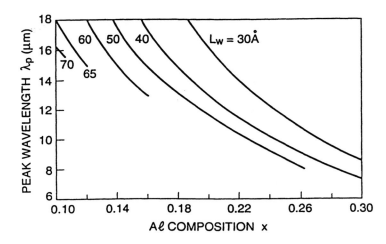

FIGURE 1. Calculated peak wavelength λ_p for a bound-to-continuum QWIP as a function of $Al_xGa_{1-x}As$ barrier composition x and GaAs quantum-well width L_w.

TABLE 1. Structure parameters for samples A-F, including quantum-well width L_w, barrier width L_b, $Al_xGa_{1-x}As$ composition x, doping density N_D, doping type, number of multiquantum-well periods, and type of intersubband transition: bound-to-continuum (B-C), bound-to-bound (B-B), and bound-to-quasicontinuum (B-QC). (Sample F has a stepped barrier.)

Sample	L_w (Å)	L_b (Å)	x	N_D (10^{18} cm^{-3})	Doping type	Periods	Intersubband transition
A	40	500	0.26	1	n	50	B-C
B	40	500	0.25	1.6	n	50	B-C
C	60	500	0.15	0.5	n	50	B-C
D	70	500	0.10	0.3	n	50	B-C
E	50	500	0.26	0.42	n	25	B-B
F	50	500	0.26	0.42	n	25	B-QC
		50	0.30				

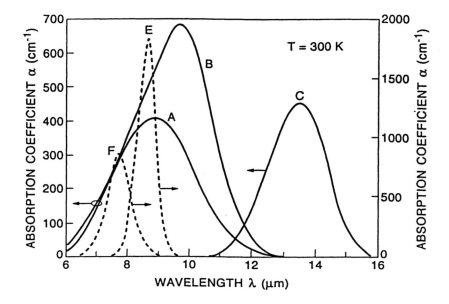

FIGURE 2. Absorption coefficient spectra $\alpha(\lambda)$ vs. wavelength measured at T = 300 K for samples A, B, C, E and F.

C   QUANTUM WELL INFRARED PHOTODETECTORS (QWIPs)

C1   Continuum Excited State

EQNS (1) and (2) relate to bound levels. However, to create a radiation detector the photoexcited carrier must escape from the well and produce a photocurrent [1]. This is advantageously arranged by narrowing the quantum well so that the second bond level is pushed into the continuum above the top of the finite barrier. Typical values are 40 Å GaAs quantum well (doped $N_D \sim 10^{18}$ cm^{-3}), $Al_{0.25}Ga_{0.75}As$ barriers (~500 Å thick to reduce the tunnelling dark current) and 50 periods of the structure to increase the intersubband absorption. This photoexcited carrier transport is schematically indicated in FIGURE 3.

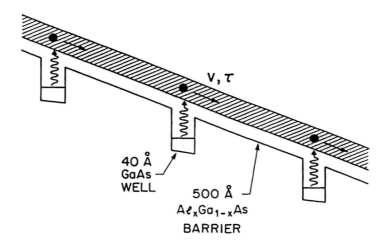

FIGURE 3. Conduction-band structure for a bound-to-continuum QWIP, showing the photoexcitation and hot-electron transport processes.

C2   Responsivity

The responsivity R (i.e. the photocurrent per incident optical power) is given by [1]

$$R = (e/h\nu)\, \eta p g \tag{3}$$

where η is the absorption quantum efficiency, p is the quantum well escape probability [9,10] for the photoexcited carrier and $g = \tau_L/\tau_T$ is the optical gain [11-13] given by the ratio of the excited carrier lifetime τ_L to the transit time τ_T. The measured normalised responsivity wavelength spectra are shown in FIGURE 4, for a variety of QWIPs having different GaAs quantum well widths and $Al_xGa_{1-x}As$ barrier heights. Shown in this figure are detectors covering the range $\lambda \sim 6 - 20\,\mu m$ but QWIPs as short as $\lambda \sim 3\,\mu m$ have also been demonstrated. The absolute values of the peak responsivities are quite large, being typically $R \simeq 0.4 - 0.7$ A/W as indicated in FIGURE 5.

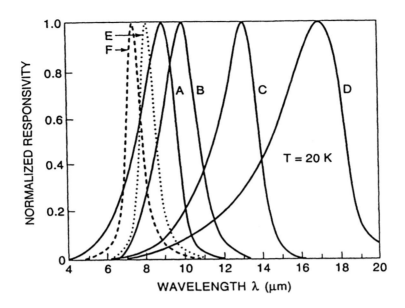

FIGURE 4. Normalised (to peak responsivity) bound-to-continuum QWIP responsivity spectra $\overline{R}(\lambda)$ vs. wavelength measured at T = 20 K, for samples A-F.

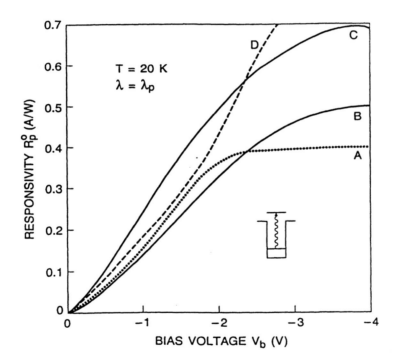

FIGURE 5. Bias-dependent peak ($\lambda = \lambda_p$) bound-to-continuum QWIP responsivity R^0_p measured at T = 20 K for samples A-D. The insert shows the conduction-band diagram.

C3 Dark Current

The dark current is produced by thermionic emission of carriers from the doped quantum wells over the top of the barriers and is thus given by [1]

$$I_d \propto e^{-(E_C - E_F)/kT} \qquad\qquad (4)$$

where E_c is the cutoff energy (i.e. the low energy for which the responsivity drops to half) and E_F is the carrier Fermi energy. This is illustrated in FIGURE 6 where the excellent straight line fit demonstrates that the dark current is dominated by thermionic emission and not defect tunnelling. Thus by lowering the temperature the dark current can be exponentially reduced. Typically for a $\lambda = 10\ \mu m$ detector (i.e. $E_c = 124\ meV$) a temperature of $T \simeq 80\ K$ is sufficiently low for high sensitivity imaging arrays.

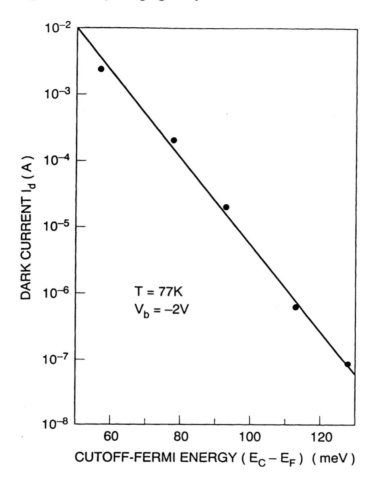

FIGURE 6. Plot of dark current I_d vs. the difference between the cutoff and Fermi energies ($E_c - E_F$) for five different QWIPs having different λ_c. The slope of the straight line corresponds to a temperature of $T = 77$ K.

C4   Detectivity

The detectivity which is a measure of the signal to noise ratio is given by [1]

$$D^* = R\sqrt{A\Delta f}/i_n \qquad (5)$$

where A is the area of the detector, $\Delta f = 1$ Hz is the noise bandwidth and i_n is the dark current noise. This is plotted in FIGURE 7 for a $\lambda = 10.7\ \mu m$ QWIP as a function of temperature. For example at $T = 77\ K$ $D^* = 10^{10}\ cm\sqrt{Hz}/W$ which is large enough to give excellent infrared imaging. Another measure of the quality of the detector is the noise equivalent temperature difference NEΔT which is the minimum temperature difference [1] which can be measured. For the device shown in FIGURE 6 at $T = 77$ K NEΔT = 10 mK (for a 50 μm square pixel with f/2 optics and an imaging bandwidth of $\Delta f = 60$ Hz). Thus an extremely small temperature change of only 10 mK can be detected. This, combined with the excellent uniformity of the GaAs technology (as compared with much lower uniformity HgCdTe technology), yields significantly higher resolution images.

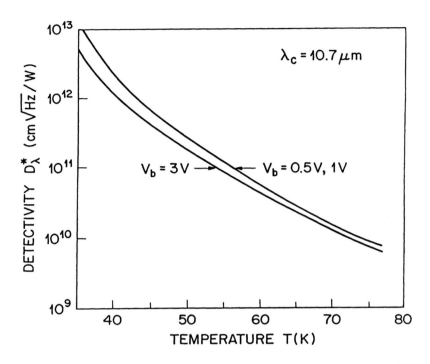

FIGURE 7. Detectivity D* vs. temperature for a bound-to-continuum QWIP having a cutoff wavelength $\lambda_c = 10.7 \,\mu m$ at $V_b = 0.5$ V. and 1 V.

C5   Imaging Arrays

Due to the mature GaAs technology and superior material uniformity, it has been possible to fabricate very large arrays (128 x 128, 256 x 256, and 640 x 480 pixels) yielding excellent high sensitivity images [14-16]. In addition, due to the 3" GaAs wafer size many arrays can be fabricated at the same time substantially lowering the cost. These arrays are processed with novel monolithic optical gratings or random scattering reflectors [7,8] and then bonded to Si multiplexers for imaging, similar to the bonding process used for HgCdTe and InSb arrays.

D   CONCLUSION

QWIPs have progressed rapidly from their first demonstration 8 years ago to practical, large area, high performance, high uniformity, and low cost arrays. Wide spectral coverage from $\lambda = 8$ to 20 μm has been demonstrated and even longer and shorter wavelengths are possible. Such high sensitivity, high resolution imaging devices are expected to find multiple uses in medicine, surveillance, factory processing and satellite earth resource mapping.

The latter area is particularly exciting and is now receiving research attention. Very long wave (VWIR) detectors ($\lambda = 14 - 20$ μm) and thermal imagers are of increasing interest for satellite spectroscopy in order to determine atmospheric composition (e.g. ozone, water, CO, CO_2 and nitrous oxide) and temperature profiles. QWIPs are well suited to this task since very long wavelength detectors have already demonstrated high detectivity at $\lambda \sim 19\,\mu m$ [1] and excellent imagery at $\lambda = 15\,\mu m$ [17] with a resolution of 128 x 128 pixels and a noise equivalent temperature difference of $NE\Delta T = 30\,mK$. For such space applications the low 1/f noise and high array uniformity are particularly advantageous.

REFERENCES

[1]   B.F. Levine [*J. Appl. Phys. (USA)* vol.74 (1993) p.R1]

[2]   L.C. West, S.J. Eglash [*Appl. Phys. Lett. (USA)* vol.46 (1985) p.1156]

[3]   B.F. Levine, K.K. Choi, C.G. Bethea, J. Walker, R.J. Malik [*Appl. Phys. Lett. (USA)* vol.50 (1987) p.1092]

[4]   E. Rosencher, B. Vinter, B. Levine (Eds) [*Intersubband Transitions in Quantum Wells*, Sept. 9-14 1991, Cargese, France (Plenum, New York, 1992)]

[5]   H.C. Liu, B.F. Levine, J.Y. Andersson (Eds) [*Quantum Well Intersubband Transition Physics and Devices* (Plenum, New York, 1994)]

[6]   K.W. Goossen, S.A. Lyon [*Appl. Phys. Lett. (USA)* vol.47 (1985) p.1257]

[7]   J.Y. Andersson, L. Lundqvist, Z.F. Paska [*Appl. Phys. Lett. (USA)* vol.58 (1991) p.2264]

[8]   G. Sarusi, B.F. Levine, S.J. Pearton, K.M.S. Bandara, R.E. Leibenguth [*Appl. Phys. Lett. (USA)* vol.64 (1994) p.960]

[9]   B.F. Levine, A. Zussman, J.M. Kuo, J. de Jong [*J. Appl. Phys. (USA)* vol.71 (1992) p.5130]

[10]   F. Luc, E. Rosencher, Ph. Bois [*Appl. Phys. Lett. (USA)* vol.62 (1993) p.2542]

[11]   G. Hasnain, B.F. Levine, S. Gunapala, N. Chand [*Appl. Phys. Lett. (USA)* vol.57 (1990) p.608]

[12]   A.G. Steele, H.C. Liu, M. Buchanan, Z.R. Wasilewski [*J. Appl. Phys. (USA)* vol.72 (1992) p.1062]

[13]   W.A. Beck [*Appl. Phys. Lett. (USA)* vol.63 (1993) p.3589]

[14]   B.F. Levine, C.G. Bethea, K.G. Glogovsky, J.W. Stayt, R.E. Leibenguth [*Semicond. Sci. Technol. (UK)* vol.6 (1991) p.C114]

[15]   L.J. Kozlowski et al [*IEEE Trans. Electron Devices (USA)* vol.38 (1991) p.1124]

[16]   T.S. Faska, J.W. Little, W.A. Beck, K.J. Ritter, A.C. Goldberg, R. LeBlanc [in *Innovative Long Wavelength Infrared Detector Workshop* Pasadena, CA, April 7-9 1992]

[17]   S.D. Gunapala et al [to be published]

CHAPTER 7

MODULATION-DOPED QUANTUM WELLS AND SUPERLATTICES

7.1 Electronic properties of GaAs/AlGaAs-based modulation doped heterojunctions, quantum wells, and superlattices

M. Shayegan

May 1995

A   INTRODUCTION

In modulation-doped semiconductors, the carriers are spatially separated from the dopant atoms to reduce the ionised impurity scattering. As a result, mobility values much higher than those in the bulk can be attained. Since the original demonstration of the modulation-doping technique in GaAs/AlGaAs heterostructures [1], the quality (as measured, for example, by the low-temperature mobility) of the electron and hole systems confined in such structures has improved dramatically and steadily. This improvement is largely a consequence of the cleanliness and vacuum integrity of the molecular beam epitaxy (MBE) chambers, availability of cleaner source and substrate materials, and the heterostructure growth procedures and design parameters. Here we provide some key features of some of the state-of-the-art GaAs/AlGaAs modulation-doped structures with an emphasis on the low-temperature mobility as a measure of quality; more details can be found in the original references and recent review articles [2].

B   HIGH-MOBILITY TWO-DIMENSIONAL ELECTRON SYSTEM AT THE GaAs/AlGaAs HETEROJUNCTION

The GaAs/AlGaAs heterojunction is the basic material structure for MODFET (modulation-doped field-effect transistor) devices which are being extensively studied and used in many laboratories world-wide. The low-temperature mobility (μ) in a typical MODFET is in the range $\sim 5 \times 10^4$ cm^2/V s to $\sim 2 \times 10^5$ cm^2/V s. During the past decade, however, major improvements have been made [3-10] and very high-quality 2D electron systems with mobilities exceeding 10^7 cm^2/V s have been reported. The structure of a state-of-the-art modulation-doped GaAs/AlGaAs sample [6] is shown in FIGURE 1 together with a schematic of the conduction-band-edge energy as a function of position after the charge transfer has taken place. The structure basically contains a two-dimensional electron system (2DES) separated from the dopant atoms (Si) by a spacer layer of undoped Al$_x$Ga$_{1-x}$As. The double-δ-doping is used to reduce the autocompensation of Si and to maximise the distance between the ionised dopants and the 2DES [4,6].

The mobility of the 2DES in such heterostructures is determined by scattering from the phonons, ionised impurities, and interface roughness. At low temperatures (T ~ 1 K), the dominant scattering is from the ionised impurities. For structures with thin spacers (≤ 400 Å thick), the remote (intentional) ionised impurities limit the mobility and typically μ depends on the areal electron density as $\mu \sim n^{3/2}$ [11]. In heterostructures with a large spacer thickness (≥ 1000 Å), a $\mu \sim n^{\gamma}$ with a $\gamma \simeq 0.6$ is observed [6,7,9]; this is the dependence expected if the dominant source of scattering is the (unintentional) residual bulk impurities in close proximity to the 2DES [6,12]. The residual impurity concentration, deduced from the mobility values for state-of-the-art 2DESs with $\mu \geq 10^6$ cm^2/V s for n ~ 5 x 10^{10} cm^{-2} is ≤ 1 x 10^{14} cm^{-3}, consistent with the residual GaAs doping expected in very clean MBE systems.

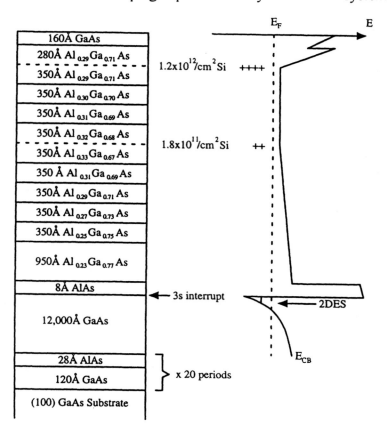

FIGURE 1. Schematic of a modulation-doped GaAs/AlGaAs heterojunction grown by MBE [6]. The conduction-band-edge energy, E_{CB}, is schematically shown on the right as a function of distance along the growth direction.

C   CONVENTIONAL ELECTRON SYSTEMS IN GaAs/AlGaAs HETEROSTRUCTURES

FIGURE 2 summarises a number of conventional modulation-doped structures containing high-quality electron systems. Unlike the structure of FIGURE 1 where the 2D electrons reside at a normal (AlGaAs grown on GaAs) interface, in most of these structures the quality of the inverted (GaAs grown on AlGaAs) interface is also important. The inverted interfaces in general have been of lower quality; this inferiority has been attributed to the interface roughness as well as impurity segregation (towards the interface) during the AlGaAs growth. There have been major recent improvements in the growth of inverted interfaces, such as the use of a GaAs/AlAs superlattice in place of an AlGaAs barrier [13], growth interruptions [13,14], and asymmetric doping [15-17] to reduce the interface roughness and also minimise the number of dopants that can reach the inverted interface. As a result, low-temperature mobility values exceeding 1 x 10^6 cm^2/V s have been achieved.

In TABLE 1 we list some of the highest low-temperature (~1 K) mobilities, for the given densities, reported for different types of structures. Some key references are given; the quoted values are from the first reference listed in each row. Note in TABLE 1 that the mobility of the superlattice structure is particularly low. This is primarily because of the thin spacer layers that are needed in such structures. As a result, the electron wavefunction overlaps substantially with the ionised impurities and this leads to a relatively low mobility.

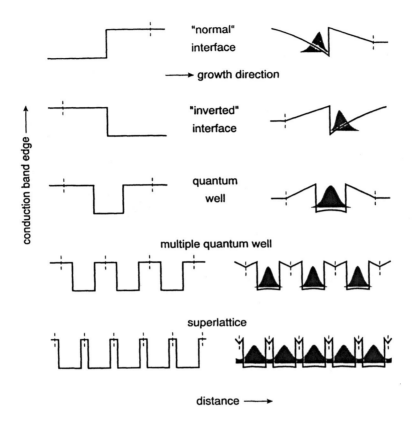

FIGURE 2. The conduction-band-edge, E_{CB}, is shown schematically as a function of distance along the growth direction for several conventional modulation-doped structures. The positions of dopant planes are shown by vertical dashed lines. On the left side, E_{CB} is shown without any electrons in the system. On the right, the self-consistent potential and charge distribution (shaded areas) are shown.

TABLE 1. Density/mobility of electron systems in some conventional modulation-doped GaAs/AlGaAs heterostructures.

Structure	Carrier density (cm^{-2})	Low-T ($\sim$1 K) mobility $(cm^2/V\ s)$	Ref
Normal interface	4.0×10^9	20×10^4	[9]
	1.0×10^{10}	4.0×10^5	[9,6]
	2.4×10^{11}	1.2×10^7	[7,8,10]
Inverted interface	5.0×10^{11}	2.1×10^6	[13a,14a]
	-	3.0×10^6	[17]
Quantum well (QW)	-	4.7×10^6	[17a]
Multiple QW	6.4×10^{10}	4.0×10^6	[18,15]
Conventional superlattice	1.4×10^{13}	6.4×10^3	[19b]

a - structure grown without back-side doping; b - total areal density in the structure which contains $\sim$30 periods of superlattice.

**D   NON-CONVENTIONAL   ELECTRON   SYSTEMS   IN   GaAs/AlGaAs
HETEROSTRUCTURES**

In FIGURE 3 a number of other modulation-doped GaAs/AlGaAs structures are shown schematically.   These systems contain two or more quantum wells, or a thick layer of electrons, and are fundamental for studying transport and many-body phenomena beyond 2D. The mobilities for these systems, listed in TABLE 2, are typically lower than those in TABLE 1 because of the presence of several interfaces and also because the electrons reside in AlGaAs in some of these structures and therefore will suffer from alloy scattering.

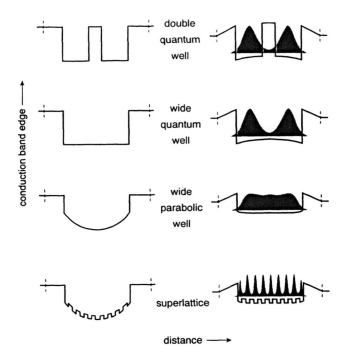

FIGURE 3. The conduction-band-edge and charge distribution for several non-conventional modulation-doped structures: the top two structures contain bilayer electron systems in either a double quantum well or a single, wide well, the third structure contains a thick electron layer system in a wide parabolic well, and the bottom structure contains an electron system with modulated charge distribution in a wide parabolic well with a superimposed periodic potential. Note that in all these structures the dopants are placed outside the entire structure.

TABLE 2. Density/mobility of electron systems in some
non-conventional GaAs/AlGaAs heterostructures.

Structure	Carrier density (cm^{-2})	Low-T (~1 K) mobility $(cm^2/V\ s)$	Ref
Double quantum well (QW)	1.5×10^{11}	1.0×10^{6}	[20,21]
Triple QW	1.4×10^{11}	7.0×10^{5}	[22,23]
Wide QW	1.6×10^{11}	1.5×10^{6}	[24,25]
Wide parabolic QW	$\sim 8 \times 10^{10}$	3.6×10^{5}	[26,27]
Superlattice in parabolic QW	1.5×10^{11}	1.1×10^{5}	[28]
Electron system in AlAs QW (X-point)	2.5×10^{11}	3.0×10^{4}	[29]

For the bottom three structures of FIGURE 3 the potential and the electron charge distribution have to be considered self-consistently (i.e. both the Schrödinger and Poisson equations have to be solved together). The repulsion between electrons gives rise to a bilayer-like system in the case of the wide quantum well, and a system with a thick, nearly-uniform distributed charge in the parabolic quantum well. In the superlattice structure, the carriers screen the long-wavelength (parabolic) part of the potential but not the short-wavelength part, and, as a result, a superlattice system with a mobility far exceeding that in the conventional superlattice (FIGURE 2 and TABLE 1) is achieved.

Finally, as listed in the last row of TABLE 2, it is possible to obtain a high-quality 2DES confined to an AlAs quantum well. Here the electron pockets are at the X-points of the Brillouin zone and thus the system has some similarities to the 2DES at the Si/SiO_2 interface.

E   HOLE SYSTEMS IN GaAs/AlGaAs STRUCTURES

Early structures containing 2D hole systems (2DHSs) were fabricated using Be as an acceptor dopant. However, Be is not an ideal dopant in GaAs/AlGaAs because of its diffusion and migration and the early samples were not of very high quality. An alternative is to grow the structure on a GaAs (311)A substrate and use Si, which is incorporated as an acceptor on the (311)A surface. Using this technique, a number of very high quality hole systems with mobility exceeding 1×10^6 cm^2/V s have been reported recently. TABLE 3 summarises some of the highest low-temperature mobilities reported for hole systems in different structures.

TABLE 3. Density/mobility of hole systems in GaAs/AlGaAs heterostructures.

Structure	Carrier density (cm^{-2})	Low-T (~1 K) mobility (cm^2/V s)	Ref
Normal interface GaAs(100), Be doping	4.6×10^{11}	4.0×10^4	[30]
Normal interface GaAs(311)A, Si doping	4.0×10^{10} 8.0×10^{10}	3.6×10^5 1.2×10^6	[32,31] [33]
Quantum well (QW) GaAs(311)A, Si doping	3.3×10^{11}	1.2×10^6	[34]
Wide parabolic QW GaAs(311)A, Si doping	1.2×10^{11}	1.2×10^5	[35]

F   CONCLUSION

Low-temperature mobilities far exceeding the bulk mobility have been achieved in modulation-doped GaAs/AlGaAs heterojunctions and quantum wells. Depending on the structure, these mobilities are still limited by ionised impurity scattering, interface roughness, or alloy scattering, rather than by the intrinsic, phonon scattering. With improvements in the purity of the grown materials and the optimisation of the growth parameters, structures with even better electronic quality and higher mobility should be achievable.

REFERENCES

[1]   R. Dingle, H.L. Stormer, A.C. Gossard, W. Wiegmann [*Appl. Phys. Lett. (USA)* vol.33 (1978) p.665]; H.L. Stormer, R. Dingle, A.C. Gossard, W. Wiegmann, M.D. Sturge [*Solid State Commun. (USA)* vol.29 (1979) p.705]

[2]   See, for example, M.R. Melloch [*Thin Solid Films (Switzerland)* vol.231 (1993) p.74]

[3]   J.H. English, A.C. Gossard, H.L. Stormer, K.W. Baldwin [*Appl. Phys. Lett. (USA)* vol.50 (1987) p.1825]

[4]   B. Etienne, E. Paris [*J. Phys. (France)* vol.48 (1987) p.2049]

[5]   M. Shayegan, V.J. Goldman, C. Jiang, T. Sajoto, M. Santos [*Appl. Phys. Lett. (USA)* vol.52 (1988) p.1086]

[6]   M. Shayegan, V.J. Goldman, M. Santos, T. Sajoto, L. Engel, D.C. Tsui [*Appl. Phys. Lett. (USA)* vol.53 (1988) p.2080]

[7]   L.N. Pfeiffer, K.W. West, H.L. Stormer, K.W. Baldwin [*Appl. Phys. Lett. (USA)* vol.55 (1989) p.1888]

[8]   C.T. Foxon, J.J. Harris, D. Hilton, J. Hewett, C. Roberts [*Semicond. Sci. Technol. (UK)* vol.4 (1989) p.582]

[9]   T. Sajoto, Y.-W. Suen, L.W. Engel, M.B. Santos, M. Shayegan [*Phys. Rev. B (USA)* vol.41 (1990) p.8449]

[10]   T. Saku, Y. Hirayama, Y. Horikoshi [*Jpn. J. Appl. Phys. (Japan)* vol.30 (1991) p.902]

[11]   F. Stern [*Appl. Phys. Lett. (USA)* vol.43 (1983) p.974]

[12]   See, for example, A. Gold [*Phys. Rev. B (USA)* vol.44 (1991) p.8818]

[13]   T. Sajoto et al [*Appl. Phys. Lett. (USA)* vol.54 (1989) p.840]

[14]   U. Meirav, M. Heiblum, F. Stern [*Appl. Phys. Lett. (USA)* vol.52 (1988) p.1268]

[15]   M. Shayegan, J.K. Wang, M. Santos, T. Sajoto, B.B. Goldberg [*Appl. Phys. Lett. (USA)* vol.54 (1989) p.27]

[16]   A.M. Lanzilotto, M. Santos, M. Shayegan [*J. Vac. Sci. Technol. A (USA)* vol.8 (1990) p.2009]

[17]   L.N. Pfeiffer, E.F. Schubert, K.W. West, C.W. Magee [*Appl. Phys. Lett. (USA)* vol.58 (1991) p.2258]

[18]   L.N. Pfeiffer, K.W. West, J.P. Eisenstein, K.W. Baldwin, P. Gammel [*Appl. Phys. Lett. (USA)* vol.61 (1992) p.1210]

[19]   H.L. Stormer, J.P. Eisenstein, A.C. Gossard, W. Wiegmann, K. Baldwin [*Phys. Rev. Lett. (USA)* vol.56 (1986) p.85]

[20]   J.P. Eisenstein, G.S. Boebinger, L.N. Pfeiffer, K.W. West, S. He [*Phys. Rev. Lett. (USA)* vol.68 (1992) p.1383]

[21]   X. Ying, S.R. Parihar, M. Shayegan [*Phys. Rev. B (USA)* vol.52 (1995) p.R11611]

[22]   T.S. Lay, X. Ying, H.C. Manoharan, M. Shayegan [*Phys. Rev. B (USA)* vol.52 (1995) p.R5511]

[23]   J. Jo, Y.W. Suen, L.W. Engel, M.B. Santos, M. Shayegan [*Phys. Rev. B (USA)* vol.46 (1992) p.9776]

[24]   Y.W. Suen, M.B. Santos, M. Shayegan [*Phys. Rev. Lett. (USA)* vol.69 (1992) p.3551]

[25]   Y.W. Suen, H.C. Manoharan, X. Ying, M.B. Santos, M. Shayegan [*Phys. Rev. Lett. (USA)* vol.72 (1994) p.3405]

[26]   M. Shayegan, J. Jo, Y.W. Suen, M. Santos, V.J. Goldman [*Phys. Rev. Lett. (USA)* vol.65 (1990) p.2916]

[27]   T. Sajoto, J. Jo, M. Santos, M. Shayegan [*Appl. Phys. Lett. (USA)* vol.55 (1989) p.1430]

[28]   J. Jo, M. Santos, M. Shayegan, Y.W. Suen, L.W. Engel, A.-M. Lanzilotto [*Appl. Phys. Lett. (USA)* vol.57 (1990) p.2130]

[29]   T.S. Lay et al [*Appl. Phys. Lett. (USA)* vol.62 (1993) p.3120]

[30]   H.L. Stormer, A.C. Gossard, W. Wiegmann, R. Blondel, K. Baldwin [*Appl. Phys. Lett. (USA)* vol.44 (1984) p.139]

[31]   A.G. Davies, R. Newburg, M. Pepper, J.E.F. Frost, D.A. Ritchie, A.G. Jones [*Phys. Rev. B (USA)* vol.44 (1991) p.13128]

[32]   M.B. Santos, Y.W. Suen, M. Shayegan, Y.P. Li, L.W. Engel, D.C. Tsui [*Phys. Rev. Lett. (USA)* vol.68 (1992) p.1188]

[33]   M. Henini, P.J. Rodgers, P.A. Crump, B.L. Gallagher, G. Hill [*Appl. Phys. Lett. (USA)* vol.65 (1994) p.2054]

[34]   J.J. Heremans, M.B. Santos, M. Shayegan [*Appl. Phys. Lett. (USA)* vol.61 (1992) p.1652]

[35]   M.B. Santos, J. Jo, Y.W. Suen, L.W. Engel, M. Shayegan [*Phys. Rev. B (USA)* vol.46 (1992) p.13639]

7.2   GaAs-based modulation doped heterostructures

K. Suzue, S. Noor Mohammad and H. Morkoc

May 1995

A   INTRODUCTION

Since the first report of a working MODFET in the open literature in 1980 [1], a fierce technological race has continued to this day to improve its properties.   Research interest remains high due to the MODFET's superlative device performance compared to the MESFET and HBT [2-4].   While the InP-based MODFET with its higher In-content InGaAs channels offers a superior material system, the GaAs-based MODFET, particularly the pseudomorphic MODFET [5], represents a longer established and inexpensive building block for MMIC and OEIC applications.   In an attempt to present a useful reference summarising recent progress (from about 1988 to the present) at a glance, the following Datareview encapsulates in tabular form vital performance characteristics of various technologies incorporating the GaAs-based MODFET.   In the following tables, the authors have made some attempt to respect the terminology used in the original referenced papers, so the terms MODFET and HEMT both appear liberally although they are equivalent (the same holds true for their pseudomorphic counterparts); this has been done to follow the dispositions of the original authors.   The majority of recent work has been devoted to the research and development of low noise amplifiers and power amplifiers: this is reflected in the tables below.

B   DISCRETE MODFET CHARACTERISTICS

The MODFET may be perceived as the compound semiconductor analogue of the widespread MOSFET [6].   It utilises a two-dimensional electron gas (2DEG) whose concentration is modulated by the gate potential.   In the most basic MODFET construction, a high bandgap material with a high doping concentration is grown above a low bandgap intrinsic material. The keystone of the MODFET's superlative performance is the dopant-free nature of the conduction channel achieved by bandgap engineering.   The conduction band discontinuity at the heterojunction interface between the two materials leads to the trapping and localisation of electrons in low bandgap material and to the formation of the 2DEG.   Two fundamental types of GaAs-based MODFET are now extant.   The older, conventional MODFET uses AlGaAs and GaAs for the high bandgap and low bandgap materials respectively.   The newer PMODFET or pseudomorphic MODFET alleviates the DX centre problem encountered by conventional MODFET by the adoption of a lower AlAs mole fraction and provides an InGaAs channel with transport properties greatly superior to GaAs.   The small gate-to-channel distance in MODFETs reduces or eliminates the enormous increase in output conductance usually observed when the gate length is scaled down.   Also, unlike a metal-semiconductor field-effect transistor (MESFET), the reduction of the gate length in the all-epitaxial MODFET is accompanied by only a minimal change in the threshold voltage.   The well known problem of the insufficiently high Schottky barrier vanishes in the MODFET because the gate formed on the large bandgap semiconductor exhibits a Schottky barrier height similar to the 0.8 eV of GaAs.   TABLE 1 is a brief survey of the device performance of MODFETs, often

TABLE 1. Measured digital and microwave properties of MODFETs.

MODFET type	Imax (mA/mm)	f_T (GHz)	f_{max} (GHz)	g_m (mS/mm)	Comments	Ref
0.25 µm AlGaAs/GaAs MODFET	100	9	15	40 115	p-channel	[7]
2 µm x 500 µm P-MODFET	250	5	4		Vertically integrated with GRINSCHSQW laser [8]	[8]
1 µm x 10 µm unstrained InGaAs/InAlAs MODFET	250	15	56	335	V_{GBr} = 23.5 V	[9]
0.2 µm x 150 µm AlGaAs/GaAs MODFET	275	72	144	330	CBE	[10]
1.8 µm AlGaAs/InGaAs	300	15	59	224	V_{BrDG} = 10 V	[11]
1.8 µm GaInP/InGaAs	400	12	42	200	V_{BrDG} = 14 V	
0.12-0.17 µm x 75 µm PHEMT	400	100	200	530		[12]
0.2 µm PHEMT	400	100	200	530	7 dB gain at 90 GHz	[13]
0.4 µm x 150 µm InAlAs/InGaAs	400	45	115	700	Metamorphic on GaAs, triangular gate	[14,15]
2 µm x 100 µm GaAs/InGaAs	500 (77 K) 690			175 (77 K) 245	δ-doped, graded channel	[16]
0.25 µm DR PMODFET	525	50	100	360	VDS = 5 V V_{BrDG} = 11 V	[17]
0.2 µm x 140 µm PMODFET	550	122	80-90	550	12 nm SQW	[18]
0.2 µm x 50 µm PMODFET	550	100	200	640	12 nm SQW	

TABLE 1 continued.

MODFET type	Imax (mA/mm)	f_T (GHz)	f_{max} (GHz)	g_m (mS/mm)	Comments	Ref
0.2 μm DR PHEMT	580	39	170	545		[19]
0.15 μm x 80 μm PHEMT 2 x heterostructure	600	90	200	500	VBrgd > 10 V	[20]
0.15 μm x 50 μm PMODFET	630	110	230	653		[21]
0.15 μm x 150 μm PMODFET		152	150			
0.3 μm PMODFET		66 66	75 75	600 450	Enhancement mode Depletion mode	[22]
1 μm x 200 μm GaInP/GaAs		17.8	23.5	163 (77 K) 213		[23,24]
0.15 μm x 200 μm N-InGaP/InGaAs/GaAs		76	191	420		[25]

with some interesting innovations. The authors have made a conscious decision to limit entry to those devices with sufficiently complete characterisation.

As mentioned above, a major advantage of MODFETs over MESFETs is the small gate-to-channel distance. The latter, along with confinement provided by the field profile in the substrate side of the channel of the MODFET, slows down the enormous increase in output conductance resulting from the scaling down of the device. In MESFETs the variation of threshold voltage ΔV_{TH} with gate length L_g is significant. As may be observed from FIGURE 1, such a variation in threshold voltage is almost absent in all-epitaxial structured MODFETs [26]. Note that these two properties guide the scalability of the device so effectively that the realisation of ultra-short submicron (≤ 0.2 μm) small-scale digital circuits has been made possible.

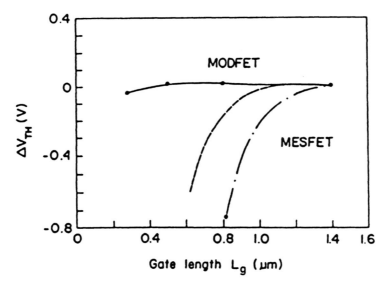

FIGURE 1. Variation of threshold voltage in MODFETs and MESFETs with gate length. The less sensitive MESFET has a buried p-buffer layer. The short-channel effect is reduced more in MODFETs than in MESFETs allowing deep-submicron devices to operate well.

The high performance of pseudomorphic MODFETs incorporating an InGaAs channel on GaAs substrates strongly demonstrates that as the InAs mole fraction of InGaAs is increased, the noise and gain performance for a given gate length and frequency improve quite remarkably. For example, the current gain cutoff frequency increases by more than a factor of two [26] as the InAs mole fraction is increased from zero to 65% (see FIGURE 2).

C   DIGITAL PERFORMANCE

MODFETs, particularly pseudomorphic MODFETs, are excellent candidates for switching elements in digital circuits. This is evident from the high intrinsic speed and large current handling capabilities demonstrated by them. Source coupled FET logic (SCFL) is based on a differential current-switching emitter-coupled-like logic used in conjunction with bipolar transistors. Employing SCFL frequency division in the range of 22 - 26 GHz has been achieved for 0.3 μm gate pseudomorphic MODFETs. SCFL is especially attractive when the highest possible speed is more important than low power consumption. FIGURE 3 summarises the performance of MODFET frequency dividers along with the competing technologies.

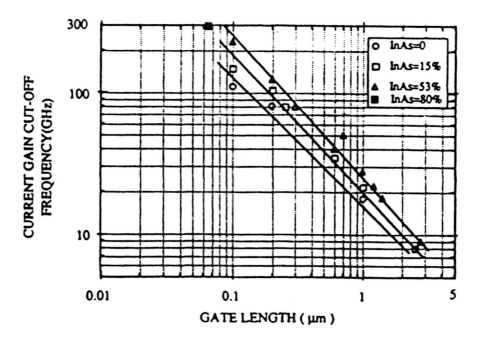

FIGURE 2. Variation of current gain cutoff frequency with gate length in InGaAs MODFETs with In mole fraction 0.0, 0.15, 0.53 and 0.8.

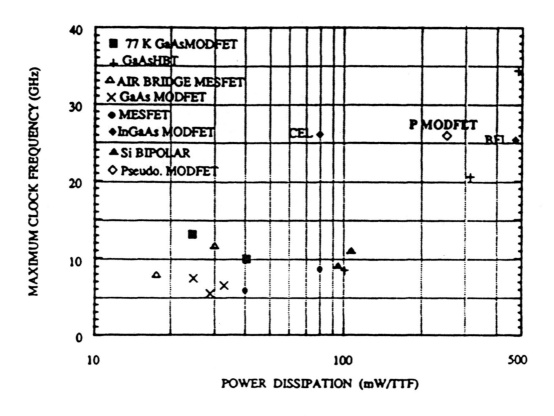

FIGURE 3. Plots of minimum frequency of division (maximum clock frequency) obtained by using pseudomorphic MODFETs and other competing technologies as a function of power consumption.

TABLE 2. Various properties of OEIC photoreceivers.

MODFET type	NRZ DR (Gbit/s)	BW (GHz)	Sensitivity (dBm)	BER	TZ (Ω)	Comments	Ref
0.3 μm AlGaAs/GaAs 2 x δ-doped	20	14.3	-16.4	10^{-9}	670	MSM PD, 115 mW power consumption, 0.84 μm	[27]
0.25 μm PMODFET	18	10			300	MSM PD, 0.85 μm	[28]
0.3 μm PMODFET	1	1 (elect.) 5.4	-25	10^{-9}		MSM PD	[29]
0.25 μm PMODFET	10	11	-20	10^{-9}	300	MSM PD, 0.85 μm tunable TZ	[30]
0.5 μm 2 x δ-doped AlGaAs/GaAs HEMT	10	8.2				MSM PD, 0.84 μm	[31,32]
0.25 μm PMODFET		4.2 (elect.) 5.6			300	MSM PD, 0.85 μm tunable TZ	[33]

TABLE 3. Various electrical properties of MMIC/MIC downconverters.

MODFET type	Frequency (GHz)	Conversion gain (dB)	NF (dB)	LO (dBm)	Comments	Ref
0.3 μm NEC NE202 HEMT	22.5-23.0	48	2.8	10	3-stage LNA, IFA, SB diode mixer/DR	[34]
75 μm wide PHEMT	Q-band	17	9	5	1-stage LNA, SE active-drain mixer 1-stage IFA, OSC	[35]
AlGaAs/GaAs	Q-band	8-11		3	2-stage LNA, SB diode mixer, 2-stage IFA	[36]
HEMT	43.5-45.5	15	6.0		3-stage LNA, 2-stage IFA, SB mixer	[37]
0.2 μm PHEMT	56-64	20	5-7	10	4-stage LNA, SB diode mixer IFA chip set also with UC and freq. mult.	[38]
0.15 μm x 100 μm AlGaAs/GaAs	60	15.5	6.8	8	4-stage RF amplifier, image rejection active-drain HEMT mixer	[39]
0.1 μm InGaAs PHEMT	90-98	5.3	6.8	10	2-stage LNA, SB diode mixer	[40]
0.1 μm PHEMT	93-95	8	6	9	3-stage LNA, image rejection >16 dB 2 SB diode mixers	[41]
0.1 μm PHEMT	94	18	4.6	10	3-stage LNA, image rejection mixer IFA hybrid IF 90° coupler	[42]

TABLE 4. Various electrical properties of MMIC upconverters.

MODFET type	IF freq. (GHz)	Gain (dB)	LO power (dBm)	LO freq. (GHz)	RF freq. (GHz)	Comments	Ref
0.2 μm PHEMT	2-10	8-11	10	54	56-64	2-stage LNA, SB diode mixer part of DC/freq. mult. chip set	[38,43,44]
0.25 μm PHEMT	6-18	3-9	5	28-40	22	3-stage RF amp, active mixer 3-stage LO amp	[45]

D   OEIC PHOTORECEIVERS

Optics is an exciting area of device integration involving MODFETs. Optics has many advantages over electronics including speed as well as immunity to electromagnetic interference, and cross-link and parasitic electronic effects such as capacitance and inductance. As the current technology is overwhelmingly electronic, a complete conversion to optics from electronics is not practical nor desirable. Rather, an interface between the optics and electronics remains more feasible. Integrated circuits employing both optics and electronics, known as OEICs, can be immensely useful for long distance fibre optic communications. Fibre optic systems convert electronic signals to optical signals for transmission, then convert the optical signals back to electronic signals at the receiving end. Thus, an effective high-speed OEIC photoreceiver is among the most pressing of prerequisites for lightwave communications systems. Successful conversion of weak optical signals to electrical signals is necessary for future optical interconnects with computers. See TABLE 2.

E   MMIC/MIC DOWNCONVERTERS

Use of millimeter-waves in new media communication designs relieves spectral overcrowding and provides sufficient bandwidth for the wireless terminal channels of Ethernet, B-ISDN, and multimedia systems. Since MMIC technologies are essential for compact, high performance, and mass-produced front-ends, the use of MMICs at millimeter-wave frequencies can provide significant advantages over the conventional hybrid integrated circuits. Downconverters are essential as the receiver's front end for numerous functions including imaging, smart munitions, and satellite broadcasting. Usually, three-stage low noise amplifiers, image rejection (IR) mixers, hybrid IF 90° couplers, and IF amplifiers are assembled together to build IR downconverters. The MODFET's laudable performance in these downconverters promises the accelerated use of low-cost and reliable MMICs. To fabricate an IR mixer using the active MODFET mixer, an active-drain mixer is suitable because the RF and LO signals can now be injected separately. Without the use of the MODFET mixer, the active-drain mixer requires a large LO power. The performances of various GaAs-based millimeter-wave downconverters built with active MODFET mixers are presented in TABLE 3.

F   MMIC UPCONVERTERS

Conversion schemes covering the full RF frequency band of 6 to 18 GHz are an essential component of many present and future electronic warfare systems. The upconversion of electrical signals complements the downconversion function. MODFETs having high-gain and low-noise capability at millimeter-wave frequencies provide an efficient means of doing so. The success of upconverter chips can stem in particular from the stable MODFET processing technology and the rigorous design methodology. The complete upconverter may consist, for example, of an active mixer, two passive baluns, an RF amplifier, and an LO amplifier. The characteristics of various MODFET upconverters fabricated in recent years are presented in TABLE 4.

G   LOW NOISE AMPLIFIERS (LNAs)

Noise figure reduction of semiconductor amplifiers is an essential element for success in microwave and millimeter-wave applications.   The key to realising low noise amplifiers is to realise high electron velocity, low source resistance, low leakage current between the gate and drain, and low gate resistance of the constituent field-effect transistors; this has been achieved by the HEMT LNAs.   HEMT LNAs possess the virtues of extremely low noise and high associated gain values in the microwave and millimeter-wave frequency bands.   Continued improvements have been seen in performance implemented by shrinking gate lengths to submicron dimensions, refinement of T-shaped gates and so on - thereby decreasing noise while minimising short-channel effects.   The state-of-the-art RF performance of HEMT LNAs is demonstrated in numerous applications, such as satellite communications, radio astronomy, airborne radar, electronic warfare, and smart munitions.   Although the InP-based discrete MODFET has demonstrated better performance at W-band or higher frequencies, the status of MMIC fabrication using GaAs-based MODFETs is much more mature than that of InP-based MODFETs.   The electrical characteristics of various GaAs-based LNAs fabricated during the past decade are summarised in TABLE 5.

H   POWER AMPLIFIERS

Highly efficient power amplifiers are essential for achieving maximum reliability, minimum size and weight, high volume, low cost, and high performance of phased array systems such as radar, satellite communications transmitters, electronic warfare, seekers, and smart munitions. In order to optimise the transmitter modules of an amplifier, its output power should be increased, which necessitates the improvement of its power performance.   To achieve this, the breakdown voltage of the device must be increased without degrading its high speed performance.   MODFET power amplifiers demonstrate superior power added efficiencies (PAE) and gain; moreover, their low cost and compact size recommend themselves well to phased array systems applications.   High linearity at high efficiency is an attractive feature that makes MODFETs especially useful for high power applications.   See TABLE 6.

I   CONCLUSION

In less than two decades, the MODFET has evolved dramatically, from an interesting research innovation to an extremely formidable yet practical device with numerous areas of application including low power communications, direct satellite broadcasting systems, millimetre-wave systems, and VLSI digital electronic systems.   Various electrical properties and application areas of MODFETs based on GaAs have been reviewed in some detail.   Most of the reports have focused on room temperature measurements of discrete devices as well as integrated circuit properties.   As is evident from the various tables presented, the MODFET has progressed to the point where the performance barriers thought to be insurmountable by three-terminal devices have been overcome with amazing dispatch.   The results emerging from various laboratories clearly indicate that the pseudomorphic MODFET has demonstrated increasingly enhanced electronic properties without compromising the breakdown voltage afforded by GaAs.   Despite this rapid progress, the day when the maximum performance of these devices has been harnessed may still be far away.   We look forward to new breathtaking results that will follow.

TABLE 5. Electrical characteristics of low noise amplifiers.

MODFET type	Frequency (GHz)	Noise figure (dB)	Associated gain (dB)	Comments	Ref
0.3 μm x 280 μm GaAs HEMT	1-13	7.8	14	2-stage hybrid lossy match	[46]
0.3 μm x 200 μm GaAs HEMT	1-20	7.5	9.5	2-stage hybrid lossy match	
0.15 μm x 300 μm PHEMT	1.5-3	1.5	21	3-stage, integrated bias + impedance	[47]
NEC 20200	2-11	5	20	Hybrid 2 x 4 matrix	[48]
0.25 μm x 150 μm AlGaAs/GaAs HEMT	2-18	5.2	11	2-stage, NF <3.5 dB @ 3-14 GHz	[49]
0.35 μm x 200 μm PHEMT	4	0.4			[50]
PHEMT	5-11	2.2	23	2-stage, broadband balanced	[51]
0.25 μm x 150 μm AlGaAs/GaAs	6-18	2.5	15.0	2-stage, 40 mW DC consumption	[52]
0.2 μm PHEMT	6-21	5.5	20	2 x 3 matrix distributed, fT = 60 GHz	[53]
0.25 μm AlGaAs/GaAs	8	0.4			[54]
0.5 μm x 300 μm AlGaAs/GaAs HEMT	8-10.5	1.5	11	2-stage	[49]
PHEMT	9-19	2.5	19	2-stage, broadband balanced	[51]
0.12-0.17 μm x 75 μm PHEMT	9-70	7	4	Distributed, ft = 100 GHz, fmax = 200 GHz, gm = 530 mS/mm, Imax = 400 mA/mm	[12]
0.3 μm PHEMT	10	0.6	13	1 shot UV lithography	[55]
0.25 μm dual gate cascode PHEMT	10	1.1	22	gm = 600 mS/mm, fT = 45 GHz	[56]
0.25 μm x 150 μm AlGaAs/GaAs HEMT	10-30	3	8	Distributed. gm = 520 mS/mm, 13 dBm @ 1 dB, 18 GHz	[57]
0.75 μm x 200 μm AlAs/GaAs SL HEMT	10	0.9	11	Vds = 2 V, 248 nm laser desorption, gm = 350-400 mS/mm	[58]
	12	1.2	10	Vds = 2 V, 248 nm laser desorption	

TABLE 5 continued.

MODFET type	Frequency (GHz)	Noise figure (dB)	Associated gain (dB)	Comments	Ref
0.25 μm x 200 μm AlGaAs/GaAs HEMT	12	0.6	11.0	VD = 2 V	[59]
0.5 μm x 200 μm AlGaAs/GaAs HEMT	12	1.04	10.8	VD = 2 V	
1.2 μm x 200 μm AlGaAs/GaAs HEMT	12	2.58	7.2	VD = 2 V	
0.2 μm x 160 μm PHEMT	12	0.68	10.4	Vds = 2 V. gm = 510 mS/mm. ft = 40 GHz off-set recess gate	[60]
	12	1.2	31	3-stage, 2nd and 3rd transistors are MESFETs	
0.25 μm x 200 μm AlGaAs/GaAs HEMT	12	0.54	12.1	WSi gate	[61]
0.5 μm x 200 μm AlGaAs/GaAs HEMT	12	1.7	15.0	2-stage. Vds = 3 V	[62]
0.2 μm x 280 μm PMODFET	12	1.25	17	2-stage, lumped element MMIC	[63]
0.15 μm x 200 μm N-InGaP/InGaAs/GaAs	12	0.41	13.0		[25]
0.25 μm x 120 μm PHEMT	12	0.65	14.5	gm = 700 mS/mm, Imax = 600 mA/mm	[64]
	18	0.82	11.5	gm = 700 mS/mm, Imax = 600 mA/mm	
0.15 μm x 50 μm DH PHEMT	18	0.55	15.2		[65]
0.15 μm x 30/50 μm PHEMT	18	0.55	15.0	gm = 650 mS/mm	[66]
0.25 μm dual gate cascode PHEMT	18	1.9	16	gm = 600 mS/mm, fT = 45 GHz	[56]
0.25 μm x 150 μm AlGaAs/GaAs HEMT	18	0.95	13	π-configured, mushroom gate	[57]
0.35 μm x 200 μm PHEMT	18	1.2			[50]

TABLE 5 continued.

MODFET type	Frequency (GHz)	Noise figure (dB)	Associated gain (dB)	Comments	Ref
0.15 μm x 30 μm PHEMT	18	0.5	15.1	ft = 135 GHz, Ids = 5.1 mA	[67]
0.1 μm x 200 μm AlGaAs/GaAs	18	0.51	10.8	Vds = 1.2 V	[68]
0.25 μm x 50/75 μm PHEMT	18	1	9	gm = 480 mS/mm, monolithic integration w/power HEMT	[69]
0.15 μm x 150 μm PHEMT	18 18	0.67 0.93	12.2 11.0	Π-gate T-gate	[70]
0.25 μm x 150 μm AlGaAs/GaAs HEMT	18-23	2.0	29	3-stage	[49]
0.1 μm x 100 μm PHEMT	18-40	3.6	5.7	Balanced cascade	[71]
0.35 μm x 75 μm AlGaAs/GaAs HEMT	20 20	2 2.5	13 18	gm = 456 mS/mm monolithic int. of SBD mixer 2-stage	[72]
0.25 μm x 200 μm AlGaAs/GaAs HEMT	20	0.96	8.6	VD = 2 V	[59]
0.5 μm x 200 μm AlGaAs/GaAs HEMT	20	1.80	8.6	VD = 2 V	
1.2 μm x 200 μm AlGaAs/GaAs HEMT	20	4.28	4.4	VD = 2 V	
0.2 μm x 120 μm PHEMT	21-23	2	33	3-stage	[73]
0.15 μm PHEMT	26	1	8	3-stage, LNA/mixer MMIC	[74]
0.1 μm x 200 μm AlGaAs/GaAs	26.5	0.9	8.8	Vds = 1.2 V	[68]
PHEMT	26-40	4.75	20	2-stage, broadband balanced	[51]
0.25 μm x 200 μm AlGaAs/GaAs HEMT	30	1.4	6.4	VD = 2 V	[59]

TABLE 5 continued.

MODFET type	Frequency (GHz)	Noise figure (dB)	Associated gain (dB)	Comments	Ref
0.5 μm x 200 μm AlGaAs/GaAs HEMT	30	2.4	5.1	VD = 2 V	
1.2 μm x 200 μm AlGaAs/GaAs HEMT	30	6.3	1.2	VD = 2 V	
0.25 μm PHEMT	33-37	3.5	17	2 x sided delta doped, 2-stage	[75]
0.2 μm PHEMT	34-36	3.9	9	Merged HEMT-HBT process	[76]
0.15 μm PHEMT	36-46 41-45	3.5 3	20 22	3-stage 3-stage	[77]
0.1 μm x 200 μm AlGaAs/GaAs	40	1.9	5.3	Vds = 1.2 V	[68]
0.1 μm x 100 μm PHEMT	43	1.32 (17 K) 0.36	6.7 (17 K) 6.9	Vds = 2.0 V	[78,79]
0.15 μm x 200 μm N-InGaP/InGaAs/GaAs	50	1.2	5.8		[25]
0.20 μm x 60 μm AlGaAs/GaAs HEMT	55-57	5	20	3-stage	[80,81]
0.3 μm PMODFET	58.5	5.2	10.5	2-stage. VD = 3.3 V, fT = 65 GHz. fmax = 130 GHz	[82,83]
0.15 μm x 30/50 μm PHEMT	60	1.6	7.6	gm = 650 mS/mm	[66]
0.25 μm AlGaAs/GaAs	60	1.8			[54]
0.15 μm x 30 μm PHEMT	60	1.6	7.6	ft = 135 GHz, Ids = 5.1 mA	[67]
0.20 μm x 60 μm AlGaAs/GaAs HEMT	60	4.5	4.0	1-stage, noise bias, 3.3 dBm @ P 1dB low-pass topology	[80,81]
		4.9	4.5	1-stage, gain bias, 3.3 dBm @ P 1 dB low-pass topology	

TABLE 5 continued.

MODFET type	Frequency (GHz)	Noise figure (dB)	Associated gain (dB)	Comments	Ref
0.20 μm x 60 μm AlGaAs/GaAs HEMT	60	5.1 5.5	4.0 4.7	1-stage, noise bias, band-pass topology 1-stage, gain bias, band-pass topology	
0.20 μm x 60 μm PHEMT	60	5.4 6.2	9.6 10.5	2-stage. noise bias, 6.4 dB @ P 1 dB. band-pass topology 2-stage. gain bias, 6.4 dBm @ P 1 dB, band-pass topology	
0.20 μm x 60 μm AlGaAs/GaAs HEMT	60	5.0 5.5	6.2 6.7	2-stage, noise bias, low-pass topology 2-stage, gain bias, low-pass topology	
0.20 μm x 60 μm AlGaAs/GaAs HEMT	60	5.3 5.8	14.2 15	3-stage, noise bias, low-pass topology 3-stage, gain bias, low-pass topology	
PHEMT	91-95 91-95 92-95	5 5.5-6 6.5-6.8	24 40-49 43-54	4-stage 8-stage. noise bias 8-stage, gain bias	[84]
0.1 μm PHEMT	92-96	4.2-4.8	12	2-stage conductor backed CPW	[85]
0.1 μm x 40 μm PHEMT	92-95 94 94	5.1 5.5 6.5	7 13.3 49	1-stage, VD = 3 V. gm = 670 mS/mm. fT = 130 GHz 2-stage 8-stage	[86,87]
0.1 μm x 40 μm PMODFET	93.5	2.1	6.3	fmax = 290 GHz	[88]
0.1 μm x 50 μm PHEMT	94	3.0	5.1	gm = 700 mS/mm, 18 dB gain at 18 GHz	[89]
0.1 μm x 40 μm PHEMT	94	5.5	13.3	2-stage	[90]
0.1 μm PHEMT	94 94	3.5 4.7	5.3 11	1-stage 2-stage	[91]
0.15 μm x 30 μm PHEMT	94 93 94	2.4 4.2 4.5	5.4 9.7 14.8	ft = 135 GHz, lds = 5.1 mA 2-stage 3-stage	[67]

TABLE 5 continued.

MODFET type	Frequency (GHz)	Noise figure (dB)	Associated gain (dB)	Comments	Ref
0.1 μm x 40 μm PHEMT	94	4.5-5.5	17	3-stage	[41]
0.1 μm x 40 μm PHEMT	94	3.5	21	3-stage. VD = 2.5 V. gm = 670 mS/mm, ft = 140 GHz	[92]
0.1 μm x 50 μm PHEMT	94	4.0	30.8	4-stage	[93]
	92	3.5	23	4-stage	
	98	4.6	26	4-stage, 1st pass	
	102	5.9	31.7	4-stage, 1st pass	
0.1 μm PHEMT	112-115	6.3	12	2-stage, 1st pass, high gain bias	[94-96]
	112-115	5.5	10	low noise bias	
	110	3.9	19.6	2-stage. 1st pass. high gain bias	
	113	3.4	15.6	low noise bias	

TABLE 6. Electrical characteristics of power amplifiers.

MODFET type	Frequency (GHz)	Gain (dB)	Power (mW)	PAE	Comments	Ref
0.32 μm x 48 μm 2-pulse doped MODFET	0.5-50	6.5	16 at 40 GHz		6-stage distributed, VD = 4 V, ft = 45 GHz, fmax = 110 GHz, NF = 4.8 dB at 0.5-26.5 GHz	[97]
0.2 μm x 50 μm recessed AlGaAs HEMT	2-52	9	2.5		Matrix distributed, 2-stage 4-section VD = 3 V	[98]
	2.45	13.5	max 12.6 10 W	63%		[99]
0.25 μm x 8 mm DR PHEMT	4	15.4	4.3 W	66%	gmmax = 430 mS/mm, Imax = 450 mA/mm. VD = 8 V	[100]
	4	12.6	5.7 W	57%	VD = 11 V	
	4	16.6	3 W	63%		
	4		4.1 W	72%	VD = 7 V	
0.25 μm x 400 μm DR PHEMT	4.5	17.2	330	63%	VD = 8 V, gmmax = 510 mS/mm, Imax = 540 mA/mm	[101]
	4.5		505	40%	VD = 14 V	
	10	11	326	62%	VD = 8 V	
0.7 μm x 3 mm 2-HJ SL AlGaAs/GaAs IHEMT	5.5	8.3	1.3 W	55%	gm = 180 mS/mm	[102]
0.25 μm x 1.12 mm recessed PHEMT	9	9.25	850	50%	VD = 5 V, gm = 428 mS/mm, fT = 50 GHz, Imax = 545 mA/mm	[103]
0.35 μm PHEMT DR 2 x pulse doped, P = 1.2 mm	10	10.4	870	59%	VD = 7 V	[104]
0.25 μm PHEMT DR. P = 1.2 mm	10	10	970	70%	Dry first recess etching, VD = 8 V	[105]
	18	6.8		48%		
0.25 μm x 1.6 mm 2-HJ PHEMT	12	14	2.2 W	39%	2-stage	[106]
	12	14	2.75 W	36%	2-stage, power tuning	
	15	10.8	600	51%	single device	

TABLE 6 continued.

MODFET type	Frequency (GHz)	Gain (dB)	Power (mW)	PAE	Comments	Ref
0.25 μm x 8 mm DR PHEMT, QW planar- and pulse-doped	12	10.8	6 W	52%	VD = 9 V, gm = 420 mS/mm. Imax = 430 mA/mm	[107]
	12	10.3	5.4 W	53%	VD = 8 V	
	12	10.8	6.0 W	52%	VD = 7 V	
	12	10	3.9 W	55.6%	VD = 6 V	
GaAs/InGaAs/GaAs PHEMT	14.25	5	4.7 W	25%	VBr = 20 V. gm = 224 mS/mm	[108]
10.5 mm wide GaAs/InGaAs/GaAs HEMT	14.25	lin. 8	4.7 W	25%	VD = 10 V. VBr = 25.7 V	[109]
PHEMT	15	12	575	50%		[110]
0.25 μm PHEMT. P = 1.2 mm	18	6.3	776	53%	Imax = 550 mA/mm, VD = 7 V. gm = 350 mS/mm. 2-pulse doped	[111]
	10	10.2	955	66%	VD = 8 V	
0.25 μm x 50 μm PHEMT	18	8	20	59%	gm = 554 mS/mm. monolithic integration w/HEMT LNA	[69]
0.33 μm x 120 μm PMODFET	25	12.9	680	45%	n⁺ GaAs supply layer. Imax = 530 mA/mm. VD = 5 V	[112]
	25	10.9	810	59%		
0.25 μm PHEMT	30	8.5	500	40%	2-stage cascaded	[113]
	31	11	141	40.3%	2-stage cascaded	
0.2 μm DR PHEMT. P = 600 μm	32	6	500	35%		[19]
	44	4.3	494	30%		
0.25 μm x 400 μm PHEMT	34-36	12	500	32%	2-sided delta-doped 2-stage	[75]
0.15 μm PHEMT, P = 1.2 mm	34-36	9	1 W	20%	VD = 5 V	[114]
	34-36	17	3 W	15%	2-stage	

TABLE 6 continued.

MODFET type	Frequency (GHz)	Gain (dB)	Power (mW)	PAE	Comments	Ref
0.25 μm x 900 μm PHEMT	35	3.2	658	24%		[115]
0.15 μm x 50 μm PHEMT	35 35	9.0 8.5	32 42	51% 37%		[65,116]
0.15 μm x 150 μm PHEMT	35 35	8.0 7.6	95 137	50% 40%		
0.2 μm PHEMT, P = 200 μm	40	19.1	41		3-stage	[73]
0.2 μm x 600 μm DR 2-pulse doped PHEMT	40-45	10-11	500-725	10-17%	1st pass, 3-stage. VD = 5 V	[117]
0.25 μm x 400 μm PHEMT	55	4.6	184	25%	gm = 500 mS/mm. Imax = 540 mA/mm	[118]
0.25 μm x 320 μm PHEMT	55	4.9	153	25%		
0.25 μm x 240 μm PHEMT	55	4.9	105	22%		
0.2 μm x 50 μm PHEMT	55	3.3	42	22.1%	gm = 760 mS/mm, Imax = 800 mA/mm	[119]
0.3 μm x 800 μm PHEMT	55	4.1	219	18%	ft = 50 GHz, fmax = 92 GHz, Imax = 420 mA, gm = 360 mSmm	[120]
0.2 μm x 80 μm PHEMT 2-HJ	59.5	8.95	313	19.9%	2-stage. VD = 5 V	[121]
0.15 μm x 320 μm DR PHEMT	59.5-63.5 59.5-63.5	7 11.7	370 740	11% 8%	V = 5 V, Isat = 500 mA/mm, fT > 75 GHz 1 amp drives 3 amps, VD = 5 V	[122,123]
0.15 μm x 50 μm PHEMT	60 60	6.0 5.9	32 42	41% 37%	NF = 1.8 dB, Ass. gain = 6.4 dB	[65,116]
0.15 μm x 150 μm PHEMT	60 60	4.7 4.5	82 125	38% 32%		

TABLE 6 continued.

MODFET type	Frequency (GHz)	Gain (dB)	Power (mW)	PAE	Comments	Ref
0.15 μm x 400 μm PHEMT 2-HJ	60	4.5	225	25.4%	VD = 5 V	[20]
0.15 μm x 400 μm	60	4.4	174	28.8%	VD = 4.5 V	
0.15 μm x 320 μm	60	5.3	170	31.1%	VD = 4.5 V	
0.15 μm x 320 μm	60	5.1	191	28.7%	VD = 5 V	
0.15 μm x 100 μm PMODFET	77	21	12		3-stage, VD = 3.5 V. fT = 110 GHz, fmax = 200 GHz	[124]
0.1 μm x 160 μm PHEMT	93.5	5.9	100	6.6%	2-stage, VD = 3.5 V	[125]
0.1 μm x 40 μm PMODFET	94	7.3	10.6	14.3%	fmax = 290 GHz	[126]
0.15 μm x 50 μm PHEMT	94 / 94	3.3 / 3.2	18 / 22	23% / 19%		[65,116]
0.15 μm x 150 μm PHEMT	94 / 94	3.0 / 2.0	45 / 57	16% / 16%		

Terminology used: 2-HJ = double heterojunction, DR = double recessed, P = periphery. VBr = breakdown voltage, gm = transconductance, VD = drain voltage.

REFERENCES

[1]   T. Mimura, S. Hiyamizu, T.Fujii, K. Nanbu [*Jpn. J. Appl. Phys. (Japan)* vol.19 (1980) p.L225-7]

[2]   H. Morkoc, H. Unlu, G. Ji [*Fundamentals and Technology of MODFETs* (Wiley and Sons, New York, USA, 1991) vol.I]

[3]   H. Morkoc, H. Unlu, G. Ji [*Fundamentals and Technology of MODFETs* (Wiley and Sons, New York, USA, 1991) vol.II]

[4]   T.J. Drummond, W.T. Masselink, H. Morkoc [*Proc. IEEE (USA)* vol.74 (1986) p.773-822]

[5]   H. Morkoc et al [U.S. Patent no.4,827,320]

[6]   H. Morkoc, P.M. Solomon [*IEEE Spectr. (USA)* vol.21 (1984) p.28-35]; see also, H. Morkoc [*IEEE Circuits Devices Mag. (USA)* vol.7 (1991) p.14-20]

[7]   H. Park, P. Mandeville, R. Saito, P.J. Tasker, W.J. Schaff, L.F. Eastman [*Proc. IEEE/Cornell Conf. on Advanced Concepts in High Speed Semiconductor Devices and Circuits* (IEEE, New York, 1989) p.101-10]

[8]   S.D. Offsey, P.J. Tasker, W.J. Schaff, L. Kapitan, J.R. Shealy, L.F. Eastman [*Electron. Lett. (UK)* vol.26 (1990) p.350-2]

[9]   N.C. Tien, J. Chen, J.M. Fernandez, H.H. Wieder [*IEEE Electron Device Lett. (USA)* vol.13 (1992) p.621-3]

[10]   R. Kempter, H. Rothfritz, J. Plauth, R. Muller, G. Trankle, G. Weimann [*Electron. Lett. (UK)* vol.28 (1992) p.1160-1]

[11]   J. Dickmann et al [*Proc. IEEE/Cornell Conf. on Advanced Concepts in High Speed Semiconductor Devices and Circuits* (IEEE, New York, 1993) p.529-38]

[12]   N. Camilleri, P. Chye, A. Lee, P. Gregory [*IEEE 1990 Microwave and Millimeter-Wave Monolithic Circuits Symp.* (IEEE, New York, 1990) p.27-30]

[13]   N. Camilleri, P. Chye, P. Gregory, A. Lee [*IEEE MTT-S Int. Microw. Symp. Dig. (USA)* (1990) p.903-6]

[14]   P. Win, Y. Druelle, A. Cappy, Y. Cordier, D. Adam, J. Favre [*Proc. IEEE Cornell Conf. on Advanced Concepts in High Speed Semiconductor Devices and Circuits* (IEEE, New York, USA, 1993) p. 511-9]

[15]   P. Win et al [*Electron. Lett. (UK)* vol.29 (1993) p.169-70]

[16]   H.-M. Shieh, W.-C. Hsu, R.-T. Hsu, C.-L. Wu, T.-S. Wu [*IEEE Electron Device Lett. (USA)* vol.14 (1993) p.581-3]

[17]   J. Perdomo et al [*IEEE 14th Annual GaAs IC Symp.* (IEEE, New York, 1992) p.203-6]

[18]   L.D. Nguyen, D.C. Radulescu, M.C. Foisy, P.J. Tasker, L.F. Eastman [*IEEE Trans. Electron Devices (USA)* vol.36 (1989) p.833-8]

[19]   J.C. Huang et al [*IEEE Electron Device Lett. (USA)* vol.14 (1993) p.456-8]

[20]   R. Lai et al [*IEEE Microw. Guid. Wave Lett. (USA)* vol.3 (1993) p.363-5]

[21]   L.D. Nguyen, P.J. Tasker, D.C. Radulescu, L.F. Eastman [*Int. Electron Devices Meet. Tech. Dig. (USA)* (1988) p.176-9]

[22]   M. Tong, K. Nummila, J.-W. Seo, A. Ketterson, I. Adesida [*Electron. Lett. (UK)* vol.28 (1992) p.1633-4]

[23]   Y.-J. Chan, D. Pavlidis [*IEEE Trans. Electron Devices (USA)* vol.41 (1994) p.637-42]

[24]   Y.-J. Chan, D. Pavlidis, M. Razeghi, F. Omnes [*IEEE Trans. Electron Devices (USA)* vol.37 (1990) p.2141-7]

[25]   M. Takikawa, K. Joshin [*IEEE Electron Device Lett. (USA)* vol.14 (1993) p.406-8]

[26]   H. Morkoc, B. Sverdlov, G.B. Gao [*Proc. IEEE (USA)* vol.81 (1993) p.492-556]

[27]   V. Hurm et al [*Electron. Lett. (UK)* vol.29 (1993) p.9-10]

[28]   A. Ketterson et al [*IEEE Trans. Electron Devices (USA)* vol.39 (1992) p.2676-7]

[29]   J.J. Morikuni et al [*IEEE Int. Solid-State Circuits Conf. (USA)* (1993) p.178-9]

[30]   A.A. Ketterson et al [*IEEE Trans. Electron Devices (USA)* vol.40 (1993) p.1406-16]

[31]   V. Hurm et al [*Electron. Lett. (UK)* vol.27 (1991) p.734-5]

[32]   V. Hurm et al [*IEEE Trans. Electron Devices (USA)* vol.38 (1991) p.2713]

[33]   A.A. Ketterson et al [*IEEE Photonics Technol. Lett. (USA)* vol.4 (1992) p.73-6]

[34]   K. Imai, H. Nakakita [*IEEE Trans. Microw. Theory Tech. (USA)* vol.39 (1991) p.993-9]

[35]   R. Ramachandran, C. Woo, N. Nijiar, D. Fisher [*IEEE 1993 Microwave and Millimeter-Wave Monolith. Circ. Symp.* (IEEE, New York, 1993) p.93-5]

[36]   T.N. Ton, T.H. Chen, G.S. Dow, K. Nakano, L.C.T. Liu, J. Berenz [*IEEE Annual GaAs Symp.* (IEEE, New York, 1990) p.185-8]

[37]   J. Berenz, M. LaCon, M. Aust [*12th Annual GaAs IC Symp.* (IEEE, New York, 1990) p.189-92]

[38]   H. Wang et al [*IEEE Trans. Microw. Theory Tech. (USA)* vol.42 (1994) p.11-7]

[39]   T. Saito, N. Hikada, Y. Ohashi, T. Shimura, Y. Aoki [*IEEE Microwave and Millimeter-Wave Monolithic Circuits Symp.* (IEEE, New York, 1993) p.77-80]

[40]   K.W. Chang et al [*IEEE 1991 Microwave and Millimeter-Wave Monolith. Cir. Symp.* (IEEE, New York, 1991) p.55-8]

[41]   H. Wang et al [*IEEE Microw. Guid. Wave Lett. (USA)* vol.3 (1993) p.281-3]

[42]   K.W. Chang et al [*IEEE Trans. Microw. Theory Tech. (USA)* vol.40 (1992) p.2332-8]

[43]   H. Wang et al [*IEEE MTT-S Int. Microw. Symp. Dig. (USA)* vol.2 (1992) p.1059-62]

[44]   H. Wang et al [*IEEE 1992 Microwave and Millimeter-Wave Monolithic Circuits Symp.* (IEEE, New York, 1992) p.197-200]

[45]   H. Fudem, S. Moghe, G. Dietz [*IEEE 14th Annual GaAs IC Symp.* (IEEE, New York, 1992) p.59-62]

[46]   Y. Ito, A. Takeda [*IEEE MTT-S Int. Microw. Symp. Dig. (USA)* (1988) vol.1 p.347-50]

[47]   H. Morkner, M. Frank, D. Millicker [*IEEE 1993 Microwave and Millimeter-Wave Monolithic Circuits Symp. Dig.* (IEEE, New York, 1993) p.13-6]

[48]   S. D'Agostino, G. D'Inezo, G. Grifoni, P. Marietti, G. Panariello [*Electron. Lett. (UK)* vol.27 (1991) p.506-7]

[49]   M.A.G. Upton et al [*IEEE 1989 Microwave and Millimeter-Wave Monolithic Circuits Symp. Dig.* (IEEE, New York, 1989) p.105-10]

[50]   R. Plana et al [*IEEE Trans. Electron Devices (USA)* vol.40 (1993) p.852-9]

[51]   B. Nelson et al [*IEEE 12th Annual GaAs IC Symp.* (IEEE, New York, 1990) p.165-8]

[52]   J. Panelli et al [*IEEE Microwave and Millimeter-Wave Monolithic Circuits Symp.* (IEEE, New York, 1992) p.21-4]

[53]   K.W. Kobayashi et al [*IEEE Microw. Guid. Wave Lett. (USA)* vol.3 (1993) p.11-3]

[54]   K.-H.G. Duh et al [*IEEE Electron Device Lett. (USA)* vol.9 (1988) p.521-3]

[55]   E.Y. Chang, K.C. Lin, E.H. Liu, C.Y. Chang, T.H. Chen, J. Chen [*IEEE Electron Device Lett. (USA)* vol.15 (1994) p.277-9]

[56]   J. Wenger, P. Narozny, H. Dambkes, J. Splettstosser, C. Werres [*IEEE Microw. Guid. Wave Lett. (USA)* vol.2 (1992) p.46-8]

[57]   C. Yuen, C. Nishimoto, M. Glenn, Y.C. Pao, S. Bandy, G. Zdasiuk [*IEEE 10th Annual GaAs IC Symp.* (IEEE, New York, 1988) p.105-8]

[58]   J.M. Dumas et al [*IEEE MTT-S Int. Microw. Symp. Dig. (USA)* (1989) p.483-6]

[59]   K. Joshin, S. Asai, Y. Hirachi, M. Abe [*IEEE Trans. Electron Devices (USA)* vol.36 (1989) p.2274-80]

[60]   O. Ishikawa et al [*IEEE MTT-S Int. Microw. Symp. Dig. (USA)* (1989) p.979-82]

[61]   I. Hanyu et al [*Electron. Lett. (USA)* vol.24 p.1327-8]

[62]   N. Ayaki et al [*IEEE 10th Annual GaAs IC Symp.* (IEEE, New York, 1988) p.101-4]

[63]   R. Bosch, P.J. Tasker, M. Schlechtweg, J. Braunstein, W. Reinert [*Electron. Lett. (UK)* vol.29 (1993) p.1394-5]

[64]   J. Wenger [*IEEE Electron Device Lett. (USA)* vol.14 (1993) p.16-8]

[65]   P.M. Smith et al [*IEEE MTT-S Int. Microw. Symp. Dig. (USA)* (1989) p.983-6]

[66]   P.C. Chao, P. Ho, K.H.G. Duh, P.M. Smith, J.M. Ballingall, A.A. Jabra [*Electron. Lett. (UK)* vol.26 (1990) p.27-8]

[67]   K.H.G. Duh et al [*IEEE MTT-S Int. Microw. Symp. Dig. (USA)* (1990) p.595-8]

[68]   H. Kawasaki, T. Shiono, M. Kawano, K. Kamei [*IEEE MTT-S Int. Microw. Symp. Dig. (USA)* (1989) p.423-6]

[69]   P. Saunier, H.Q. Tserng, H.D. Shih, K. Bradshaw [*Electron. Lett. (UK)* vol.25 (1989) p.583-4]

[70]   K.C. Hwang, P. Ho, F.R. Bardsley [*Electron. Lett. (UK)* vol.29 (1993) p.1116-7]

[71]   M. Kimishima, T. Ashizuka [*IEEE MTT-S Int. Microw. Symp. Dig. (USA)* vol.2 (1993) p.523-6]

[72]   W.-J. Ho et al [*IEEE 10th Annual GaAs IC Symp.* (IEEE, New York, 1988) p.301-4]

[73]   J.A. Lester, W.L. Jones, P.D. Chow [*IEEE MTT-S Int. Microw. Symp. Dig. (USA)* (1991) p.433-6]

[74]   H.C. Huang et al [*IEEE Microwave and Millimeter-Wave Monolithic Circuits Symp.* (IEEE, New York, 1994) p.37-40]

[75]   A. Kurdoghlian et al [*IEEE Microwave and Millimeter-Wave Monolithic Circuits Symp.* (IEEE, New York, 1993) p.97-8]

[76]   D.K. Umemoto, D.C. Streit, K.W. Kobayashi, A.K. Oki [*IEEE Microw. Guid. Wave Lett. (USA)* vol.4 (1994) p.361-3]

[77]   K.H.G. Duh, S.M.J. Liu, S.C. Wang, P. Ho, P.C. Chao [*IEEE 1993 Microwave and Millimeter-Wave Monolithic Circuits Symp.* (IEEE, New York, 1993) p.99-102]

[78]   R.E. Lee, R.S. Beaubien, R.H. Norton, J.W. Bacon [*IEEE Trans. Microw. Theory Tech. (USA)* vol.37 (1989) p.2086-92]

[79]   R.E. Lee, R.S. Beaubien, R.H. Norton, J.W. Bacon [*IEEE MTT-S Int. Microw. Symp. Dig. (USA)* (1989) p.975-8]

[80]   M. Aust, J. Yonaki, K. Nakano, J. Berenz, G.S. Dow, L.C.T. Liu [*IEEE 11th Annual GaAs IC Symp.* (IEEE, New York, 1989) p.95-8]

[81]   M. Aust, J. Yonaki, K. Nakano, J. Berenz, G. Dow, L.C.T. Liu [*IEEE Milcom Symp. Proc. (USA)* vol.3 (1989) p.739-43]

[82]   M. Schlechtweg et al [*IEEE Trans. Microw. Theory Tech. (USA)* vol.40 (1992) p.2445-51]

[83]   M. Schlechtweg et al [*IEEE 1992 Microwave and Millimeter-Wave Monolithic Circuits Symp.* (IEEE, New York, 1992) p.29-32]

[84]   T.N. Ton, B. Allen, H. Wang, G.S. Dow, E. Barnachea, J. Berenz [*IEEE Microw. Guid. Wave Lett. (USA)* vol.2 (1992) p.63-4]

[85]   T.-N. Ton et al [*Electron. Lett. (UK)* vol.29 (1993) p.1804-5]

[86]   H. Wang et al [*IEEE Trans. Microw. Theory Tech. (USA)* vol.40 (1992) p.417-28]

[87]   H. Wang et al [*IEEE MTT-S Int. Microw. Symp. Dig. (USA)* vol.3 (1991) p.943-6]

[88]   K.L. Tan et al [*IEEE Electron Device Lett. (USA)* vol.11 (1990) p.585-7]

[89]   P.C. Chao, K.H.G. Duh, P. Ho, P.M. Smith, J.M. Ballingall, A.A. Jabra [*Electron. Lett. (UK)* vol.25 (1989) p.504-5]

[90]   K.L. Tan et al [*Electron. Lett. (UK)* vol.27 (1991) p.1166-7]

[91]   H. Wang et al [*Int. Electron Devices Meet. Tech. Dig. (USA)* (1991) p.939-42]

[92]   H. Wang et al [*IEEE MTT-S Int. Microw. Symp. Dig. (USA)* (1992) p.803-6]

[93]   D.-W. Tu et al [*IEEE 1993 Microwave and Millimeter-Wave Monolithic Circuits Symp.* (IEEE, New York, 1993) p.29-32]

[94]   H. Wang et al [*IEEE J. Solid-State Circuits (USA)* vol.28 (1993) p.988-93]

[95]   H. Wang et al [*IEEE MTT-S Int. Microw. Symp. Dig. (USA)* vol.2 (1993) p.783-5]

[96]   H. Wang et al [*IEEE 14th Annual GaAs IC Circuit Symp.* (IEEE, New York, 1992) p.23-6]

[97]   J. Perdomo, M. Mierzwinski, H. Kondoh, C. Li, T. Taylor [*IEEE 11th Annual GaAs IC Symp.* (IEEE, New York, 1989) p.91-4]

[98]   R. Heilig, D. Hollman, G. Baumann [*IEEE MTT-S Int. Microw. Symp. Dig. (USA)* (1994) p.459-62]

[99]   L. Aucoin et al [*IEEE 15th Annual GaAs IC Symp.* (IEEE, New York, 1993) p.351-3]

[100]   S.T. Fu et al [*IEEE MTT-S Int. Microw. Symp. Dig. (USA)* (1993) p.1469-72]

[101]   M.-Y. Kao et al [*Int. Electron Devices Meet. Tech. Dig. (USA)* (1992) p.319-21]

[102]   J.M. Van Hove et al [*IEEE Electron Device Lett. (USA)* vol.9 (1988) p.530-2]

[103]   C.S. Wu, F. Ren, S.J. Pearton, M. Hu, C.K. Pao, R.F. Wang [*Electron. Lett. (UK)* vol.30 (1994) p.1803-5]

[104]   S. Shanfield et al [*IEEE 14th Annual GaAs IC Symp.* (IEEE, New York, 1992) p.207-10]

[105]   S. Shanfield et al [*IEEE MTT-S Int. Microw. Symp. Dig. (USA)* (1992) p.639-41]

[106]   D. Helms et al [*IEEE MTT-S Int. Microw. Symp. Dig. (USA)* (1991) p.819-21]

[107]   S.T. Fu, L.F. Lester, T. Rogers [*IEEE MTT-S Int. Microw. Symp. Dig. (USA)* (1994) p.793-6]

[108]   T. Fujii et al [*IEEE 13th Annual GaAs IC Symp.* (IEEE, New York, 1991) p.109-12]

[109]   T. Sonoda et al [*Electron. Lett. (UK)* vol.27 (1991) p.1304-5]

[110]   P.M. Smith et al [*Electron. Lett. (UK)* vol.27 (1991) p.270-1]

[111]   D. Danzilio, L.K. Hanes, B. Lauterwasser, B. Ostrowski, F. Rose [*IEEE14th Annual GaAs IC Symp.* (IEEE, New York, 1992) p.255-7]

[112]   G.-G. Zhou, K.T. Chan, B. Hughes, M. Mierzwinski, H. Kondoh [*Int. Electron Devices Meet. Tech. Dig. (USA)* (1989) p.109-12]

[113]   H.Q. Tserng, P. Saunier, Y.-C. Kao [*IEEE Microwave and Millimeter-Wave Monolithic Circuits Symp.* (IEEE, New York, 1992) p.51-4]

[114]   M.V. Aust et al [*IEEE Microwave and Millimeter-Wave Monolithic Circuits Symp.* (IEEE, New York, 1993) p.45-8]

[115]   P.M. Smith et al [*Electron. Lett. (UK)* vol.25 (1989) p.639-40]

[116]   A.W. Swanson, P.M. Smith, P.C. Chao, K.H. Duh, J.M. Ballingall [*IEEE Milcom Symp. Proc. (USA)* vol.3 (1989) p.744-8]

[117]   W. Boulais et al [*IEEE MTT-S Int. Microw. Symp. Dig. (USA)* vol.2 (1994) p.649-52]

[118]   K.L. Tan et al [*IEEE Electron Device Lett. (USA)* vol.12 (1991) p.213-4]

[119]   P. Saunier, R.J. Matyi, K. Bradshaw [*IEEE Electron Device Lett. (USA)* vol.9 (1988) p.397-8]

[120]   S. Arai, H. Kojima, K. Otsuka, M. Kawano, H. Ishimura, H. Tokuda [*IEEE 13th Annual GaAs IC Symp.* (IEEE, New York, 1991) p.105-8]

[121]   R.E. Kasody et al [*IEEE Microw. Guid. Wave Lett. (USA)* vol.4 (1994) p.303-4]

[122]   A.K. Sharma et al [*IEEE MTT-S Int. Microw. Symp. Dig. (USA)* vol.2 (1994) p.813-6]

[123]   A.K. Sharma, G. Onak, R. Lai, K.L. Tan [*IEEE Microwave and Millimeter-Wave Monolithic Circuits Symp.* (IEEE, New York, 1994) p.73-6]

[124]   M. Schlechtweg et al [*Electron. Lett. (UK)* vol.29 (1993) p.1119-20]

[125]   T.H. Chen et al [*IEEE 14th Annual GaAs IC Symp.* (IEEE, New York, 1992) p.71-4]

[126]   D.C. Streit et al [*IEEE Electron Device Lett. (USA)* vol.12 (1991) p.149-50]

7.3 Properties and applications of InP-based modulation doped heterostructures

L.D. Nguyen

January 1996

A INTRODUCTION

The InP-based modulation doped heterostructures have been a subject of great interest to the electronic materials community during the past decade. In the most basic form, this type of heterostructure consists of a single-sided modulation-doped $Al_{0.48}In_{0.52}As/Ga_{0.47}In_{0.53}As/Al_{0.48}In_{0.52}As$ quantum well grown lattice-matched to an InP substrate. Exceptionally high electron sheet concentrations and room-temperature mobilities have been achieved from such quantum wells, leading to the eventual successful development of a new class of InP-based high electron mobility transistor (HEMT) for high speed, high frequency applications. This Datareview provides a set of basic material properties associated with InP-based modulation-doped heterostructures and devices.

B MATERIAL PROPERTIES

The InP-based heterostructure has generated great interest in the device research community because it possesses some of the most desirable material properties for device applications. TABLE 1 lists a set of material parameters for a typical lattice-matched AlInAs/GaInAs/AlInAs modulation-doped heterostructure [1-5]. Of these, the high conduction band offset (0.52 eV) and room-temperature mobility (>10,000 cm^2/V s) are the two most important parameters that differentiate the InP-based modulation-doped heterostructures from other material systems.

TABLE 1. Material properties for a typical InP-based modulation doped heterostructure.

	$Al_{0.48}In_{0.52}As$	$Ga_{0.47}In_{0.53}As$	$Al_{0.48}In_{0.52}As/$ $Ga_{0.47}In_{0.53}As$
Band gap (E_g)	1.4 eV	0.75 eV	
Electron effective mass ($m*$)	0.078 m_0	0.047 m_0	
Schottky barrier height (ϕ)	0.6 eV	0.2 eV	
Conduction band offset (ΔE_c)			0.52 eV
300 K mobility (μ)			10,000 cm^2/V s
Electron sheet concentration (n_s)			2.6×10^{12} cm^{-2}
Dielectric constant (e)	12.9	12.5	

Furthermore, the room-temperature mobility of $AlInAs/Ga_{1-x}In_xAs$ modulation-doped heterostructures increases with x, reaching a value as high as 16,000 cm^2/V s for x = 0.80 [6]. In recent years, significant effort has been devoted to the study of the 'pseudomorphic' $AlInAs/Ga_{1-x}In_xAs$ modulation-doped heterostructures in the quest to improve further the performance of InP-based HEMTs [7-10].

C DEVICE PROPERTIES

While the material properties for InP-based modulation-doped heterostructures were determined in early studies, the establishment of a set of basic device properties for InP-based HEMTs has been a subject of much debate in the past decade. Central to the debate are what form does the 'velocity field' curve for InP-based HEMTs take and what value should be used for the 'saturation velocity'. For short gate length (<0.25 μm) devices, the simplest and arguably most successful approach has been one that assumed a two-part velocity-field curve, in which the saturation velocity is taken to be the peak velocity (v_{peak}) in the bulk GaInAs material [11,12]. Proponents of this approach argue that in short gate length devices and at sufficiently low drain voltages (<1.0 V) the electrons quickly attain the peak velocity and traverse the depletion region at the peak velocity before being scattered into the heavy-mass valleys. Thus, the usual formula for the transconductance of a HEMT must be written in terms of the peak, and not the saturation, velocity:

$$g_m = C_s v_{peak} \eta \tag{1}$$

where g_m and C_s are the transconductance and gate capacitance per unit area, respectively, and η, the modulation efficiency, is a less-than-unity factor that characterises the efficiency of the high mobility electrons to be modulated [13]. Since η approaches 1 for a well-designed InP HEMT [14], it is convenient to characterise such a device by its average velocity, $v_{ave} = v_{peak} \eta \cong v_{peak}$. Thus EQN (1) becomes

$$g_m = C_s v_{ave} \tag{2}$$

The advantage of this approach is that there is no longer any ambiguity regarding the velocity-field curve (the velocity-field curve is embedded in the modulation efficiency, which approaches unity for a well-designed HEMT). TABLE 2 lists the typical aspect ratio, average velocity, and transconductance for a contemporary 0.1 μm InP-based HEMT and FIGURE 1 shows its equivalent circuit model [15].

TABLE 2. Typical parameters for a 0.1 μm gate-length InP-based HEMT.

Parameter	Value
Aspect ratio (L_s/a)	5
Average velocity (v_{ave}), (cm/s)	2.6×10^7
Transconductance (g_m), (mS/mm)	800 - 1000

f (GHz)	Fmin (dB)	Γopt (Mag)	Γopt (Ang)	Rn/50
30	1.2	0.571	82.2	0.081
32	1.2	0.551	86.3	0.081
34	1.3	0.533	90.3	0.081
36	1.3	0.518	94.2	0.081
38	1.3	0.504	98.1	0.081
40	1.3	0.492	101.8	0.081
42	1.3	0.482	105.5	0.081
44	1.4	0.473	109.1	0.081
46	1.4	0.465	112.6	0.081
48	1.4	0.458	116.1	0.081
50	1.4	0.453	119.5	0.081

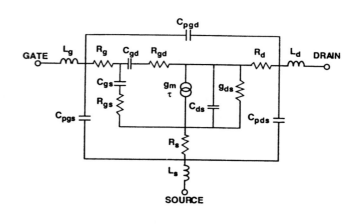

PARAMETER	Cgs	Rgs	Cgd	Rgd	gm	Tau	Cds	gds
Min.	100 fF	—	19 fF	—	130 mS	—	—	7.5 mS
Typ.	110 fF	1 Ω	21 fF	5 Ω	140 mS	0.2 ps	18 fF	10 mS
Max.	120 fF	—	23 fF	—	150 mS	—	—	12.5 mS

PARAMETER	Rg	Rd	Rs	Cpgs	Cpgd	Cpds	Lg	Ld	Ls
Typ.	1 Ω	3 Ω	2 Ω	7.5 fF	3.5 fF	22 fF	17 pH	18 pH	5 pH

FIGURE 1. Equivalent circuit model for a 0.1 x 150 μm InP HEMT at V_{ds} = 1.0 V. I_{ds} = 21 mA. The HEMT has six gate fingers.

D   APPLICATIONS

The InP HEMT technology has emerged as a leading candidate for ultra-low-noise and high frequency applications. Currently, InP HEMTs are being developed for use in advanced satellite communications systems, radio astronomy telescopes, wideband instrumentation, missile seekers, and automotive electronics [16-18]. TABLE 3 lists the potential system applications of InP-based HEMTs.

TABLE 3. Potential system applications of InP-based HEMTs.

System	Frequency (GHz)	Volume
Space-based communications	30, 44, 60	Low
High-speed wireless communications	20, 60	High
Automotive	77	High
Missile seekers	35, 94	Moderate

E   CONCLUSION

After years of intense research and debate, much is known about this material system. Today, the InP-based modulation-doped heterostructures and devices are critical components for advanced millimetre-wave satellite communications and remote sensing systems. In the near future, these heterostructures may enable a number of new applications at millimetre-wave frequencies, such as high-speed wireless communications and automotive electronics.

REFERENCES

[1]   T.P. Pearsall et al [*Inst. Phys. Conf. Ser. (UK)* no.45 (1979) ch.1 p.94-102]

[2]   R. People, K.W. Wecht, K. Alavi, A.Y. Cho [*Appl. Phys. Lett. (USA)* vol.43 (1983) p.118-20]

[3]   W.P. Hong, G.I. Ng, P. Bhattacharya, D. Pavlidis, S. Willing, B. Das [*J. Appl. Phys. (USA)* vol.64 no.4 (1988) p.1945-9]

[4]   J.L. Thobel, L. Baudry, A. Cappy, P. Bourel, R. Fauquembergue [*Appl. Phys. Lett. (USA)* vol.56 (1990) p.346-8]

[5]   L.P. Sadwick, C.W. Kim, K.L. Tan, D.C. Streit [*IEEE Electron Devices Lett. (USA)* vol.12 no.11 (1991) p.626-8]

[6]   A.S. Brown, A.E. Schmitz, L.D. Nguyen, J.A. Henige, L.E. Larson [in *1994 Indium Phosphide and Related Materials (USA)* p.263-6]

[7]   U.K. Mishra, A.S. Brown, S.E. Rosenbaum [in *1988 IEDM Tech. Digest (USA)* p.180-3]

[8]   L.D. Nguyen, A.S. Brown, M.A. Thompson, L.M. Jelloian [*IEEE Trans. Electron Devices (USA)* vol.39 no.9 (1992) p.2007-14]

[9]   T. Enoki, K. Arai, T. Akazaki, Y. Ishii [*IEICE Trans. Electron. (Japan)* vol.E76-C (1993) p.1402-11]

[10]   M. Wojtowicz et al [*IEEE Electron Devices Lett. (USA)* vol.15 no.11 (1994) p.477-9]

[11]   N. Moll, M.R. Hueschen, A. Fischer-Colbrie [*IEEE Trans. Electron Devices (USA)* vol.35 (1988) p.878-86]

[12]   L.D. Nguyen, P.J. Tasker, D.C. Radulescu, L.F. Eastman [*IEEE Trans. Electron Devices (USA)* vol.36 (1989) p.2243-8]

[13]   M.C. Foisy, P.J. Tasker, B. Hughes, L.F. Eastman [*IEEE Trans. Electron Devices (USA)* vol.35 (1988) p.871-8]

[14]   L.D. Nguyen, L.E. Larson, U.K. Mishra [*Proc. IEEE (USA)* vol.80 no.4 (1992) p.494-518]

[15]   L.D. Nguyen et al [*Solid-State Electron. (UK)* vol.38 no.9 (1995) p.1575-9]

[16]   R. Isobe et al [*IEEE MTT-S Int. Microw. Symp. Dig. (USA)* (1995) p.1133-6]

[17]   M.W. Pospieszalski et al [*IEEE MTT-S Int. Microw. Symp. Dig. (USA)* (1995) p.1121-4]

[18]   J. Pusl et al [*IEEE MTT-S Int. Microw. Symp. Dig. (USA)* (1995) p.1661-4]

CHAPTER 8

DEVICE APPLICATIONS

8.1 Lattice-matched GaAs- and InP-based QW lasers

N.K. Dutta

January 1996

A   INTRODUCTION

A double heterostructure laser consists of an active layer sandwiched between two higher gap cladding layers. The active layer thickness is typically in the range of 0.1 to 0.3 μm. Over the last decade, double heterostructure lasers with an active layer thickness ~100 Å have been fabricated. In these structures the carrier (electron or hole) motion normal to the active layer is restricted. As a result, the kinetic energy of the carriers moving in that direction is quantised into discrete energy levels similar to the well-known quantum mechanical problem of the one-dimensional potential well, and hence these lasers are called quantum well lasers.

When the thickness of the active region (or any low gap semiconductor layer confined between higher gap semiconductors) becomes comparable to the de Broglie wavelength ($\lambda \sim h/p$), quantum mechanical effects are expected to occur. These effects are observed in the absorption and emission (including laser action) characteristics and transport characteristics including phenomena such as tunnelling. The optical characteristics of semiconductor quantum well double heterostructures were initially studied by Dingle et al [1]. Since then extensive work on GaAlAs quantum well lasers has been reported [2-6]. Quantum well lasers fabricated using the InGaAsP material system have been extensively studied [7-10].

One advantage of quantum well DH lasers over regular DH lasers is that the emission wavelength of the former can be varied simply by varying the width of the quantum wells which form the active region. The restriction of the carrier motion normal to the well leads to a modification of the density of states in a quantum well which changes the radiative and nonradiative recombination rates of electrons and holes in a quantum well heterostructure compared to those of a regular DH. These changes may result in several desirable characteristics such as lower threshold current density, higher efficiency and lower temperature dependence of threshold current of suitably designed quantum well lasers compared to regular DH lasers.

B   ENERGY LEVELS

A carrier (electron or hole) in a double heterostructure is confined in a three-dimensional potential well. The energy levels of such carriers are obtained by separating the Hamiltonian into three parts, corresponding to the kinetic energies in the x-, y- and z-directions, each of which forms a continuum of states. When the thickness of the heterostructure (L_z) is comparable to the de Broglie wavelength, the kinetic energy corresponding to the particle motion along the z-direction is quantised. The energy levels can be obtained by separating the Hamiltonian into energies corresponding to x-, y- and z-directions. For the x-, y-directions, the energy levels form a continuum of states given by

$$E = \frac{\hbar^2}{2m}\left(k_x^2 + k_y^2\right) \tag{1}$$

where m is the effective mass of the carrier and k_x and k_y are the wave vectors along the x- and y-directions respectively. Thus the electrons or holes in a quantum well may be viewed as forming a two-dimensional Fermi gas.

The energy levels in the z-direction are obtained by solving the Schrödinger equation for a one-dimensional potential well. For the limiting case of an infinite well, the energy levels and the wavefunctions are

$$E_n = \frac{\hbar^2}{2m}\left(\frac{n\pi}{L_z}\right)^2 \text{ and } \Psi_n = A \sin\frac{n\pi z}{L_z} \ (n = 1, 2, 3) \tag{2}$$

where A is a normalisation constant. For very large L_z, EQN (2) yields a continuum of states and the system no longer exhibits quantum effects.

The potential well for electrons and holes in a double heterostructure depends on the materials involved. The commonly used relationships between conduction-band (ΔE_c) and valence-band (ΔE_v) discontinuities for GaAs/AlGaAs double heterostructures are as follows [11,12]:

$$\Delta E_c/\Delta E = 0.65 \pm 0.03, \quad \Delta E_v/\Delta E = 0.35 \pm 0.03 \tag{3}$$

where ΔE is the bandgap difference between the confining layers and active region. A knowledge of ΔE_c and ΔE_v is necessary in order to accurately calculate the energy levels. For an InGaAsP/InP double heterostructure, the values obtained by Forrest et al [12] are

$$\Delta E_c/\Delta E = 0.39 \pm 0.01, \quad \Delta E_v/\Delta E = 0.61 \pm 0.01 \tag{4}$$

The energy eigenvalues for a particle confined in the quantum well are

$$E(n, k_z, k_y) = E_n + \frac{\hbar^2}{2m^*_n}\left(k_x^2 + k_y^2\right) \tag{5}$$

where E_n is the nth confined-particle energy level for carrier motion normal to the well and m^*_n is the effective mass. FIGURE 1 shows schematically the energy levels E_n of the electrons and holes confined in a quantum well. The confined-particle energy levels (E_n) are denoted by E_{1c}, E_{2c}, E_{3c} for electrons, E_{1hh}, E_{2hh} for heavy holes and E_{1lh}, E_{2lh} for light holes. These quantities can be calculated by solving the Schrödinger equations for a given potential barrier (ΔE_c or ΔE_v).

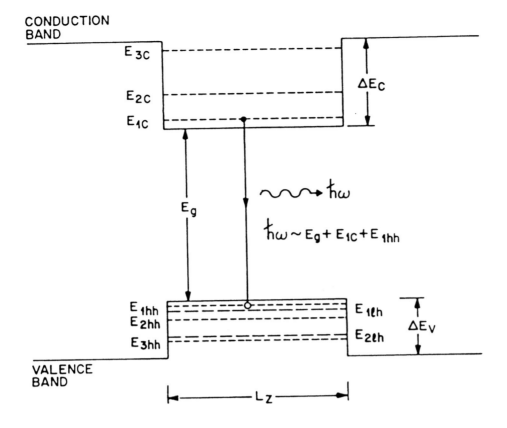

FIGURE 1. Energy levels in a quantum well structure.

Electron-hole recombination in a quantum well follows the selection rule $\Delta n = 0$, i.e. the electrons in states E_{1c} (E_{2c}, E_{3c}, etc.) can combine with the heavy holes E_{1hh} (E_{2hh}, E_{3hh}, etc.) and with light holes E_{1lh} (E_{2lh}, E_{3lh}, etc.). Note, however, that since $E_{1lh} > E_{1hh}$ the light-hole transitions are at a higher energy than the heavy-hole transitions. Since the separation between the lowest conduction-band level and the highest valence-band level is given by

$$E_q = E_g + E_{1c} + E_{1hh} \cong E_g + \frac{h^2}{8L_z^2} \left(\frac{1}{m_c} + \frac{1}{m_{hh}} \right) \tag{6}$$

we see that in a quantum well structure the energy of the emitted photons can be varied simply by varying the well width L_z. FIGURE 2 shows the experimental results of Temkin et al [13] for InGaAs quantum well lasers with different well thicknesses bounded by InP cladding layers. For low well thicknesses the laser emission shifts to higher energies.

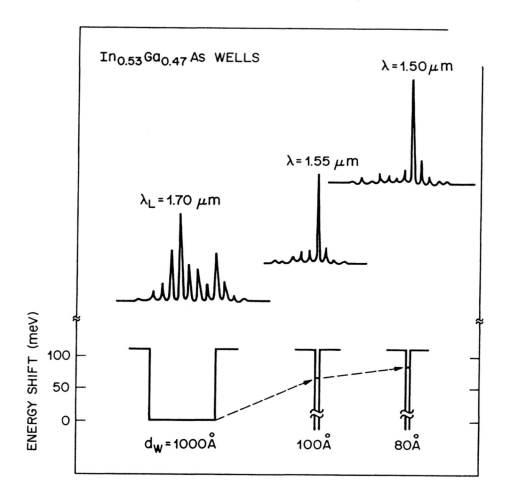

FIGURE 2. Lasing wavelength for different well thicknesses of an InGaAs/InP MQW structure.

C   DENSITY OF STATES AND THRESHOLD CURRENT

The basis of light emission in semiconductors is the recombination of an electron in the conduction band with a hole from the valence band and the excess energy is emitted as a photon (light quantum). The process is called radiative recombination. Sufficient numbers of electrons and holes must be excited in the semiconductor for stimulated emission or net optical gain. The condition for net gain at a photon energy E is given by

$$E_{fc} + E_{fv} = E - E_g \tag{7}$$

where E_{fc} and E_{fv} are the quasi-Fermi levels of electrons and holes, respectively, measured from the respective band edges (positive into the band) and E_g is the bandgap of the semiconductor. The quasi-Fermi levels can be calculated as follows from a knowledge of the density of states of electrons and holes in the quantum well heterostructure.

In a quantum well structure, the kinetic energy of the confined carriers (electrons or holes) for velocities normal to the well (z-direction) is quantised into discrete energy levels. This modifies the density of states from the well-known three-dimensional case. Using the principle of box quantisation of kinetic energies along the x- and y-directions, the number of electron states per unit area in the x-y direction for the ith subband within an energy interval dE is given by

$$D_i(E) \, dE = 2\frac{d^2k}{(2\pi)^2} \qquad (8)$$

where the factor 2 arises from two spin states and $\mathbf{k} = (k_x, k_y)$ is the momentum vector. Using the parabolic band approximation, i.e. $E = \hbar^2 k^2/2m_{ci}$, EQN (8) may be written as

$$D_i = \frac{m_{ci}}{\pi \hbar^2} \qquad (9)$$

where m_{ci} is the effective mass of the electrons in the ith subband of the conduction band. Thus the density of states per unit volume is given by

$$g_{ci} = \frac{D_i}{L_z} = \frac{m_{ci}}{\pi \hbar^2 L_z} \qquad (10)$$

A similar equation holds for the holes in the valence band. For the regular three-dimensional case, the density of states is given by

$$\rho(E) = 2(2\pi m_c k_B T/h^2)^{3/2} E^{1/2} \qquad (11)$$

where k_B is the Boltzmann constant and T is the temperature. A comparison of EQNS (10) and (11) shows that the density of states in a quantum well is independent of the carrier energy and temperature. The modification of the density of states in a quantum well is sketched in FIGURE 3. This modification can significantly alter the recombination rates in a quantum well double heterostructure compared to those for a regular double heterostructure. The simple model for the density of states described above neglects the effect of collisions between carriers which would broaden the discrete energy levels in the z-direction. The step-like density of states in a QW allows better utilisation of carriers for gain at a given temperature in a QW compared to that for a regular DH. This results in higher gain at a lower current in a

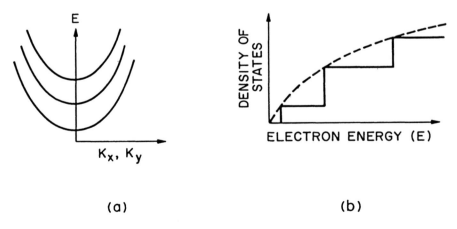

(a)   (b)

FIGURE 3. (a) Energy vs. wavevector for each subband. (b) Schematic representation of the density of states in a quantum well.

QW compared to that for a regular DH. The gain vs. current density in a QW structure has been calculated [14]. It is approximately given by

$$g = a(J - J_0) \qquad (12)$$

where a is the gain constant and J_0 is the transparency current density. At threshold, theoretical gain equals the total optical loss in the cavity. The condition for threshold is

$$\Gamma g_{th} = \alpha_a \Gamma + \alpha_c (1 - \Gamma) + \frac{1}{L} \ln\left(\frac{1}{R}\right) \qquad (13)$$

where Γ is the fraction of the lasing mode confined in the active region (generally known as the confinement factor), g_{th} is the optical gain at threshold, α_a and α_c are the optical losses such as free carrier absorption and scattering in the active region and cladding layer, respectively, L is the cavity length and R is the reflectivity of the two cleaved facets so that the last term represents the distributed mirror loss.

The evaluation of Γ is, in general, quite tedious. For the fundamental mode, however, a remarkably simple expression,

$$\Gamma \cong D^2/(2 + D^2) \text{ with } D = k_0(\mu_a^2 - \mu_c^2)^{1/2}d \qquad (14)$$

is found to be accurate to within 1.5% [15]. In the above expression, d is the active layer thickness, $k_0 = 2\pi/\lambda$ where λ is the wavelength in free space, and μ_a and μ_c are the refractive indices of the active and the cladding layer, respectively. For a single quantum well laser (d < 200 Å), EQN (14) reduces to a simpler expression

$$\Gamma \cong \frac{1}{2} D^2$$

The calculated confinement factor (Γ) for a 250 Å thick GaAs/Al$_{0.52}$Ga$_{0.48}$As quantum well is 0.04. Using an internal loss $\alpha = 10$ cm^{-1}, R = 0.3 and L = 380 μm, we get $g_{th} \cong 10^3$ cm^{-1} from EQN (13). The threshold gain for single quantum well lasers is an order of magnitude larger than that for regular double heterostructure lasers. This difference is principally due to the smaller confinement factor for a single quantum well. From EQNS (12) and (13), the threshold current density J_{th} is given by

$$J_{th} = J_0 + \left(\alpha + \frac{1}{L} \ln\left(\frac{1}{R}\right)\right)/\alpha\Gamma \qquad (15)$$

where $\alpha = \alpha_a \Gamma + \alpha_c(1 - \Gamma)$.

EQN (15) shows that J_{th} decreases with increasing cavity length and for small losses it equals the transparency current density J_0 as $L \rightarrow \infty$. The two terms that constitute the threshold current density in EQN (15) are (i) transparency current density, and (ii) current needed to overcome losses. The latter is small if α is small, as seen in QW lasers, and can be further reduced using high reflectivity coatings. The low absorption loss of QW structures allows the fabrication of very low threshold lasers.

D SINGLE QUANTUM WELL AND MULTI-QUANTUM WELL LASERS

Quantum well injection lasers with both single and multiple active layers have been fabricated. Quantum well lasers with one active region are called single quantum well (SQW) lasers and those with multiple active regions are called multi-quantum well (MQW) lasers. The layers separating the active layers in an MQW structure are called barrier layers. The energy band diagrams of these laser structures are schematically shown in FIGURE 4. The MQW lasers where the bandgap of the barrier layers is different from that of the cladding layers are sometimes referred to as modified multi-quantum well lasers.

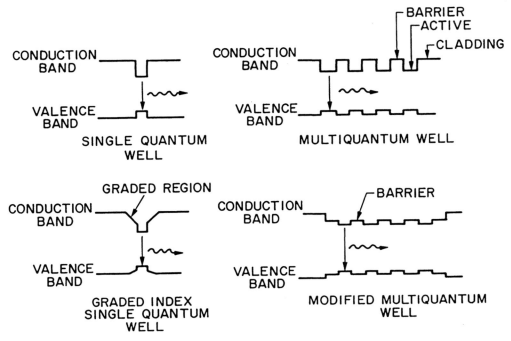

FIGURE 4. Single quantum well and multiquantum well laser structures.

One of the main differences between the SQW and MQW lasers is that the confinement factor (Γ) of the optical mode is significantly smaller for the former than for the latter. This can result in higher threshold carrier density and in higher threshold current density of SQW lasers when compared with MQW lasers. The confinement factor of an SQW heterostructure can be significantly increased using a graded-index cladding layer (see FIGURE 4). This allows the use of the intrinsic advantage of the quantum well structure (high gain at low carrier density) without the penalty of a small mode-confinement factor. Threshold current densities as low as 200 A/cm^2 have been reported for a GaAs graded-index (GRIN) SQW laser [16].

The confinement factor of the fundamental mode of a double heterostructure is approximately given by EQN (14). For small active layer thickness, such as that of a single quantum well, the expression for Γ can be further simplified since the normalised waveguide thickness D < 1 and we obtain [17]

$$\Gamma \cong 2\pi^2(\mu_a^2 - \mu_c^2)d^2/\lambda_0^2 \qquad (16)$$

where μ_a and μ_c are the refractive indices of the active and cladding layer, respectively, d is the active layer thickness and λ_0 is the free space wavelength.

The mode confinement in MQW structures can be analysed by solving the electromagnetic wave equations for each of the layers with appropriate boundary conditions. This procedure is quite tedious because of the large number of layers involved. Streifer et al [18] have shown that the following simple formula gives reasonably accurate results

$$\Gamma(MQW) = \gamma \frac{N_a d_a}{N_a d_a + N_b d_b} \tag{17}$$

where

$$\gamma = 2\pi^2 \, (N_a d_a + N_b d_b)^2 (\bar{\mu}^2 - \mu_c^2)/\lambda_0^2$$

and

$$\bar{\mu} = \frac{N_a d_a \mu_a + N_b d_b \mu_b}{N_a d_a + N_b d_b}$$

N_a, N_b are the number of active and barrier layers in the MQW structure and d_a, d_b (and μ_a, μ_b) are the thicknesses (and refractive indices) of the active and barrier layers, respectively. For most current applications, a graded index separate confinement MQW structure is used.

E   GaAs/AlGaAs QUANTUM WELL LASERS

A considerable amount of experimental work has been reported on quantum well lasers fabricated using the AlGaAs material system. The epitaxial layers are grown using molecular beam epitaxy (MBE) or organometallic vapour phase epitaxy (OMVPE or MOCVD) growth techniques. These results show that AlGaAs quantum well lasers exhibit lower threshold current, somewhat higher differential quantum efficiency, and a weaker temperature dependence of threshold current when compared with regular double heterostructure lasers.

Some of the results that demonstrate the high performance of AlGaAs quantum well lasers are: fabrication of GRIN SQW lasers with the lowest reported threshold current density of 200 A/cm^2 [6] (this compares with values in the range 0.6-1 kA/cm^2 for regular DH lasers), fabrication of modified MQW lasers with threshold current densities of 250 A/cm^2 [4], and fabrication of laser arrays with MQW active regions which have been operated to very high CW power [19].

Tsang [4] has reported measurements of threshold current densities of AlGaAs multi-quantum well lasers with different barrier heights (FIGURE 5). The cladding layers were $Al_{0.35}Ga_{0.65}As$. The results show that there is an optimum value for the barrier height ($Al_xGa_{1-x}As$ with x ~ 0.19) for lowest threshold current. For larger x, the barrier height is too large for sufficient current injection and for very small x, the effect of the potential well which gives rise to confined two-dimensional-like states is reduced [20].

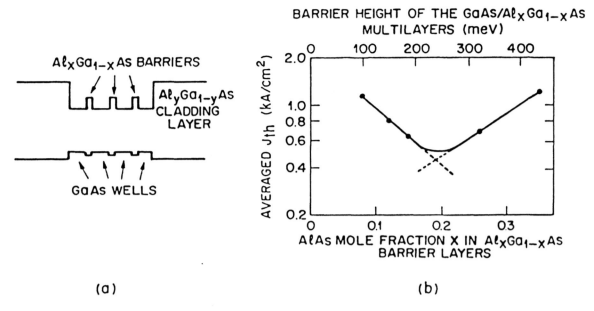

FIGURE 5. (a) Schematic of an AlGaAs MQW structure. (b) Threshold current density as a function of barrier energy of an AlGaAs MQW laser.

E1 Low Threshold Operation

For low threshold laser designs, it is important to have lateral carrier and optical confinement. This is accomplished using a buried heterostructure (BH) design shown in FIGURE 6 [21]. The active region is a graded index separate confinement heterostructure design with a single quantum well. The latter allows low transparency current density. The measured light vs. current characteristics of a BH laser for different facet reflectivities is shown in FIGURE 7. The typical threshold current of ~4 mA with uncoated facets is reduced to ~0.5 mA when the facets are coated with high reflectivity dielectric coatings [21].

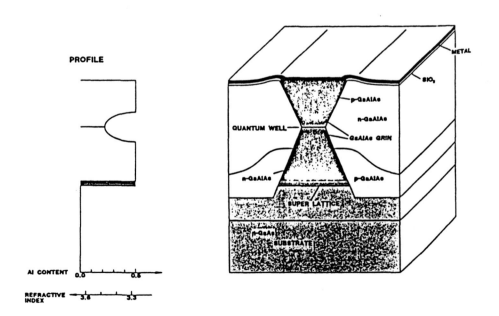

FIGURE 6. Buried heterostructure AlGaAs QW laser. Also shown is the GRIN SCH QW design.

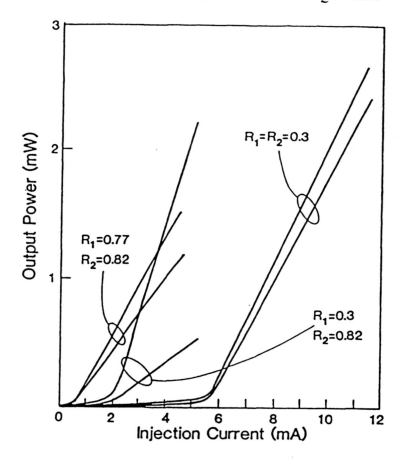

FIGURE 7. Light vs. current characteristics of a BH laser with different facet coatings.
Cavity length = 120 μm.

E2   High Speed Operation

The quantum well laser has a higher differential gain coefficient (rate of change of gain with carrier density) than a regular DH laser. Since the modulation bandwidth of a laser is proportional to the square root of the gain coefficient, QW lasers exhibit higher bandwidth than regular DH lasers. Modulation bandwidths of ~10 GHz have been reported for AlGaAs BH QW lasers [8].

E3   VCSEL Design

The GaAs/AlGaAs QW active layer design has also been used extensively for the fabrication of high performance vertical cavity surface emitting lasers (VCSELs). These lasers are typically grown by MBE or MOCVD epitaxial growth techniques. A schematic of a low threshold VCSEL design is shown in FIGURE 8. In this design, selective oxidation of a thin AlAs layer grown on top of the active region is used for lateral optical and current confinement [22]. The distributed Bragg reflectors (DBR) on the p- and n-side of the active region provide reflectivities of >99% to the lasing mode. Threshold currents of 70 μA have been reported using the above design.

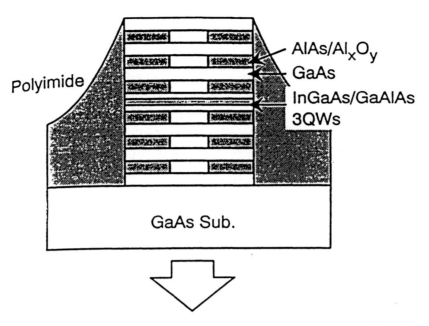

FIGURE 8. Schematic of a low threshold vertical cavity surface emitting laser.

F   InGaAsP/InP QW LASERS

The quaternary alloy $In_xGa_{1-x}As_yP_{1-y}$ has been grown epitaxially over crystalline InP using liquid phase epitaxy (LPE), vapour phase epitaxy (VPE), chemical beam epitaxy (CBE) and metal organic chemical vapour deposition (MOCVD) growth techniques. The material grown is of high quality to allow highly reliable low threshold operation for lasers fabricated using this material. The bandgap (E_g) of $In_xGa_{1-x}As_yP_{1-y}$ lattice matched to InP varies with composition. It is given by

$$E_g \text{ (in eV)} = 1.35 - 0.72y + 0.12y^2 \qquad (18)$$

at 300 K with $x \sim 0.45y$.

InGaAsP/InP QW lasers have been fabricated using the MOCVD growth technique. The parameters for optimisation of a quantum well laser structure are the well and barrier widths, barrier composition and grading for a graded index separate confinement heterostructure (GRIN-SCH) design. The schematic of QW laser structures with different barrier layer composition is shown in FIGURE 9. Best results are obtained for lasers with high energy barriers.

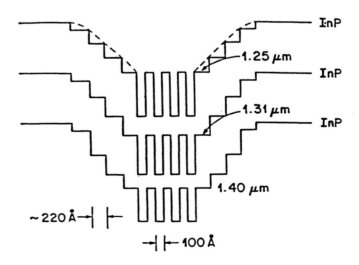

FIGURE 9. Schematic of three GRIN MQW designs. The bandgap of the barrier layer is varied from 1.25 μm to 1.40 μm.

F1   Index Guided Lasers

Index guiding along the junction plane is necessary in order to control the light emission pattern of a semiconductor laser. The schematic of an index guided buried heterostructure (BH) laser is shown in FIGURE 10. The active region width is ~1 μm. The laser utilises Fe doped InP layers for confinement of the current to the active region and for providing lateral index guiding. The light vs. current characteristics of an MQW BH laser at different temperatures are shown in FIGURE 11.

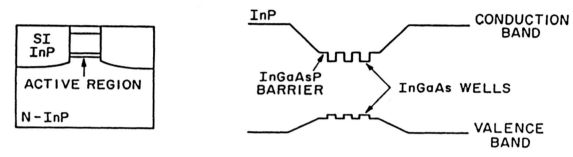

FIGURE 10. Schematic of a buried heterostructure laser utilising a semi-insulating Fe-doped InP current confining layer.

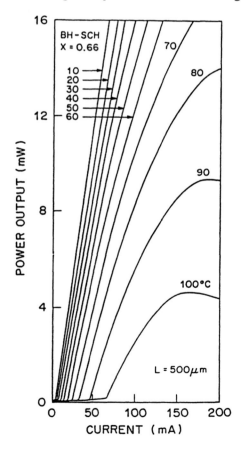

FIGURE 11. Light vs. current characteristics of an MQW BH laser at different temperatures.

Low threshold InGaAsP/InP QW lasers have also been reported using high reflectivity facet coatings. The light vs. current characteristics of an MQW BH laser with 90% and 70% facet coatings are shown in FIGURE 12. The laser has a threshold current of 11 mA at 25°C and 0.9 mA at 10°C.

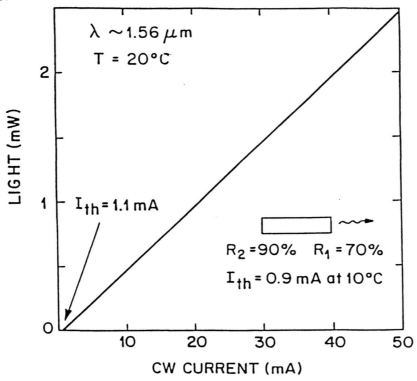

FIGURE 12. Light vs. current characteristics of an InGaAsP/InP MQW BH laser with 90% and 70% facet coatings. The cavity length is 200 μm.

F2   High Speed Performance

Due to higher gain coefficient, QW lasers are expected to have higher bandwidth than regular DH lasers. However, in order to demonstrate this it is necessary to have a low capacitance parasitic free device structure. An example of such a structure is shown in FIGURE 13. In

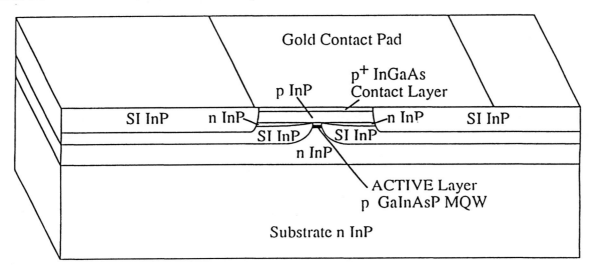

FIGURE 13. Schematic of a high speed laser design.

this structure, most of the laser device has semi-insulating InP material which reduces both leakage current and device capacitance. A modulation bandwidth of 25 GHz has been reported using this design [23].

Another important parameter related to high speed operation in fibre optic transmission systems is spectral width under modulation also known as chirp. If the chirp is too large, the light pulse spreads due to fibre dispersion as it propagates through the fibre. Hence a single frequency laser (such as a distributed feedback laser) with low chirp is needed for a high speed transmission system. Semiconductor lasers with an MQW active region have lower chirp than lasers with a regular DH active region. The measured full width at half maximum of the spectral width (chirp) is shown as a function of modulation current in FIGURE 14. Note that the MQW lasers exhibit smaller chirp than regular DH lasers [24] and strained MQW lasers exhibit even lower chirp. The lower chirp is due to a smaller linewidth enhancement factor for MQW and strained MQW lasers compared to that for regular DH lasers. The same mechanism is also responsible for lower CW linewidth of MQW and strained MQW lasers.

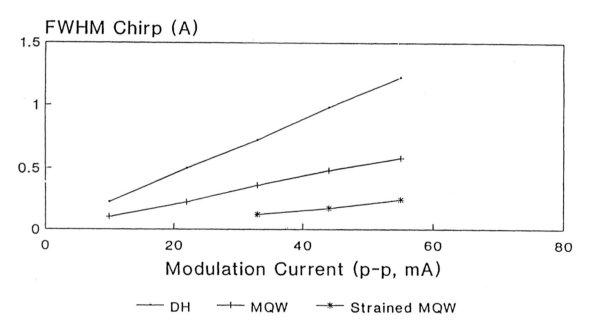

FIGURE 14. Measured full width at half maximum of the spectral width
(chirp) plotted as a function of modulation current.

F3   Unipolar Laser

The injection lasers described so far operate on the principal of population inversion near a p-n junction. A new type of unipolar injection laser was reported in 1994 [25]. In this device, the lasing action takes place between the conduction band levels shown in FIGURE 15. The laser emits in the far infrared (4 μm to 5 μm) region of the spectrum. The active region is the InGaAs quantum well and the barrier region is InAlGaAs. The structure is grown by MBE over an InP substrate. The light vs. current characteristics of such a laser at three different temperatures are shown in FIGURE 16. Improvements to this laser are being pursued.

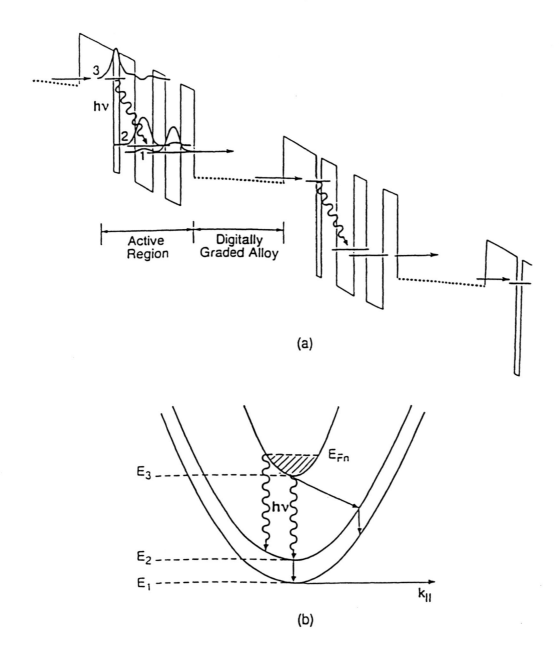

(a)

(b)

FIGURE 15. Schematic band diagram of a unipolar laser.

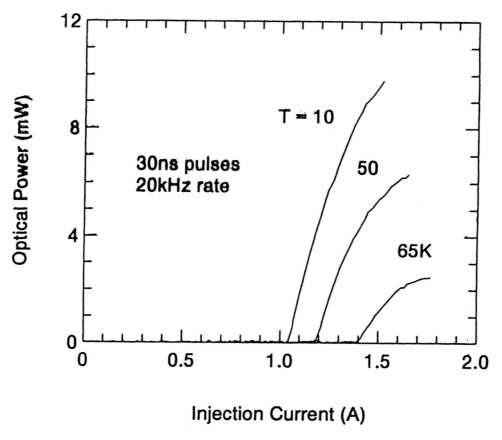

FIGURE 16. Light vs. current characteristics of an MQW unipolar laser.
The laser wavelength is 4.2 μm.

REFERENCES

[1]   R. Dingle, W. Wiegmann, C.H. Henry [*Phys. Rev. Lett. (USA)* vol.33 (1974) p.827]; R. Dingle, C.H. Henry [U.S. Patent no.3-982-207, 21 Sept. 1976]

[2]   N. Holonyak Jr., R.M. Kolbas, R.D. Dupuis, P.D. Dapkus [*IEEE J. Quantum Electron. (USA)* vol.16 (1980) p.170]

[3]   N. Holonyak Jr. et al [*J. Appl. Phys. (USA)* vol.51 (1980) p.1328]

[4]   W.T. Tsang [*Appl. Phys. Lett. (USA)* vol.39 (1981) p.786]

[5]   W.T. Tsang [*IEEE J. Quantum Electron. (USA)* vol.20 (1986) p.1119]

[6]   S.D. Hersee, B. DeCremoux, J.P. Duchemin [*Appl. Phys. Lett. (USA)* vol.44 (1984) p.476]

[7]   N.K. Dutta et al [*Appl. Phys. Lett. (USA)* vol.46 (1985) p.19; *Appl. Phys. Lett. (USA)* vol.46 (1985) p.1036]

[8]   See, for example, P. Zory (Ed.) [*Quantum Well Lasers* (Academic Press, New York, 1993)]

[9]   P.J.A. Thijs, L.F. Tiemeijer, P.I. Kuindersma, J.J.M. Binsma, T. van Dougen [*IEEE J. Quantum Electron. (USA)* vol.27 (1991) p.1426]

[10]   U. Koren, M. Oron, B.I. Miller, J.L. DeMiguel, G. Raybon, M. Chien [*Electron. Lett. (UK)* vol.26 (1990) p.465]

[11]   T. Ishikawa, J. Bowers [*IEEE J. Quantum Electron. (USA)* vol.30 (1994) p.562]

[12]   S.R. Forrest, P.H. Schmidt, R.B. Wilson, M.L. Kaplan [*Appl. Phys. Lett. (USA)* vol.45 (1984) p.1199]; S.R. Forrest [in *Heterojunction Band Discontinuities: Physics and Applications* Eds F. Capasso, G. Margaritondo (North Holland, Amsterdam, 1987) ch.8]

[13]   H. Temkin, M.B. Panish, P.M. Petroff, R.A. Hamm, J.M. Vandenberg, S. Sumski [*Appl. Phys. Lett. (USA)* vol.47 (1985) p.394]

[14]   N.K. Dutta [*J. Appl. Phys. (USA)* vol.53 (1982) p.7211]

[15]   D. Botez [*IEEE J. Quantum Electron. (USA)* vol.17 (1981) p.178]

[16]   W.T. Tsang [*Appl. Phys. Lett. (USA)* vol.39 (1981) p.134]

[17]   W.P. Dumke [*IEEE J. Quantum Electron. (USA)* vol.11 (1975) p.400]

[18]   W. Streifer, D.R. Scifres, R.D. Burnham [*Appl. Opt. (USA)* vol.18 (1979) p.3547]

[19]   D.R. Scifres, R.D. Burnham, C. Lindstrom, W. Streifer, T.L. Paoli [*Appl. Phys. Lett. (USA)* vol.42 (1983) p.645]

[20]   N.K. Dutta [*IEEE J. Quantum Electron. (USA)* vol.19 (1983) p.794]

[21]   K.Y. Lau [in *Quantum Well Lasers* Ed. P. Zory (Academic Press, New York, 1993) ch.4]

[22]   Y. Hayashi, T. Mukaihara, N. Hatori, N. Ohnoki, A. Matsutani, F. Koyama [*IEEE Photonics Technol. Lett. (USA)* vol.7 (1995) p.1234]

[23]   P. Morton, R.A. Logan, T. Tanbun-Ek, P.F. Sciortino, A.M. Segent [*Electron. Lett. (UK)* vol.28 (1992) p.2156]

[24]   D.T. Nichols, P. Bhattacharya et al [*IEEE J. Quantum Electron. (USA)* vol.8 (1992) p.1239]

[25]   F. Capasso et al [*Science (USA)* vol.264 (1994) p.553]

8.2 Quantum well modulation and switching devices

G. Parry, C.F. Tang, R.I. Killey, P.N. Stavrinou, M. Whitehead and C.C. Button

June 1996

A INTRODUCTION

Quantum well modulators are devices which modify the amplitude or phase of an optical signal propagating through the device, in response to an applied voltage. The physical phenomenon exploited in these devices is known as the quantum confined Stark effect [1-4]. The effect is essentially the electric field induced modification of the confinement energies and wavefunctions of the electrons and holes in a quantum well. Specifically, the electric field reduces the energy of confined electron and hole states so that the energy separation between an electron and hole is reduced with increasing electric field. Consequently the optical absorption edge shifts to lower energies with electric field. The effect occurs when an electric field is applied perpendicular to the wells so that the electron and hole wavefunctions become polarised to opposite sides of the well. The overlap of the wavefunctions and consequently the absorption coefficient of the exciton reduces with applied field. FIGURE 1 shows the effect in GaAs/AlGaAs quantum well structures and FIGURE 2 shows the absorption spectra and the field induced change in absorption coefficient for an InGaAs/InGaAsP quantum well structure.

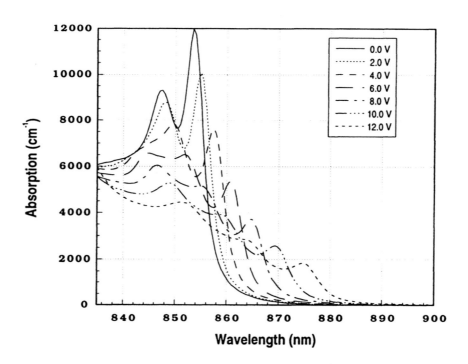

FIGURE 1. Absorption spectra of a sample of GaAs/AlGaAs QWs as a function of applied voltage. The QWs are 10 nm wide and are located in the intrinsic region of a p-i-n structure. The average applied electric field at 12 V is 120 kV/cm.

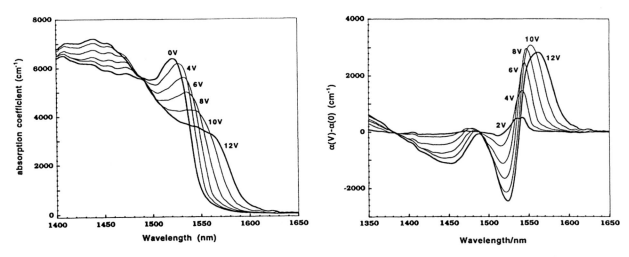

FIGURE 2. Absorption spectra of a sample of InGaAs/InGaAsP as a function of applied voltage. The sample contains 60 QWs of width 6.5 nm separated by barriers of width 10 nm. The quantum wells are located in the intrinsic region of a p-i-n structure.

It is clear from FIGURE 1 and FIGURE 2 that, even with no applied electric field, a quantum well structure exhibits rather different features in its absorption spectrum when compared with bulk material. This occurs primarily as a result of exciton formation during optical absorption. The binding energy of excitons in quantum well structures increases as a result of the confinement of the electron and hole and very sharp exciton features are observed even at room temperature. The resonantly enhanced absorption clearly results in larger changes in absorption coefficient than would be apparent in bulk semiconductor structures [5,6].

It is these two features, the enhanced excitonic absorption and the quantum confined Stark effect, which are crucial to the operation of optical modulators.

B   DEVICE GEOMETRY

Several distinct device geometries have been used. It is convenient to consider them under two headings: waveguide geometries [7,8] where light is launched into the structure parallel to the surface of the semiconductor, and normally incident geometries in which light propagates perpendicular to the surface of the semiconductor. The normal incidence geometry includes transmission devices [2], reflection devices [9] and micro-cavity or Fabry-Perot devices [10]. FIGURE 3 shows all four geometries. In all cases the electric field is applied across the quantum wells by locating the wells in the intrinsic region of a 'p-i-n' diode structure. When a reverse bias voltage is applied across the diode a near uniform electric field is produced across the quantum wells. In the reflection and Fabry-Perot geometries the mirrors required are usually produced using Bragg reflectors. These are produced using two distinct but lattice matched semiconductor materials selected so that the difference between the refractive indices is as large as possible. Each semiconductor layer is a quarter wavelength in thickness, and multiple stacks of the layers are used to achieve high reflectivities. In the Fabry-Perot device the mirror reflectivities are usually chosen to be lower for the front and maximum for the back reflectors and the devices are then generally referred to as asymmetric Fabry-Perot modulators [11-13].

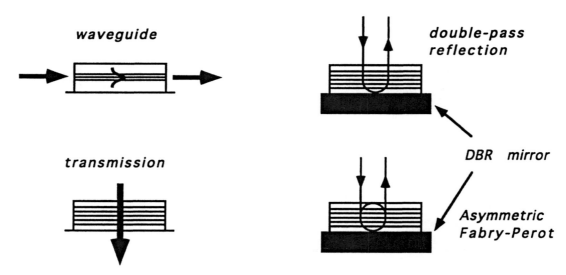

FIGURE 3. Four possible device geometries for modulators.

C   DEVICE OPERATION

Modulators can operate in either the electro-absorptive or electro-refractive mode. In the electro-absorptive mode the width of the quantum wells is selected so that the peak exciton absorption is located on the high energy side of the operating wavelength. Application of an applied field shifts the exciton and increases the absorption at the operating wavelength. Alternatively the quantum wells can be chosen so that the exciton absorption occurs at the operating wavelength and in this case the absorption decreases with applied field. Almost all waveguide electro-absorption devices operate in the former mode as otherwise the insertion loss becomes prohibitively high for most applications. The latter mode is essential for some switching devices such as self-electro-optic effect devices (SEEDs) [14] and for normal incidence devices which have low or zero reflectivity with no applied voltage [15]. For Fabry-Perot devices there is an additional requirement on the location of the optical resonance. Various modes of operation occur, one example being the asymmetric Fabry-Perot modulator in which the device is normally reflecting and the exciton peak is located on the low energy side of the optical resonance. Application of the electric field then moves the strongly absorbing exciton into the optical resonance causing a reduction in reflectivity at the resonance - even achieving zero reflectivity for an appropriate voltage.

In the electro-refractive mode modulators impose phase changes on an incident optical signal since there is a change in refractive index of the quantum well material associated with the QCSE. The magnitude of the refractive index change can be estimated by carrying out a Kramers-Kronig transformation of the absorption spectrum with and without an applied field. Refractive index changes of the order of 1% can be obtained apparently offering very large phase changes in light propagating through the structure but these large changes are usually accompanied by significant (and often unwanted) changes in intensity of the light. Early results on phase modulation were reported in [16,17].

D   MATERIALS

Material choice dictates the range of wavelength over which a modulator can operate. TABLE 1 lists many of the III-V materials which have been studied and the wavelength range over which modulation is achieved.

TABLE 1.

Material	Approximate wavelength (nm)	Ref
InGaAs/InP	1550	[12,21,16]
InGaAsP/InGaAsP	1550	[19]
InGaAs/InAlAs	1550	[18]
InGaAsP/InP	1300-1560	[26]
InAsP/GaInP	1300	[20,25]
InGaAs/GaAs	920-1070	[24,13]
InAsP/InP	1060	[25]
InGaAs/AlGaAs	920-980	[13]
GaAs/AlGaAs	820-850	[1-4]
GaAs/AlGaAs on silicon	850	[23]

E   APPLICATIONS

There has been a substantial effort directed towards developing high speed low drive voltage, and low chirp quantum well waveguide modulators for external modulation of 1.55 μm lasers. Electrical bandwidths of over 40 GHz have been achieved with InGaAs/InAlAs [18] and InGaAsP [19] quantum well electro-absorption modulators. Some designs show a high tolerance to input polarisations.

In contrast normal incidence devices have been used in applications where moderate speed, one and two dimensional arrays of modulators have been required. Arrays integrated by flip-chip packaging methods to silicon electronics have been reported. Uniform 8x8 arrays of InGaAs/InP multiquantum well asymmetric Fabry-Perot modulators have been demonstrated [21], and recently ring oscillators with optical and electrical readout have been produced using GaAs MQW modulators hybrid bonded to 0.8 μm silicon VLSI [22].

F   SWITCHING DEVICES

Modulators can form the basis of switching elements either in the waveguide or normal incidence geometry. Waveguide switching elements are usually formed using directional couplers or Mach Zehnder interferometers to produce NxN space switches [26] although it should be noted that the polarisation sensitivity of quantum well waveguides can lead to the requirement of more complex designs and some of the best switches have involved bulk rather than quantum well structures.

Normal incidence devices can achieve switching using the concept of the self electro-optic effect which essentially combines together a quantum well modulator with a photodetector. There are a number of possible implementations: R-SEED (resistor biased), D-SEED (diode biased), S-SEED (symmetric configuration), F-SEED (FET biased) and bipolar transistor based switches [27,28]. FIGURE 4 shows the optical arrangement and a typical characteristic achieved with a Fabry-Perot S-SEED configuration [29]. References [30] and [41] provide excellent reviews of SEED devices.

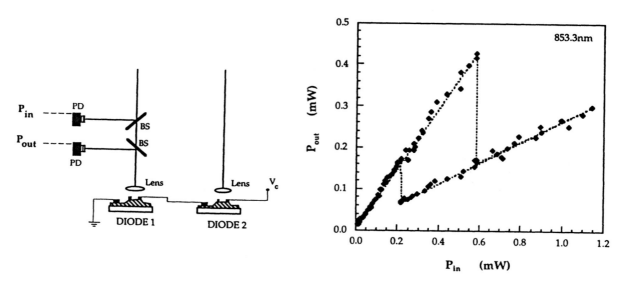

FIGURE 4. Experimental arrangement and input-output characteristics of a SEED device. The input-output characteristics are obtained from diode 1 and a constant optical bias is applied to diode 2. As the input intensity increases from 0 mW the output power increases until it reaches the maximum at approximately 0.42 mW after which it switches to the lower value of 0.15 mW. Bistable behaviour is clear from the hysteresis loop.

The most impressive technology at present is the F-SEED which has formed the basis of small circuits operating with 22 fJ input energies and at speeds of up to 650 Mb/s. More complex multistage circuits (96 optical beams and 400 transistors) operating with 100 fJ optical input energies at 155 Mb/s have been reported [31]. Alternative approaches to integration of modulators and electronic devices involving epitaxial or hybrid integration with silicon have also been demonstrated [21-23].

G CONCLUSION

The discovery of the QCSE has led to the development of a new family of optoelectronic modulators and switching devices which will undoubtedly have a major impact on advanced systems for communication and computing.

REFERENCES

[1]   D.A.B. Miller et al [*Phys. Rev. B (USA)* vol.32 (1985) p.1043-60]
[2]   T.H. Wood et al [*Appl. Phys. Lett. (USA)* vol.44 (1984) p.16-8]
[3]   M. Whitehead, G. Parry, J.S. Roberts, P. Mistry, P. Li Kam Wa, J.P.R. David [*Electron. Lett. (UK)* vol.23 (1987) p.1048]
[4]   M. Whitehead et al [*Appl. Phys. Lett. (USA)* vol.53 (1988) p.956]
[5]   R. Dingle, W. Wiegmann, C.H. Henry [*Phys. Rev. Lett. (USA)* vol.33 (1974) p.827-30]

[6]   D.S. Chemla, D.A.B. Miller, P.W. Smith, A.C. Gossard, W. Wiegmann [*IEEE J. Quantum Electron. (USA)* vol.20 (1984) p.265-75]

[7]   J.S. Weiner et al [*Appl. Phys. Lett. (USA)* vol.47 (1985) p.664-7]

[8]   T.H. Wood [*J. Lightwave Technol. (USA)* vol.6 (1988) p.743-57]

[9]   G.D. Boyd, D.A.B. Miller, D.S. Miller, S.L. McCall, A.C. Gossard, J.H. English [*Appl. Phys. Lett. (USA)* vol.50 (1987) p.1119-21]

[10]   M. Whitehead, G. Parry, P. Wheatley [*IEE Proc. J (UK)* vol.136 (1989) p.52-8]

[11]   M. Whitehead, A. Rivers, G. Parry, J.S. Roberts, C. Button [*Electron. Lett. (UK)* vol.25 (1989) p.984]

[12]   A.J. Moseley, J. Thompson, M.Q. Robbins, M.J. Goodwin [*Electron. Lett. (UK)* vol.26 (1990) p.913-5]

[13]   L. Buydens, P. Demeester, Z. Yu, P. Van Daelle [*IEEE Photonics Technol. Lett. (USA)* vol.3 (1991) p.1104-6]

[14]   D.A.B. Miller [*Opt. Quantum Electron. (UK)* vol.22 (1990) p.S61-98]

[15]   M. Whitehead, A. Rivers, G. Parry, J.S. Roberts, C. Button [*Electron. Lett. (UK)* vol.26 (1990) p.1588-9]

[16]   U. Koren, T.L. Koch, H. Presting, B.I. Miller [*Appl. Phys. Lett. (USA)* vol.50 (1987) p.368-70]

[17]   J.E. Zucker, T.L. Hendrickson, C.A. Burrus [*Electron. Lett. (UK)* vol.24 (1988) p.112-3]

[18]   O. Mitomi et al [*Appl. Opt. (USA)* vol.31 (1992) p.2030-5]

[19]   K. Satzke et al [*Electron. Lett. (UK)* vol.31 (1995) p.2030-1]

[20]   K.K. Loi, X.B. Mei, C.W. Tu, W.S. Chang [*IEEE Photonics Technol. Lett. (USA)* vol.8 (1996) p.626-8]

[21]   A.J. Moseley, M.Q. Kearley, R.C. Morris, D.J. Robbins, J. Thompson, M.J. Goodwin [*Electron. Lett. (UK)* vol.28 (1992) p.12-4]

[22]   A.V. Krishnamoorthy et al [*Electron. Lett. (UK)* vol.31 (1995) p.1917-8]

[23]   P. Barnes et al [*Opt. Quantum Electron. (UK)* vol.24 (1992) p.S177-92]

[24]   T.K. Woodward, T. Sizer II, D.L. Sivco, A.Y. Cho [*Appl. Phys. Lett. (USA)* vol.57 (1990) p.548-50]

[25]   T.K. Woodward, T.-H. Chiu, T. Sizer II [*Appl. Phys. Lett. (USA)* vol.60 (1992) p.2846-8]

[26]   M.N. Kahn, J.E. Zucker, L.L. Buhl, B.I. Miller, C.A. Burrus [*Proc. Euro. Conf. Opt. Commun. (ECOC'95)* Brussels, Belgium, 1995, p.103-7]

[27]   S. Goswami, S.-C. Hong, D. Biswas, P.K. Bhattacharya, J. Singh, W.-Q. Li [*IEEE J. Quantum Electron. (USA)* vol.27 (1991) p.760-8]

[28]   W.-Q. Li, S. Goswami, P.K. Bhattacharya, J. Singh [*Electron. Lett. (UK)* vol.27 (1991) p.31-3]

[29]   P. Zouganelli, A.W. Rivers, G. Parry, J.S. Roberts [*Opt. Quantum Electron. (UK)* vol.25 (1993) p.S935-51]

[30]   D.A.B. Miller et al [*IEEE J. Quantum Electron. (USA)* vol.21 (1985) p.1462-76]

[31]   A.L. Lentine et al [*Conf. on Lasers and Electro-Optics* Baltimore, Maryland, May 2-7 1993 (Optical Soc. of America) post deadline paper CPD24]

8.3 Quantum well and superlattice photodetectors

J.C. Campbell

May 1995

A INTRODUCTION

Avalanche photodiodes have the advantage, relative to PIN photodiodes, of internal gain. The gain is achieved through carrier multiplication that results from impact ionisation at high electric fields. This is usually achieved by biasing the photodiode near its breakdown voltage. The net effect of the avalanche gain is an improvement in sensitivity compared to PINs. For many applications the added performance more than compensates for the higher cost and more complex bias circuitry of the APDs. Two of the crucial performance characteristics of APDs, the gain-bandwidth product and the excess noise arising from the random nature of the multiplication process, are determined primarily by the electron and hole ionisation coefficients (α and β, respectively) or, more specifically, by k which is defined as the ratio of the ionisation coefficients (either α/β or β/α such that $k < 1$). For low noise and high gain-bandwidth products, a large difference in the ionisation rates, i.e. $k \ll 1$ is necessary [1,2].

Conventional APD structures utilise a relatively thick multiplication region. For short-wavelength applications ($\lambda < 1.0 \ \mu m$), Si is usually the material of choice [3,4] and in the wavelength range from $1.0 \ \mu m$ to $1.6 \ \mu m$ the APD structure that has demonstrated the best performance is the $InP/In_{0.53}Ga_{0.47}As$ SAM (separate absorption and multiplication regions) APD [5-9]. The value of k for these APDs is a characteristic of the material used for the multiplication layer and little can be done to alter it. For Si, $\alpha \gg \beta$ which results in high gain-bandwidth products and low multiplication noise [1]. For the III-V compounds, on the other hand, $k \approx 1$ [10,11]. A great deal of research has been devoted to the development of novel, low-noise APD structures in III-V compounds that have very low values of k.

One approach to achieving low multiplication noise in avalanche photodiodes is the use of heterojunctions to artificially enhance the ionisation rate of either the electrons or the holes. Several heterojunction APD structures have been proposed and analysed theoretically [12-17]. While there has been some diversity of approach, all of these structures utilise a multicell concept with ionisation-rate enhancement via heterojunctions. The most successful APD of this type is the multiple-quantum-well (MQW) APD. The band structure of an MQW APD is shown in FIGURE 1. The operating principle has been described as follows. As electrons emerge from the wide-bandgap portion of each cell into the narrow-bandgap portion of the adjacent cell, the discontinuity in the conduction band provides

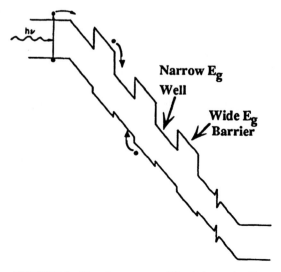

FIGURE 1. Band structure (Γ minimum) of a multiquantum-well avalanche photodiode [21].

sufficient additional energy to initiate ionisation. This enhances the ionisation rate of the electrons. The ionisation rate for holes, on the other hand, is not enhanced to the same degree since the valence band discontinuity is smaller than that of the conduction band.

There are some aspects to this explanation that would suggest that it is perhaps too simplistic. For example, Monte Carlo calculations have shown that when the fields are high enough to produce significant gain through impact ionisation, the carriers are distributed relatively uniformly throughout the Brillouin zone [18]. Electrons in satellite valleys do not experience the same conduction discontinuity as those in the Γ valley. Also, the model in FIGURE 1 does not explain the dependence of the ionisation rate ratio on the Al content in the barrier layers of GaAs/Al$_x$Ga$_{1-x}$As MQW APDs [19,20]. Nevertheless, there is incontrovertible evidence of favourable modification of the ionisation coefficients in GaAs/Al$_x$Ga$_{1-x}$As [19-25] and InAlAs/InGaAs(P) [29-35] MQW APDs.

B   GaAs-BASED PHOTODETECTORS

Chin et al [12] first predicted that the difference between the conduction band and valence band discontinuities at GaAs/Al$_x$Ga$_{1-x}$As interfaces could be used to enhance the ionisation rates of electrons in GaAs/Al$_x$Ga$_{1-x}$As MQWs compared to bulk GaAs. Capasso et al [21] subsequently reported experimental confirmation of enhancement of the electron ionisation rate using a multiplication region consisting of 25 pairs of GaAs(450 Å)/Al$_{0.45}$Ga$_{0.55}$As (550 Å). Since then there have been numerous reports on GaAs/Al$_x$Ga$_{1-x}$As MQW APDs [19,20,23-25]. Initial work on GaAs/Al$_x$Ga$_{1-x}$As multi-quantum-well APDs relied on differences in the multiplication for hole injection and electron injection to determine the ionisation rates. Kagawa and co-workers [19,20] successfully demonstrated that the modified ionisation coefficients result in lower multiplication noise and a high gain-bandwidth product (126 GHz). An interesting aspect of this work was the observation that the quantum-well effects were strongly dependent on the Al content of the wide-bandgap layers. For alloy compositions for which the bandgap of Al$_x$Ga$_{1-x}$As was direct ($x \leq 0.45$), α/β decreased with increasing x, e.g. $\alpha/\beta = 0.5$ at $x = 0.3$ and $\alpha/\beta = 0.14$ at $x = 0.45$. However, the electron ionisation rate was drastically reduced when the Al$_x$Ga$_{1-x}$As layers became indirect ($x > 0.45$). This is illustrated in FIGURE 2 which contrasts the measured excess noise factor for $x = 0.45$ and $x = 0.55$.

C   InP-BASED PHOTODETECTORS

Multiple-quantum-well APDs have also been fabricated in the InP/In$_{0.53}$Ga$_{0.47}$As [26-28] and InAlAs/InGaAs(P) materials systems. These APDs were designed for operation near $\lambda = 1.55\ \mu m$, the wavelength of minimum attenuation in optical fibres. For the InP/In$_{0.53}$Ga$_{0.47}$As structures, β was observed to become slightly larger than α since the valence band discontinuity is larger than that of the conduction band but no significant change in the ratio of the ionisation coefficients was achieved. For the In$_{0.52}$Ga$_{0.48}$As/In$_{0.53}$Ga$_{0.47}$As system, it has been shown with PIN structures that the large conduction band offset [29] results in an electron ionisation rate that is 10 to 20 times that of the holes [30]. In order to achieve pure electron injection, and thus to take advantage of this disparity in ionisation rates, it is necessary to use an SAM structure with a p-type In$_{0.53}$Ga$_{0.47}$As absorption layer and an

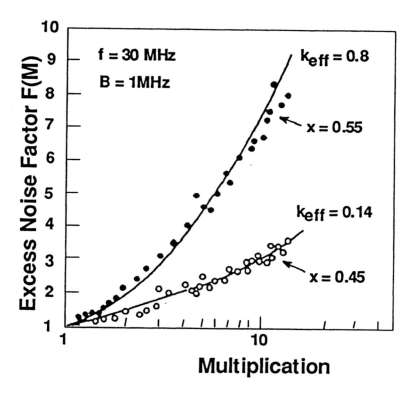

FIGURE 2. Excess noise factor for GaAs/Al$_x$Ga$_{1-x}$As MWQ APDs for
x = 0.45 and x = 0.55 in the barrier layers [19].

n-type Al$_{0.48}$In$_{0.52}$As/In$_{0.53}$Ga$_{0.47}$As multi-quantum well multiplication region [31]. The excess noise factor of these APDs is consistent with a k factor between 0.1 and 0.2 which is a significant improvement over the conventional InP/In$_{0.53}$Ga$_{0.47}$As APDs where k is typically ≈0.4. Hanatani et al [32] have obtained a gain-bandwidth product of 130 GHz with the Al$_{0.48}$In$_{0.52}$As/In$_{0.53}$Ga$_{0.47}$As multiple quantum well structure. Kagawa et al [33,34] have obtained excellent performance by utilising a quaternary alloy of InGaAsP in the narrow-bandgap portion of the MQW multiplication region. The addition of P effectively eliminates the valence band discontinuity, ΔE_v, without significantly lowering the conduction band discontinuity, ΔE_c. At low gains (M$_o$ < 10) the response was transit-time limited and the bandwidth was 17 GHz. At higher multiplication values a gain-bandwidth-limited response of 110 GHz was obtained [35]. One of the best measures of the performance of an APD is the sensitivity that can be achieved in an optical receiver. These Al$_{0.48}$In$_{0.52}$As/In$_{0.53}$Ga$_{0.47}$As multi-quantum well APDs have been successfully deployed in optical receivers operating at 10 Gbit/s [36-38].

It is well known that strain can be used to improve the performance of MQW lasers. Gutierrez-Aitken and co-workers [39,40] have measured the electron and hole ionisation coefficients in biaxially strained In$_x$Ga$_{1-x}$As/In$_y$Al$_{1-y}$As (0.44 < x < 0.62, 0.44 < y < 0.62) MQWs. They reported a strain-induced enhancement in β/α for tensile strain in the quantum well and compressive strain in the barrier. This effect was attributed to changes in the bandgap, the band offsets, and the band structure.

D   CONCLUSION

MQW APDs have proved to be an effective approach to enhance the ionisation rate of electrons relative to that of holes. $Al_xGa_{1-x}As/GaAs$ multiquantum-well structures have demonstrated suppression of the multiplication noise compared to bulk APDs and gain-bandwidth products as high as 126 GHz have been achieved. $Al_{0.48}In_{0.52}As/In_{0.53}Ga_{0.47}As$ MQW APDs have been demonstrated for long-wavelength transmission systems. A gain-bandwidth product of 130 GHz has been achieved with low multiplication noise. There is, however, still much to be learned about impact ionisation in MQW structures. Further improvements in the performance of APDs will undoubtedly follow closely behind new developments in materials fabrication and processing.

REFERENCES

[1]   R.J. McIntyre [*IEEE Trans. Electron Devices (USA)* vol.13 (1966) p.164-8]

[2]   R.B. Emmons [*J. Appl. Phys. (USA)* vol.38 (1967) p.3705-14]

[3]   P.P. Webb, R.J. McIntyre, J. Conradi [*RCA Rev. (USA)* vol.35 (1974) p.234-78]

[4]   H. Melchior, A.R. Hartman, D.P. Schinke, T.E. Seidel [*Bell Syst. Tech. J. (USA)* vol.57 (1978) p.1791-807]

[5]   K. Nishida, K. Taguchi, Y. Matsumoto [*Appl. Phys. Lett. (USA)* vol.35 (1979) p.251-3]

[6]   J.C. Campbell, A.G. Dentai, W.S. Holden, B.L. Kasper [*Electron. Lett. (UK)* vol.18 (1983) p.818-20]

[7]   J.C. Campbell, W.T. Tsang, G.J. Qua, J.E. Bowers [*Appl. Phys. Lett. (USA)* vol.51 (1987) p.1454-6]

[8]   L.E. Tarof [*Electron. Lett. (UK)* vol.27 (1991) p.34-6]

[9]   L.E. Tarof, J. Yu, R. Bruce, D.G. Knight, T. Baird, B. Oosterbrink [*IEEE Photonics Technol. Lett. (USA)* vol.5 (1993) p.672-4]

[10]   G.E. Bulman, V.M. Robbins, K.F. Brennan, K. Hess, G.E. Stillman [*IEEE Electron Device Lett. (USA)* vol.4 (1983) p.181-3]

[11]   F. Osaka, T. Mikawa, O. Wada [*IEEE J. Quantum Electron. (USA)* vol.21 (1986) p.1326-38]

[12]   R. Chin, N. Holonyak Jr., G.E. Stillman, J.Y. Tang, K. Hess, [*Electron. Lett. (UK)* vol.16 (1980) p.467-9]

[13]   K. Brennan [*IEEE Trans. Electron Devices (USA)* vol.32 (1985) p.2197-205]

[14]   H. Blauvelt, S. Margalit, A. Yariv [*Electron. Lett. (UK)* vol.18 (1982) p.375-6]

[15]   J.S. Smith, L.C. Chiu, S. Margalit, A. Yariv, A.Y. Cho [*J. Vac. Sci. Technol. B (USA)* vol.1 (1983) p.376-8]

[16]   S.L. Chuang, K. Hess [*J. Appl. Phys. (USA)* vol.59 (1986) p.2885-94]

[17]   M.C. Teich, K. Matsuo, B.E.A. Saleh [*IEEE Trans. Electron Devices (USA)* vol.33 (1986) p.1475-88]

[18]   V. Chandramouli, C.M. Maziar [*Solid-State Electron. (UK)* vol.36 (1993) p.285-90]

[19]   T. Kagawa, H. Iwamura, O. Mikami [*Appl. Phys. Lett. (USA)* vol.54 (1989) p.33-5]

[20]   T. Kagawa, K. Mogi, H. Iwamura, O. Mikami [*Int. Conf. Integrated Opt. and Opt. Fiber Communication* Kobe, Japan, 1989]

[21]   F. Capasso, W.T. Tsang, A.L. Hutchinson, G.F. Williams [*Appl. Phys. Lett. (USA)* vol.40 (1982) p.38-40]

[22]   F. Capasso, W.T. Tsang, G.F. Williams [*IEEE Trans. Electron Devices (USA)* vol.30 (1983) p.381-90]

[23]   F.-Y. Juang, U. Das, Y. Nashimoto, P.K. Bhattacharya [*Appl. Phys. Lett. (USA)* vol.47 (1989) p.972-4]

[24]   J. Allam, F. Capasso, K. Alavi, A.Y. Cho [*Proc. 13th Int. Symp. on GaAs and Related Compounds* Ed. W.T. Lindley (Institute of Physics, Bristol, 1986) p.405-10]

[25]   M. Toivonen, A. Salokatve, M. Hovinen, M. Pessa [*Electron. Lett. (UK)* vol.28 (1992) p.32-4]

[26]   F. Osaka, T. Mikawa, O. Wada [*IEEE J. Quantum Electron. (USA)* vol.22 (1986) p.1986-91]

[27]   F. Beltram, J. Allam, F. Capasso, U. Koren, B. Miller [*Appl. Phys. Lett. (USA)* vol.50 (1987) p.1170-2]

[28]   A.J. Moseley, J. Urquhart, J.R. Riffat [*Electron. Lett. (UK)* vol.24 (1988) p.313-5]

[29]   R. People, K.W. Wecht, K. Alavi, A.Y. Cho [*Appl. Phys. Lett. (USA)* vol.43 (1983) p.118-20]

[30]   T. Kagawa, Y. Kawamura, H. Asai, M. Naganuma, O. Mikami [*Appl. Phys. Lett. (USA)* vol.55 (1989) p.993-5]

[31]   T. Kagawa, Y. Kawamura, H. Asai, M. Naganuma [*Appl. Phys. Lett. (USA)* vol.57 (1990) p.1895-7]

[32]   S. Hanatani, H. Nakamura, S. Tanaka, C. Notsu, H. Sano, K. Ishida [*Tech. Digest Conf. on Opt. Fiber Comm./Int. Conf. on Integrated Opt. and Opt. Fiber Comm.* San Jose, CA, USA, Feb. 1993 (Opt. Soc. America, 1993) p.187]

[33]   T. Kagawa, Y. Kawamura, H. Iwamura [*IEEE J. Quantum Electron. (USA)* vol.28 (1992) p.1419-23]

[34]   T. Kagawa, Y. Kawamura, H. Iwamura [*Tech. Digest Conf. on Opt. Fiber Comm./Int. Conf. on Integrated Opt. and Opt. Fiber Comm.* San Jose, CA, USA, Feb. 1993 (Opt. Soc. America, 1993)]

[35]   T. Kagawa, Y. Kawamura, H. Asai, M. Naganuma, O. Mikami [*Int. Electron Devices Meet. Tech. Dig. (USA)* (1989) p.725-8]

[36]   I. Watanabe et al [*Tech. Digest Conf. on Opt. Fiber Comm./Int. Conf. on Integrated Opt. and Opt. Fiber Comm.* San Jose, CA, USA, Feb. 1993 (Opt. Soc. America, 1993) p.184-5]

[37]   Y. Miyamoto et al [*Tech. Digest Conf. on Opt. Fiber Comm./Int. Conf. on Integrated Opt. and Opt. Fiber Comm.* San Jose, CA, USA, Feb. 1993 (Opt. Soc. America, 1993) p.13-5]

[38]   Y. Miyamoto, K. Hagimoto, T. Kagawa [*IEEE Photonics Technol. Lett. (USA)* vol.3 (1991) p.372-4]

[39]   A.L. Gutierrez-Aitken, S. Goswami, Y.C. Chen, P.K. Bhattacharya [*Int. Electron Devices Meet. Tech. Dig. (USA)* (1992) p.647-50]

[40]   A.L. Gutierrez-Aitken, P.K. Bhattacharya [*J. Appl. Phys. (USA)* vol.73 (1993) p.5014-6]

8.4 Quantum well tunnelling devices

S. Luryi and A. Zaslavsky

August 1995

A INTRODUCTION

Advances in epitaxy have permitted the design of semiconductor devices with sufficient precision to make use of quantum mechanical tunnelling. The potential advantages of tunnelling-based devices consist of high-speed operation and higher functionality inherent in their strongly nonlinear current-voltage characteristics, while the disadvantages include the relative weakness of quantum effects at room temperature and the difficult integration into mainstream semiconductor technology. To date, these difficulties have largely impeded tunnelling devices from evolving beyond the research stage, but continuing progress in epitaxy, fabrication and novel device concepts may render these devices suitable for some applications. Section B of this Datareview will cover the extensively researched double-barrier resonant tunnelling structure (RTS) as a discrete device, as well as a two-terminal nonlinear component in simple circuits, and discuss the prospects of three-terminal operation. Optical sources and detectors based on intersubband transitions in quantum wells populated or emptied by tunnelling are discussed in Section C. Tunnelling devices sufficiently small to employ single-carrier or Coulomb blockade effects that offer distant prospects of general-purpose single-electron logic or more immediate possibilities of high-precision current sources are discussed in Section D. Finally, expanding the definition of a device to include tunnelling structures that provide experimental probes of the electronic and structural properties of quantum wells, we discuss double-barrier and double-well RTSs as spectroscopic systems for subband dispersions, carrier coherence lifetimes, and strain distributions.

B RESONANT TUNNELLING DEVICES

B1 Two-Terminal Double-Barrier Resonant Tunnelling Diodes

A schematic band diagram of a double-barrier RTS in the tunnelling regime, the device itself, and the corresponding nonlinear current-voltage I(V) characteristic are shown in FIGURE 1. Originally proposed [1] and demonstrated [2] in the 1970s, this general structure has attracted much scientific interest. As shown in FIGURE 1(a), confinement of carriers in the well quantises their electronic states into two-dimensional (2D) subbands E_i and transport between the doped electrodes cladding the double-barrier regions proceeds mainly by tunnelling (as long as the temperature is low or the barriers high enough to suppress thermally activated above-barrier transport). Originally tunnelling was analysed in terms of coherent quantum-mechanical transmission through the double-barrier potential [1-3], which predicts very sharp resonances when the incident energy matches E_i. A more realistic description in the presence of scattering is the sequential tunnelling model [4]: carriers tunnel from the emitter into a 2D subband conserving energy E and transverse momentum $k_\perp$, then lose coherence and eventually tunnel out to the collector in a separate tunnelling event. The E and

$k_{\perp}$ conservation rules determine the number of carriers that can tunnel into the well for a given alignment, set by the applied bias V, of the subband E_i and the emitter Fermi level E_F: current can flow only when $E_F > E_i$ (E conservation, leading to a threshold bias V_{th}) and $E_i > E_c$ ($k_{\perp}$ conservation, leading to a peak bias V_p). The sequential tunnelling model [4] describes very well the experimental low-temperature I(V) of a high-quality GaAs/AlGaAs RTS shown in FIGURE 1(c) [5], except for the non-zero valley current for $V > V_p$ which requires explicit evaluation of phonon and impurity scattering-assisted tunnelling that need not conserve E and $k_{\perp}$.

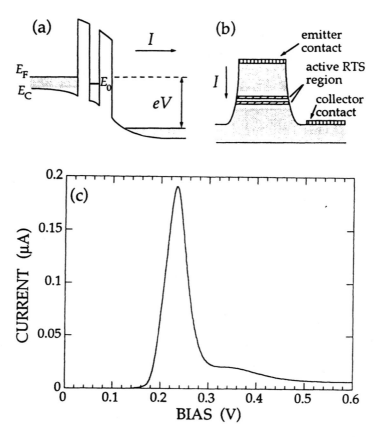

FIGURE 1. (a) Schematic band diagram of an n-type double-barrier resonant tunnelling structure under bias. (b) Cross-sectional diagram of a two-terminal RTS device pillar, doped regions shaded. (c) Low-temperature I(V) characteristic of a GaAs/AlGaAs double-barrier RTS. The valley current for $V > V_p$ is due to scattering or phonon-assisted tunnelling.

The double-barrier RTS is a two-terminal device with a highly nonlinear I(V) characteristic. The strong negative differential resistance (NDR) region for $V > V_p$, with the peak-to-valley ratio of the current reaching more than 50 even at room temperature in appropriately designed RTSs [6], is useful in the design of solid-state oscillators. The principle of operation is analogous to the classic p-n junction-based tunnel diode [7] and the maximum oscillation frequency is limited by the RC time constant. In a tunnel diode both sides of the junction are doped heavily and the tunnelling barrier is set by the bandgap, while optimised RTSs can have tailored barriers leading to very high peak current densities (lower R) without high doping on both sides of the tunnel barriers (leading to lower C). Microwave oscillators operating at frequencies up to 420 GHz in InGaAs/AlAs/InAs and 712 GHz in InAs/AlSb RTS diodes have been fabricated [8], as well as circuit-compatible InAs/AlSb RTS diodes with 1.7 ps switching speeds [9].

B2 Three-Terminal Resonant Tunnelling Structures

As a discrete two-terminal device the double-barrier RTS has limited utility beyond solid-state oscillators. There has been considerable research into the fabrication of three-terminal RTS devices, either by separately contacting the well of a double-barrier RTS or by adding a gate electrode. The former approach is analogous to a heterojunction bipolar transistor: the schematic band diagram of such a device is shown in FIGURE 2, which combines the bipolar [10] and unipolar [11] versions. The bipolar version permits the separate contacting of the narrow well by doping it p-type. Negative transconductance is observed in the collector current I_c characteristic whenever the hole current in the quantum well biases the RTS into the NDR region [10]. However, the narrow quantum well necessary for large 2D subband separation leading to strong NDR also results in high base resistance unless the well is very heavily doped, but then scattering tends to wash out the tunnelling characteristics. This difficulty might be overcome in polytype GaSb/AlSb/InAs RTSs where bandgap blocking of the tunnelling current gives excellent peak-to-valley ratios even for wide quantum wells [12]. Structures with wider n-type quantum wells have also been fabricated into unipolar three-terminal tunnelling hot-electron transistors, where the injected electrons traverse the well ballistically and contribute to the collector current [11]. The current gain in these devices depends on the fraction of the ballistic electrons reaching the collector rather than scattering into the quantum well; it is typically smaller than in bipolar transistors. An interesting variation on this scheme is the lateral tunnelling transistor [13], where the two barriers are

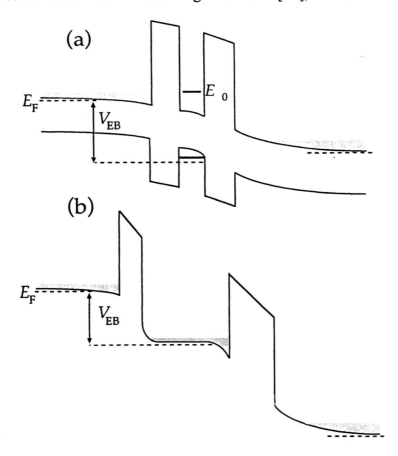

FIGURE 2. (a) Bipolar three-terminal RTS (after [10]); p-type doping permits a separate contact to the narrow quantum well. (b) Unipolar n-type three-terminal RTS, with a separate contact to a wider quantum well (analogous to the base in a heterojunction bipolar transistor, after [11]).

produced by narrow electrostatic gates deposited on top of a modulation-doped heterostructure with a 2D electron gas [14]. The high mobility of the 2D electron gas greatly enhances the fraction of ballistic electrons reaching the collector and reduces the base resistance, leading to current gains of over 100, but only at cryogenic temperatures [13].

An alternative route to a three-terminal RTS device is the fabrication of a gated structure. FIGURE 3 illustrates the cross-section and schematic band diagram of a 2D RTS, where a gate controls the density of the 2D electrons that tunnel through 1D subbands in the quantum well [15]. Negative transconductance was predicted for this device since the fringing electric fields generated by gate bias lower the subband energies E_i with respect to the emitter E_F, shifting the current peaks towards lower drain bias values. Large-scale fabrication of such devices requires the regrowth of modulation-doping heterostructures on etched surfaces, but the negative transconductance characteristic has been confirmed [16] in a similar structure produced by cleaved-edge overgrowth [17]. The same functionality can in principle be obtained from a lateral tunnelling transistor [13] if the electrostatically confined quantum well can be made sufficiently narrow for well-resolved subband quantisation. Finally, narrow vertical

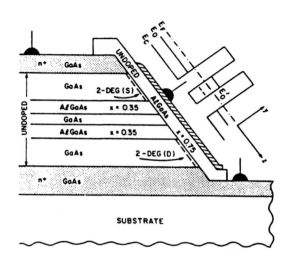

FIGURE 3. Gated two-dimensional RTS structure, after [15]. The electrons tunnel from the 2D electron gas (E_0 labels the bottom of the 2D subband) at the modulation-doped interface through the 1D subbands of the quantum wire (only the lowest 1D subband E_0' is shown). The gate potential controls both 2D gas density and E_0' [15]; for experimental realisation see [16].

double-barrier RTSs can be controlled laterally by a Schottky gate [18], although this approach involves great fabrication complexity. It should be stressed that while it has been suggested that three-terminal RTS devices with negative transconductance can in principle perform complementary functions for logic applications [19], no RTS circuit analogous to a CMOS inverter has ever been demonstrated. The difficulty lies in the fact that complementary CMOS transistors have more than just a negative transconductance; their current flow is effected by carriers of opposite polarity, which makes it possible to connect the drains of two transistors in series, rather than the source of one to the drain of the other.

B3   Integration of Resonant Tunnelling Diodes in Devices and Circuits

The highly nonlinear I(V) of an RTS makes it a potentially useful circuit element when inserted in another device or integrated with other devices. Much research has been done on resonant hot-electron transistors (RHETs) where a double-barrier RTS is inserted into the emitter of a hot-electron transistor as shown in FIGURE 4 [20]. The RHET exhibits negative transconductance together with current gain, as the I(V) characteristic of the emitter RTS is reproduced in the collector current. Adding two logic inputs at the base of the RHET provides exclusive NOR functionality in a single device [20]. Various logic gates and circuits have been fabricated by combining RHETs and resistors with reduced transistor counts compared to standard bipolar or CMOS circuits. However, the reduced transistor count in such circuits as a full adder [21] typically comes at the expense of fabricating a large number of resistors.

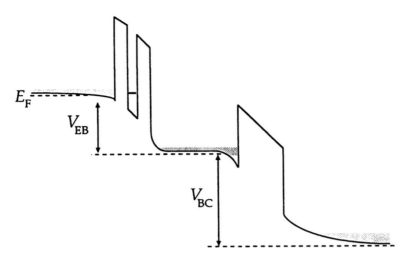

FIGURE 4. Schematic resonant hot-electron transistor (RHET) band diagram under bias. The resonant peak current of the RTS emitter structure is reproduced in the collector current (after [20]).

The epitaxy of several decoupled double-barrier RTSs separated by wide doped regions can produce a two-terminal device with a multipeak I(V) characteristic as each of the constituent RTSs is biased through the resonance similar to FIGURE 1(c). When such a structure is inserted in the emitter of a bipolar transistor, the multipeak collector current provides frequency multiplication with current gain [22]. In an optimised structure of decoupled RTSs in series, a regular sawtooth I(V) characteristic with a large number of resonant peaks can be produced, as shown in FIGURE 5. When biased with a constant current from a field effect transistor, such a device has a number of stable voltage operating points that can be used for multistate memory [23]: after an input

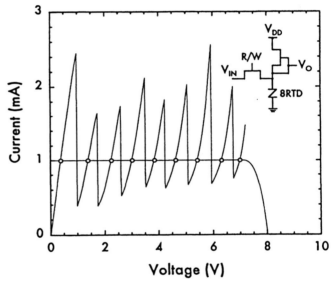

FIGURE 5. Multipeak I(V) characteristic of a cascaded RTS structure with a field-effect transistor load employed for a nine-state memory at room temperature [23]. Inset shows the schematic memory circuit: the field-effect transistor load provides a constant current to set the various stable points.

voltage value is applied, the circuit moves to the nearest stable point and maintains that value indefinitely.

A variant of static random access memory can be constructed by connecting two double-barrier RTSs in series with an additional connection to the middle node, either by fabricating a double-emitter RHET with a common floating base [24] or by fabricating two RTSs with a common collector on top of a single-barrier tunnel diode [25]. When two devices with NDR I(V) characteristics are connected in series and the total applied bias exceeds twice the peak voltage V_p, the middle node becomes bistable. Most of the total applied bias drops over one of the devices, while the other takes less than V_p and current continuity is preserved. The potential of the middle node (for example, the floating common base of a double-emitter RHET device) can be flipped between the two stable points by the biasing of a third terminal (the collector). As a result, very compact static memory cells consisting of two vertical RTS devices and associated contact lines have been fabricated [24,25]. The difficulty with these RTS-based memories is their relatively large standby power consumption. The bistable operating points of two RTSs in series biased just beyond $2V_p$ pass a current that is at least as large as the valley current value in the I(V) characteristic of a single RTS. Yet the valley current cannot be reduced indefinitely by reducing the RTS device size because in large-scale memory circuits sizable peak resonant currents are required to charge up the interconnect capacitances. Consequently, significant improvements in the peak-to-valley ratios of the RTS elements will be required before these tunnelling-based static memories become competitive with low-power CMOS.

C   OPTOELECTRONIC TUNNELLING DEVICES

If many identical double-barrier potentials are repeated epitaxially, the result is a superlattice (SL) in which the quantised subbands broaden into minibands of width ΔE_i separated by minigaps. Consider transport in the presence of an electric field F applied along the SL. If F is weak, such that the potential drop $eFd \ll \Delta E$, where d is the SL period, current will flow by miniband conduction. Given sufficiently long scattering times τ, the electric field would accelerate carriers into the region of negative curvature in the miniband dispersion, giving rise to NDR in the I(V) characteristic [26]. For even greater τ, the carriers would reach the minizone boundary and experience Bragg reflection, giving rise to Bloch oscillations [26,27]. These effects have proven to be very difficult to observe in transport because of scattering, Zener tunnelling between different minibands, and especially the charge-driven break-up of the superlattice into high- and low-field domains [28]. For this reason, despite much research into the physics of low-field superlattice transport [29], SL I(V) nonlinearities have not been used in devices to date.

In the strong F limit, where $eFd > \Delta E$, the miniband breaks up into 2D subbands localised within one well and a resonant tunnelling current flows when the ground subband in one well is exactly aligned with the first (or higher) excited subband in the adjacent well: $eFd = (E_i - E_0)$, i = 1, 2 ... [30,31]. As in ordinary double-barrier resonant tunnelling, this energy conservation rule can be altered by an inelastic process, such as phonon or photon emission. In particular, when $eFd > (E_i - E_0)$ a carrier can tunnel into the adjacent well by emitting a photon of energy $\hbar\omega = eFd - (E_i - E_0)$, implying the exciting possibility of an infrared laser tunable by an applied electric field, an idea dating back to 1971 [30]. A conceptually similar device can also be operated at resonance, $eFd = (E_2 - E_0)$ and $\hbar\omega = (E_2 - E_1)$, provided a population inversion between the subbands E_2 and E_1 is maintained by some means, such as a longer nonradiative lifetime in the E_2 subband compared to the E_1 subband. In this case, the radiation frequency cannot be tuned by the electric field but is set by the intersubband energy separation determined by the quantum well parameters.

The fabrication of such a device faces two fundamental obstacles. First, efficient amplification of a given frequency requires F to be uniform over many SL periods, whereas the current-carrying charge tends to break up the SL into high- and low-field domains [28]. Second, the higher-lying states of a quantum well are confined by lower tunnelling barriers, which works against population inversion. Recently, intersubband laser action was demonstrated in a quantum cascade laser (QCL) [32], illustrated in FIGURE 6. The QCL is a periodic structure alternating between a short-period SL and a double-well active region in which population inversion is established. As shown in FIGURE 6, under operating bias carriers flow through a superlattice miniband and tunnel into the highest E_2 subband of the double-well active region. Tunnelling out of the E_2 subband is impeded by the minigap of the downstream SL, so the carriers relax by radiative and nonradiative processes down to the E_1 and E_0 subbands, which can tunnel out into the downstream SL. Active region parameters are chosen to fix the $(E_1 - E_0)$ energy separation close to the optical phonon energy, leading to a much shorter lifetime of the E_1 subband and establishing population inversion and laser action at $\hbar\omega = (E_2 - E_1)$. Finally, the QCL retains overall charge neutrality under bias by proper doping of the SL regions, hence avoiding the domain formation problem. As a result, infrared lasers at $\lambda \approx 4.5$ and 8.4 μm have been demonstrated at operating temperatures above 100 K [32], leading to prospects of devices competitive with other sources in that frequency range.

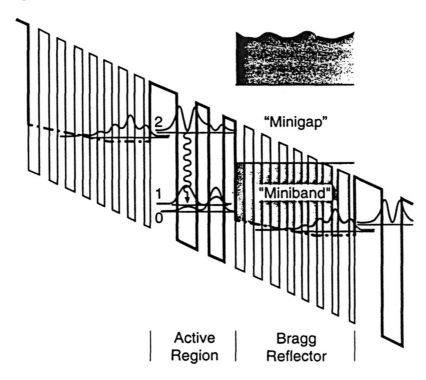

FIGURE 6. Schematic diagram of one period in the quantum cascade layer (QCL). The radiative transition frequency is indicated; it can be varied by adjusting the double quantum well parameters [32].

Another application of tunnelling-based optoelectronic devices comes in the area of quantum well intersubband IR photodetectors. Although in most of these devices IR radiation excites a transition from the lowest quantum well subband E_0 to the above-barrier continuum final states and current flows by hot-electron transport rather than tunnelling [33], there have been versions where the final state is a higher-lying subband that is emptied by tunnelling into an adjacent SL miniband [34]. In this case the photodetector structure is essentially the inverse of the QCL in FIGURE 6, except that the active region consists of a single quantum well and the SL blocks tunnelling out of the ground-state subband E_0.

D   TUNNELLING NANOSTRUCTURES AND COULOMB BLOCKADE

If a double-barrier RTS diode of FIGURE 1(b) is made sufficiently narrow in one or both lateral directions, the electronic states in the quantum well will be further quantised, leading to 1D quantum wire subbands or fully quantised, atomic-like states. The I(V) characteristics of such RTS nanostructures are still governed by the same physics of tunnelling into reduced dimensionality states [4], but with another effect coming into play: the tunnelling of a single electron can tangibly alter the electrostatic field distribution over the entire structure. Modelling the double-barrier RTS as a small parallel-plate capacitor, the addition of a single electron to the quantum well population requires the energy of $e^2/2C$, where C is the total capacitance between the quantum well and the rest of the structure. This discrete energy barrier arises from charge quantisation and must be overcome for a carrier to tunnel into the well - a phenomenon known as the Coulomb blockade [35]. At low temperatures and in sufficiently small RTSs, $e^2/2C$ can be the largest parameter in the system and, instead of a smooth rise in the I(V) characteristic above V_{th}, the current increases in discrete steps corresponding to the opening of additional single-electron tunnelling channels [36]. These I(V) measurements, as well as low-frequency capacitance measurements on double-barrier RTSs with impenetrable collector barriers [37], have been employed to probe the energy spectrum of artificial few-electron atoms in and out of magnetic fields.

Given an additional gate electrode to control the potential of the quantum well, the tunnelling of single electrons into the well can be controlled in a transistor-like manner. Effective gate control is easier in the planar geometry, where electrostatically biased gates on top of a modulation-doped heterostructure [14] deplete the 2D electron gas and create a small 2D island isolated from the rest of the electron gas by potential barriers. An additional gate electrode can alter the effective size and capacitance of the island. As additional single-electron tunnelling channels are opened by changing the gate voltage, very regular conductance peaks as a function of V_g have been observed [38,39] at low temperatures. There have been numerous proposals of single-electron transistors (SETs) and other devices [40,41] as the ultimate limit of miniaturisation-driven semiconductor technology. It should be emphasised, however, that in addition to the extremely stringent fabrication requirements faced by large-scale SET circuitry at non-cryogenic temperatures, it is not clear that semiconductor SET realisations have any advantage over metal tunnel junctions, where the Coulomb-blockade phenomena were originally observed [42]. Thus, the first SET with voltage gain was realised in small Al tunnel junction capacitors [43].

One advantage of semiconductor-based Coulomb-blockade RTSs with barriers created by electrostatic gating of a 2D electron gas [39] is the separate tunability of barriers. Given small emitter-collector biasing in the Coulomb-blockade regime, only one electron at a time can pass through the device. If the emitter and collector barriers are lowered and then raised sequentially, a single electron will tunnel onto the island when the emitter barrier is lowered and tunnel out when the collector barrier is lowered, resulting in the transport of a single electron through the device with every barrier biasing cycle. A very accurate current source that supplies a current $I = ef$, where f is the barrier biasing cycle frequency, has been constructed [44,45]. Such devices may find metrological applications as current standards.

E   TUNNELLING SPECTROSCOPY OF QUANTUM WELL PARAMETERS

Insofar as an experimental test structure for semiconductor characterisation may be termed a device, tunnelling structures have found a number of applications in the study of quantum well parameters. Here we briefly describe two quantities of interest accessible via tunnelling measurements - quantum well coherent lifetimes and in-plane dispersions.

An optically pumped double-well system for measuring coherent lifetimes is illustrated in FIGURE 7 [46]. By adjusting the quantum well parameters and applying an external electric field it is possible to bring the electron subbands into resonance (with the electron eigenstates of the double-well system becoming symmetric and antisymmetric combinations of the single-well states) without aligning the hole subbands. In this case, radiative recombination of electrons and holes occurring in different wells will produce luminescence at different frequencies. If one of the wells is selectively populated with a short optical pulse of the appropriate wavelength, the photoexcited electrons will execute Rabi oscillations between the quantum wells with a frequency ω corresponding to the symmetric-antisymmetric energy splitting. The time-resolved photoluminescence from the double-well system then contains two oscillating signal frequencies that are out-of-phase and decay on a time scale of the coherent lifetime in the quantum well [46] - an

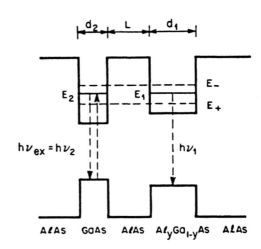

FIGURE 7. Schematic diagram of the double-well tunnelling structure for measuring coherent lifetimes in a quantum well implemented in GaAs/AlGaAs. Electrons oscillate between the quantum wells giving rise to out-of-phase luminescence signals at frequencies ν_1 and ν_2. The decay of the oscillating luminescence gives the coherent lifetime [46,47].

effect that was observed in high-quality GaAs/AlGaAs quantum wells by femtosecond spectroscopy [47]. Similar information on the coherent lifetimes in a quantum well can be extracted from a modulation-doped double-well system with separate contacts to the two quantum wells and a gate V_g to control subband alignment [48]. At low temperatures, tunnelling transport between the two quantum wells separated by a thick tunnelling barrier is governed by the same E and $k_\perp$ conservation laws as in standard RTS, with the difference that tunnelling occurs between fully quantised 2D subbands. Hence tunnelling is possible only when the subbands are perfectly aligned and in an ideal system the tunnelling $I(V_g)$ peak would be a delta function. The measured width of the tunnelling peak, combined with independently measured in-plane 2D electron gas mobility, provides an experimental measure of the coherent lifetimes in the 2D subband.

Magnetotunnelling measurements in p-type double-barrier RTSs have been used to extract the complex, anisotropic valence subband dispersions in quantum wells. If the tunnelling I(V) characteristics are measured in a transverse magnetic field $B_\perp$, the $k_\perp$ conservation rule for tunnelling into the well is modified by $\Delta k_\perp = eB_\perp \langle z \rangle / \hbar$, where $\langle z \rangle$ is the distance traversed in the direction of tunnelling [49]. As a result, the in-plane dispersion of the 2D subbands $E_i(k_\perp)$ can be calculated from the measured shifts in the magnetotunnelling $I(V,B_\perp)$ peak positions [50]. Furthermore, since the $B_\perp$-induced $\Delta k_\perp$ shift affects the $k_\perp$ component perpendicular to $B_\perp$, magnetotunnelling measurements as a function of $B_\perp$ orientation should reveal the dispersion anisotropy between different in-plane crystallographic directions [51]. While measurements on GaAs/AlGaAs RTSs did not show a significant anisotropy at attainable $B_\perp$ values, measurements on strained InGaAs/InAlAs p-type RTSs [52] have been successful in extracting the in-plane band structure anisotropy (and stronger effects have been observed by this technique in the highly anisotropic Si/SiGe quantum wells [53]).

As a final note, another application of tunnelling spectroscopy is the determination of strain redistribution in small strained heterostructures. As semiconductor devices push deep into the submicron regime, their size becomes comparable to strain relaxation length scales. Since the energy separation between the 2D subbands in RTSs is partially determined by strain, I(V) peak positions and lineshapes of small RTS devices should reveal the strain distributions in the quantum well. In this way, measurements of submicron p-Si/SiGe RTSs have been employed to extract the strain relaxation in quantum wells of small devices [54].

F   CONCLUSION

Intensive research over the past two decades has examined a number of interesting quantum-well structures in which carrier tunnelling phenomena are harnessed to perform potentially useful electronic or optoelectronic functions. While to date none of these structures has progressed to real technological implementation because of fabrication and high-temperature operation problems, as well as competition from ever-improving standard transistor-based technology, further advances in epitaxial and lithographic techniques should enable tunnelling-based devices to occupy niche applications (such as metrological current standards or infrared lasers). The possibility of tunnelling-based memories or logic circuitry finding widespread technological application appears less likely, unless major breakthroughs in materials and device performance are accomplished.

REFERENCES

[1]   R. Tsu, L. Esaki [*Appl. Phys. Lett. (USA)* vol.22 (1973) p.562-4]

[2]   L.L. Chang, L. Esaki, R. Tsu [*Appl. Phys. Lett. (USA)* vol.24 (1974) p.593-5]

[3]   B. Ricco, M.Ya. Azbel [*Phys. Rev. B (USA)* vol.29 (1984) p.1970-81]

[4]   S. Luryi [*Appl. Phys. Lett. (USA)* vol.47 (1985) p.490-2]

[5]   A. Zaslavsky, Y.P. Li, D.C. Tsui, M. Santos, M. Shayegan [*Phys. Rev. B (USA)* vol.42 (1990) p.1374-80]

[6]   J.H. Smet, T.P.E. Broekart, C.G. Fonstad [*J. Appl. Phys. (USA)* vol.71 (1992) p.2475-7]

[7]   S. Sze [*Physics of Semiconductor Devices, 2nd Edition* (Wiley-Interscience, New York, 1981) ch.9 p.516-37]

[8]   E.R. Brown, T.C.G.L. Sollner, C.D. Parker, W.D. Goodhue, C.L. Chen [*Appl. Phys. Lett. (USA)* vol.55 (1989) p.1777-9]; E.R. Brown et al [*Appl. Phys. Lett. (USA)* vol.58 (1991) p.2291-3]

[9]   E. Özbay, D.M. Bloom, D.H. Chow, J.N. Schulman [*IEEE Electron Device Lett. (USA)* vol.14 (1993) p.400-2]

[10]   M.A. Reed, W.R. Frensley, R.J. Matyi, J.N. Randall, A.C. Seabaugh [*Appl. Phys. Lett. (USA)* vol.54 (1989) p.1034-6]

[11]   M. Heiblum, I.M. Anderson, C.M. Knoedler [*Appl. Phys. Lett. (USA)* vol.49 (1986) p.207-9]; M. Heiblum et al [*Phys. Rev. Lett. (USA)* vol.56 (1986) p.2854-7]

[12]   R. Beresford, L.F. Luo, K.F. Longenbach, W.I. Wang [*Appl. Phys. Lett. (USA)* vol.56 (1990) p.551-3]

[13]   A. Palevski, C.P. Umbach, M. Heiblum [*Appl. Phys. Lett. (USA)* vol.55 (1989) p.1421-3; *Phys. Rev. Lett. (USA)* vol.62 (1989) p.1776-9]

[14]   [Datareview in this book: 7.2 GaAs-based modulation doped heterostructures]

[15]   S. Luryi, F. Capasso [*Appl. Phys. Lett. (USA)* vol.47 (1985) p.1437-9]; erratum [*Appl. Phys. Lett. (USA)* vol.48 (1986) p.1693]

[16]   Ç. Kurdak et al [*Appl. Phys. Lett. (USA)* vol.64 (1994) p.610-2]

[17]   L.N. Pfeiffer et al [*Appl. Phys. Lett. (USA)* vol.58 (1991) p.269-71]; A. Zaslavsky, D.C. Tsui, M. Santos, M. Shayegan [*Appl. Phys. Lett. (USA)* vol.58 (1991) p.1440-2]

[18]   T.K Woodward, T.C. McGill, R.D. Burnham [*Appl. Phys. Lett. (USA)* vol.50 (1987) p.451-3]; P. Guéret, N. Blanc, R. Germann, H. Rothuizen [*Phys. Rev. Lett. (USA)* vol.68 (1992) p.1896-9]

[19]   F. Capasso, K. Mohammed, A.Y. Cho [*IEEE J. Quantum Electron. (USA)* vol.22 (1986) p.1853-69]

[20]   N. Yokoyama, K. Imamura, H. Ohnishi, T. Mori, S. Muto, A. Shibatomi [*Jpn. J. Appl. Phys. (Japan)* vol.24 (1985) p.L853-4]; N. Yokoyama et al [*Solid State Electron. (UK)* vol.31 (1988) p.577-82]

[21]   M. Takatsu et al [*IEEE J. Solid-State Circuits (USA)* vol.27 (1992) p.1428-30]

[22]   S. Sen, F. Capasso, D. Sivco, A.Y. Cho [*IEEE Electron Device Lett. (USA)* vol.9 (1988) p.402-4]

[23]   A.C. Seabaugh, Y.-C. Kao, H.-T. Yuan [*IEEE Electron Device Lett. (USA)* vol.13 (1992) p.479-81]

[24]   T. Mori, S. Muto, H. Tamura, N. Yokoyama [*Jpn. J. Appl. Phys. (Japan)* vol.33 (1994) p.790-3]

[25]   J. Shen, G. Kramer, S. Tehrani, H. Goronkin, R. Tsui [*IEEE Electron Device Lett. (USA)* vol.16 (1995) p.178-80]

[26]   L. Esaki, R. Tsu [*IBM J. Res. Dev. (USA)* vol.14 (1970) p.61-5]

[27]   H.M. James [*Phys. Rev. (USA)* vol.76 (1949) p.1611-24]; L.V. Keldysh [*Zh. Eksp. Teor. Fiz. (USSR)* vol.43 (1962) p.661-6], translated in [*Sov. Phys.-JETP (USA)* vol.16 (1963) p.471-4]

[28]   L. Esaki, L.L. Chang [*Phys. Rev. Lett. (USA)* vol.33 (1974) p.495-8]; K.K. Choi, B.F. Levine, R.J. Malik, J. Walker, C.G. Bethea [*Phys. Rev. B (USA)* vol.35 (1987) p.4172-5]

[29]   A. Sibille, J.F. Palmier, H. Wang, F. Mollot [*Phys. Rev. Lett. (USA)* vol.64 (1990) p.52-5]; H.T. Grahn, K. von Klitzing, K. Ploog, G.H. Döhler [*Phys. Rev. B (USA)* vol.43 (1991) p.12094-7]

[30]   R.F. Kazarinov, R.A. Suris [*Fiz. Tekh. Poluprovodn. (USSR)* vol.5 (1971) p.797-800], translated in [*Sov. Phys.-Semicond. (USA)* vol.5 (1971) p.707-9]

[31]   F. Beltram et al [*Phys. Rev. Lett. (USA)* vol.64 (1990) p.3167-70]

[32]   J. Faist, F. Capasso, D.L. Sivco, C. Sirtori, A.L. Hutchinson, A.Y. Cho [*Science (USA)* vol.264 (1994) p.553-6]; J. Faist et al [*Appl. Phys. Lett. (USA)* vol.66 (1995) p.538-40]; C. Sirtori et al [*Appl. Phys. Lett. (USA)* vol.66 (1995) p.3242-4]

[33]   [Datareview in this book: 8.3 Quantum well and superlattice photodetectors]

[34]   B.F. Levine et al [*J. Appl. Phys. (USA)* vol.72 (1992) p.4429-43]

[35]   H. van Houten, C.W.J. Beenakker [*Phys. Rev. Lett. (USA)* vol.63 (1989) p.1893-6]

[36]   Bo Su, V.J. Goldman, J.E. Cunningham [*Science (USA)* vol.255 (1992) p.313-6]

[37]   R.C. Ashoori et al [*Phys. Rev. Lett. (USA)* vol.71 (1993) p.613-6]

[38]   U. Meirav, M.A. Kastner, S.J. Wind [*Phys. Rev. Lett. (USA)* vol.65 (1990) p.771-4]

[39]   L.P. Kouwenhoven et al [*Z. Phys. B (Germany)* vol.85 (1991) p.367-73]

[40]   M.A. Kastner [*Rev. Mod. Phys. (USA)* vol.64 (1992) p.849-58]; K.K. Likharev, T. Claeson [*Sci. Am. (USA)* vol.266 (1992) p.80-5]

[41] D.V. Averin, K.K. Likharev [in *Single Charge Tunneling: Coulomb Blockade Phenomena in Nanostructures* Eds H. Grabert, M.H. Devoret (Plenum Press, New York, 1992) p.371]

[42] T.A. Fulton, G.J. Dolan [*Phys. Rev. Lett. (USA)* vol.59 (1987) p.109-12]

[43] G. Zimmerli, R.L. Kautz, J.M. Martinis [*Appl. Phys. Lett. (USA)* vol.61 (1992) p.2616-8]

[44] L.J. Geerligs et al [*Phys. Rev. Lett. (USA)* vol.64 (1990) p.2691-4]

[45] L.P. Kouwenhoven et al [*Phys. Rev. Lett. (USA)* vol.67 (1991) p.1626-9]

[46] S. Luryi [*Solid State Commun. (USA)* vol.65 (1988) p.787-9]; S. Luryi [*IEEE J. Quantum Electron. (USA)* vol.27 (1991) p.54-60]

[47] K. Leo et al [*Phys. Rev. Lett. (USA)* vol.66 (1991) p.201-4]

[48] J.P. Eisenstein, L.N. Pfeiffer, K.W. West [*Phys. Rev. Lett. (USA)* vol.69 (1992) p.3804-7]

[49] R.A. Davies, D.J. Newson, T.G. Powell, M.J. Kelly, H.W. Myron [*Semicond. Sci. Technol. (UK)* vol.2 (1987) p.61-4]

[50] R.K. Hayden et al [*Phys. Rev. Lett. (USA)* vol.66 (1991) p.1749-52]

[51] J.P. Eisenstein, T.J. Gramila, L.N. Pfeiffer, K.W. West [*Phys. Rev. B (USA)* vol.44 (1991) p.6511-4]

[52] S.Y. Lin, A. Zaslavsky, K. Hirakawa, D.C. Tsui, J.F. Klem [*Appl. Phys. Lett. (USA)* vol.60 (1992) p.601-3]

[53] U. Gennser, V.P. Kesan, D.A. Syphers, T.P. Smith III, S.S. Iyer, E.S. Yang [*Phys. Rev. Lett. (USA)* vol.67 (1991) p.3828-31]

[54] A. Zaslavsky, K.R. Milkove, Y.H. Lee, B. Ferland, T.O. Sedgwick [*Appl. Phys. Lett. (USA)* vol.67 (1995) p.3921-3]

8.5 Properties and device applications of antimony-based quantum wells and superlattices

R.A. Stradling

March 1996

A   OVERVIEW OF MATERIALS AND ELECTRONIC PROPERTIES

The most developed semiconductor heterostructures are the lattice-matched systems consisting of GaAs (lattice constant = 5.654 Å) with $Ga_{1-x}Al_xAs$ and InP (lattice constant = 5.868 Å) with $Ga_{1-x}In_xAs$. However, neither of these systems provides a bandgap suitable for long-wavelength optical sources, or mid-infrared detectors. There are also potential applications for narrow gap materials for low power and fast electronic devices. These developments have produced demands for new materials based on InAs, InSb, AlSb and GaSb and alloy combinations of these binaries. The lattice constants of InAs (6.058 Å at room temperature), GaSb (6.095 Å) and AlSb (6.135 Å) are quite similar but the lattice constant of InSb (6.479 Å) is much greater and all these materials are poorly matched to GaAs or InP. Reasonable quality substrates are available of InAs, GaSb and InSb but all of these materials are quite conducting at room temperature. For reasons of cost and electrical isolation, GaAs substrates are frequently employed. In this case an $InAs/In_{0.7}Ga_{0.3}As$ superlattice can be useful as a buffer to prevent the threading dislocations from reaching the surface [1].

The band structure of InAs, and InSb, is characterised by the small band direct bandgap at the centre of the Brillouin zone and the large separation in energy between the conduction band minima at the Γ-X and Γ-L points. The primary electronic properties arise from the band structure which results in high mobilities and saturation drift velocities for the electrons. In the case of InSb the room temperature electron mobility is nearly ten times that of GaAs and the saturation drift velocity exceeds that of silicon by a factor of five. The low effective masses of InAs and InSb give high quantum confinement energies. The latter characteristic opens up the possibility of the observation of mesoscopic and Coulomb blockade effects at higher temperatures. The heterostructure combinations of particular interest are $InAs/In(As_{1-x}Sb_x)$, $InSb/In(Al_{1-x}Sb_x)$, InAs/GaSb and InAs/AlSb.

InAs has a particularly low lying conduction band which leads to the formation of type II band alignments at heterojunctions and to very large conduction band offsets which can be exploited in such devices as tunnel diodes. The type II alignment can drastically modify the electronic properties, e.g. by leading to the suppression of Auger recombination. Another special property is that the deep lying conduction band causes native defect levels to lie about 200 meV above the conduction band edge rather than in the middle of the forbidden gap [2-4]. Consequently electron accumulation layers form naturally at the surfaces of bulk layers and the Fermi energy at a metal semiconductor contact is pinned within the conduction band at a similar energy. Thus a Schottky barrier is not formed and contacts which are extremely transparent to electron flow are readily fabricated.

The defect levels which cause the surface pinning also act to stabilise the Fermi level which provides an electronic reference level for the defect annihilation energies. In the case of amphoteric impurities this determines the maximum free carrier concentration which can be obtained from doping [2]. Consequently InAs can be doped very heavily with Si donors (the preferred dopant in MBE growth) where concentrations as high as 5×10^{19} cm^{-3} can be achieved. With GaAs, where the Fermi level is pinned mid-gap, the donor doping limit is $\cong 10^{19}$ cm^{-3}. With InSb the pinning energy is near to the valence band so Si acts amphoterically but almost complete activation of the silicon as a donor up to concentrations of $\sim 3 \times 10^{18}$ cm^{-3} can be obtained by reducing the temperature to 350°C. With GaSb and AlSb silicon only acts as an acceptor. These trends can be understood qualitatively in terms of the amphoteric native defect model introduced in [2] where the defects act to stabilise the Fermi level. The position of the defect levels with respect to the band edges therefore determines the maximum free carrier concentration which can be obtained by silicon doping. The defect levels lie in the conduction band of InAs but close to the valence band edge in GaSb.

Doping with elemental tellurium (and other group VI atoms) is discouraged in MBE as Te has a very high vapour pressure and severe long term memory effects are found. Congruent evaporation using PbTe or GaTe works well with little contamination [5,6] but ties up an additional cell in the MBE chamber as Si will also be required for n-type doping of InAs or InSb. Surface segregation appears to be a problem in AlSb [7]. The untreated GaSb surface is known to produce donor-like levels which act to pin the Fermi energy at the surface about 0.2 eV above the valence band edge [7,8]. The surface donors provide an additional source of electrons for the InAs quantum well above the intrinsic concentration arising from the semimetallic band alignment with the concentration of extra electrons varying approximately inversely with the thickness of the GaSb cap [7]. Apart from the question of segregation the use of group VI elements for doping presents less of a problem with MOVPE.

Remote doping of InAs/GaSb or InAs/AlSb quantum wells presents a problem with molecular beam epitaxy (MBE). Use of a group VI dopant risks long term memory effects. Silicon, which is the preferred MBE dopant for most other III-V systems, acts as an acceptor with GaSb and in AlSb.

An alternative technology has been developed [9] for remote doping of the InAs quantum wells where a double well structure is employed. The second InAs well is thin (~ 2 nm) and doped with silicon where concentrations as high as 5×10^{19} cm^{-3} can be employed. Because of the high confinement energy this well acts as a source of electrons for the first well.

GaSb grown without deliberate doping always turns out to be p-type with a hole concentration at room temperature of about 10^{16} cm^{-3}. The native defect responsible for the residual p-type conductivity is thought to be a double acceptor with levels 33 and 80 meV above the valence band edge [10,11]. The defect has been variously identified including the possibilities that an antisite point defect formed by a Ga atom on the Sb site or a gallium vacancy complexed with a Ga atom on the Sb site may be responsible. The lower the growth temperature the lower is the density of the native defects and the higher the hole mobility, and room temperature mobilities of 900 cm^2/V s together with peak mobilities at 50 K of 15,000 cm^2/V s have been reported.

AlSb grown by MBE is generally p-type with a hole concentration of $\sim 10^{16}$ cm^{-3} at room temperature. Material grown by MOVPE generally shows very large carbon acceptor contamination although the use of new precursors such as tritertiarybutylaluminium have dropped the carbon level down from 10^{19} cm^{-3} to 10^{18} cm^{-3} [12]. A favourable feature of AlSb is its relatively low refractive index compared with its near lattice-matched partner GaSb. The ratio of the two refractive indices at 2 μm wavelength is 1.24 which is considerably larger than the corresponding value for GaAs/AlAs. Thus a simple ten period AlSb/GaSb distributed Bragg reflector has been shown to have a reflectance of over 98% at 1.92 μm wavelength and a 12 period $Al_{0.2}Ga_{0.8}Sb$/AlSb DBR had a reflectivity of 99% at 1.38 μm wavelength [13].

The $InAs_{1-x}Sb_x$ alloy system with $x \sim 0.65$ has the narrowest direct bandgap of any thick-film III-V material. Remarkably, when InSb is mixed together with the wider bandgap material InAs at low compositions, the bandgap decreases, reaching a minimum value of 0.145 eV at 10 K and a value of about 0.10 eV at room temperature. Unfortunately the alloy is prone to metallurgical problems such as ordering and phase separation in the mid alloy range and even 'natural' superlattices can be grown when material is supplied at constant composition [14]. Strain-layer superlattices of both $InAs/InAs_{1-x}Sb_x$ and $InSb/InAs_{1-x}Sb_x$ have been grown with high mobility despite the large mismatch. Interband magneto-optics and luminescence [15-17] show that the photon energy emitted falls extremely rapidly with increasing x with the bandgap decreasing at 10 K from 0.44 eV for x = 0 to 0.10 eV for x = 0.4.

The reason for this rapid narrowing of the superlattice energy gap has been controversial with different groups suggesting conflicting schemes [18]; (i) 'type II' band alignments with the valence band offsets changing rapidly with x (in which case the photon emission is spatially indirect) with the conduction band minimum being either in the alloy [17] or in the InAs [19] or alternatively (ii) a spatially direct bandgap in which the alloy bandgap is anomalously narrowed by microstructural effects such as atomic ordering [15,16].

The band alignments at the $InSb/InAl_{1-x}Sb_x$ interface are not known accurately. In contrast the offsets for the InAs/GaSb/AlSb system have been determined quite precisely. In the case of InAs/GaSb, the InAs conduction band minimum lies about 150 meV lower than the top of the GaSb valence band for the [001] orientation (type III alignment) with differences of about 20 meV [20] depending on whether the bonding at the interfaces is InSb-like or GaAs-like [21]. For the [111]A orientation the overlap is found to be 200 meV and the difference in energy compared with [001] is attributed to the dipole expected at the interface [22]. Thus the system is naturally semimetallic.

The InAs/AlSb system has an exceptionally high conduction band offset (1.36 eV at room temperature) particularly suited to tunnel and other microelectronic devices. InAs/GaSb/AlSb combinations are attractive for both infrared and electronic applications.

The semimetallic band alignments with InAs/GaSb heterostructures lead to the simultaneous presence of electrons and holes in undoped structures. There has been considerable speculation for some time that such a system could lead to the formation of stable excitonic states at low temperatures. Experimental evidence for such states has been lacking until recently when Cheng et al [23] reported the presence of an additional line approximately 2 meV above the cyclotron resonance line in a number of InAs/(Ga,Al)Sb quantum wells which disappeared with increasing temperature. In this experiment the persistent photoconductive effect was used to control the carrier concentration and the line was only present when the electron and hole concentrations were approximately equal.

A metal insulator transition occurs with single InAs/GaSb quantum wells when the well width is reduced below about 10 nm [24]. When short period superlattices are grown the material can remain closely intrinsic down to the lowest temperatures provided that a thick surface cap is grown to reduce the influence of the surface [25]. A novel type of quantum Hall effect appears for these near intrinsic samples where the normal plateaux are replaced by minima which approach zero Hall resistance [26]. InAs/AlSb quantum wells and superlattices on the other hand appear much more n-type and it has been speculated that antisite donors in the bulk consisting of As substituting on the Al site and similar native defects such as As on the Al site at AlAs interfaces can be responsible for these residual donors [27]. Donor planes can be formed in AlSb by an interrupted growth technique where the growth surface is first made aluminium rich and then 'soaked' with As for 60 seconds, after which further aluminium is supplied. By this means As antisite defects are formed [7].

A particular problem arises with InAs/AlSb devices because of the instability in air of AlSb. In order to prevent corrosion, the final AlSb layer has to be capped by either a thin GaSb or InAs layer. When GaSb is employed the Fermi energy is pinned by the surface donors [8,28]. Alternatively a $Ga_{1-x}Al_xSb$ alloy with $x > 0.5$ can be used as an air-stable cap.

There is no common anion or cation across the interface between InAs and GaSb (or AlSb). It is therefore possible to induce two different types of bonding, InSb-like or AlAs (or GaAs)-like at the interface by careful control of the shutter sequences [7,26]. Without such control the bonding at the interfaces will be random. The band offsets, the local vibrational properties and the electronic mobilities will depend on the nature of the interfaces.

Generally samples grown with AlAs (GaAs)-like interfaces have inferior structural quality and much lower mobility than those with InSb interfaces [7]. The problem appears to be the roughening or intermixing of the surface during the growth of the first AlAs (GaAs) interface due to the exposure of the AlSb (GaSb) surface to excess As [29,30]. The optimum temperature in MBE growth for obtaining high mobility with InSb-interfaces is about 450°C [31] for InAs/GaSb quantum wells. However InAs/GaSb quantum wells with even higher structural quality and mobility can be grown at 400°C by the use of minimum As overpressure to reduce the surface roughening [31].

The well width dependence of the electrical properties of InAs/AlSb quantum wells has been studied in an extensive series of experiments by the Santa Barbara group [32,33]. For studies of the well width dependence all samples had a top barrier consisting of 50 nm of AlSb capped by a 5 nm GaSb layer and InSb interfaces. The bottom barrier consisted of 20 nm of AlSb grown on top of a ten period (2.5 nm + 2.5 nm) GaSb/AlSb smoothing superlattice. The growth temperature was 500°C. The optimum mobility was found for a well thickness of about 15 nm [32]. The decrease in mobility with decreasing thickness found for well widths below 10 nm is thought to be due to interface roughness [33]. In contrast to the rapid decrease of carrier concentration with decreasing well thickness found with InAs/GaSb wells and explained by the metal-insulator transition [24], the carrier concentration in the InAs/AlSb wells increased rapidly with decreasing thickness below 8 nm [32]. This behaviour has yet to be explained.

Similar structures to those employed in [32,33] were employed in [8] and [34] for a study of the effect of the barrier thickness on the mobility except that the well width was kept the same for all samples at 15 nm and the GaSb surface cap was increased to 10 nm. The carrier concentration was found to increase rapidly with decreasing distance between the quantum well and the surface (L) in the manner expected in the surface pinning model. An increased mobility was also observed with decreasing L. This was assumed to be correlated with the increasing carrier concentration as would be expected with a fixed concentration of scattering centres and transfer of carriers from the surface. For very small values of L (or large values of carrier concentration), however, the mobility fell sharply, either because of surface roughness or because of the onset of intersubband scattering [34].

The compatibility of two apparently disparate materials has been demonstrated with the growth of high-quality heterostructures between metallic Sb and GaSb [35,36]. It has proved possible to perform multilayer growth on [111] surfaces using MBE and migration enhanced epitaxy despite the fact that the crystal structure is different with GaSb being zinc blende while Sb is rhombohedral. Regrowth of one on the other is possible because on a [111] surface there is near perfect lattice match (to 0.06%) and the interface atomic nets are both hexagonal.

B   INFRARED EMITTERS

A major area of development concerns infrared LEDs and lasers where there is an urgent requirement to provide low cost and sensitive systems for pollution monitoring where trace gases are to be detected by their fundamental vibrational-rotational absorption bands. There has been great progress recently in the development of III-V antimonide-based lasers operating between 2 and 4 μm wavelength. The first report of operation with a III-V antimonide materials system was with bulk InSb where lasing was observed at a wavelength of 5.2 μm but only below a temperature of 10 K [37]. TABLE 1 lists the characteristics of mid IR laser systems which have been developed recently with III-V materials [37-50] and compares the results with the recently developed cascade lasers [51-54] which employ the AlInAs/GaInAs materials system and with II-VI ($Hg_{1-x}Cd_xTe$) [55] and IV-VI (PbSe/PbSrSe) laser systems [56]. Column one gives the institute and first author concerned and column two the materials system involved. Columns three to seven list the operating parameters at a particular temperature, column eight the structure and column nine the mode of excitation and the maximum operating temperature.

The longest wavelength operation yet reported for a III-V laser at room temperature is at 2.78 μm where pulsed operation was achieved with a quaternary multiquantum well system with $In_{0.24}Ga_{0.76}As_{0.16}Sb_{0.84}$ wells and $Al_{0.25}Ga_{0.75}As_{0.02}Sb_{0.98}$ barriers [49]. The main problem preventing CW operation seems to have been an unexpectedly large series resistance, possibly from poor conductivity in the cladding layers.

Liquid nitrogen temperature operation of antimonide-based lasers has been achieved at 5.2 μm wavelength with an InSb [50] laser using a relatively simple pseudo double heterostructure using an $In_{1-x}Al_xSb$ barrier with x = 0.14.

TABLE 1. Laser system characteristics.

Group	Material	λ (micron)	Power	T	Threshold	To	Structure	Mode of operation
MIT Mengalis [37]	InSb						Bulk	
Choi [38-39]	(GaInAl)(AsSb)	2	1.3 W (CW)	300 K	140 A/cm^2	110 K	Broad stripe	
		2	100 mW	300 K	20 mA		Ridge waveguide	
		2	200 mW	300 K			Tapered	
		3	90 mW	100 K	9 A/cm^2 at 40 K		DH	255 K pulsed/CW to 170 K
Le [40]	InAsSb/GaSb	3.9	0.8 W	85 K	Optic pulse			210 K optically pumped
Eglash [41]	InAsSb/AlInAs	3.9	30 mW				DH	170 K pulsed/105 K CW
	InAsSb/AlInAs	4.5						85 K pulsed
	InAsSb/AlAsSb	4	200 mW mean				DH/diode pumped	155 K pulsed/80K CW
Choi and Turner [42]	InAsSb/InAlAsSb	3.9	60 mW	80 K	78 A/cm^2	30 K	SLS	165 K pulsed/128 K CW
Sandia Biefeld/Kurtz [43]	InAsSb/InGaAs	3.9		<100 K			SLS	Optically pumped
Biefeld/Kurtz	InAs/InAsSb/InPSb	3.5					SL MQW	135 K pulsed/77 K CW
Ioffe Baranov [44]	InAsSbP/InAsSb	3.2	8 mW	80 K	40 mA	30 K	DH (LPE grown)	80 K CW
				180 K	6 A			180 K pulsed
Hughes Miles [45] Chow [46]	GaInSb/InAs	3.3		<170 K			Type II SL	170 K pulsed
		3.2						5 K optically pumped
		3.8						
Zhang [47,48]	InAs/AlAsSb	3		95 K		32 K	DH	CW
	InAsSb/InAs	3.4					Diode pumped SLS	

DH = double heterostructure; SLS = strained layer superlattice; SL = superlattice; MQW = multiple quantum well.

TABLE 1 continued.

Group	Material	λ (micron)	Power	T	Threshold	To	Structure	Mode of operation
Sarnoff Research Center Lee [49]	AlGaAsSb/InGaAsSb	2.78	30 mW	288 K	10 kA/cm²	58 K	MQW	333 K pulsed
DRA Ashley [50]	InSb/AlInSb	5.1	1 mW	80 K	2.6 kA/cm²	17 K	DH	Pulsed
ATT Capasso [51-54]	AlInAs/GaInAs	4.5	80 mW	10 K	1.7 kA/cm²		Cascade layer	Pulsed
		4.5		80 K				Pulsed
		4.6	20 mW	100 K	3 kA/cm²			CW
		4.6	2 mW	200 K	5 kA/cm²			Pulsed
		8.4	40 mW	85 K	1.7 kA/cm²			CW
	AlInAs/GaInAs	8 to 13	6 nW	10 K	2.1 kA/cm²		Cascade LED (blue Stark shift)	CW
				10-200 K				
MIT Le [55]	CMT/CdZnTe	3.2	1.3 W peak 105 mW mean	88 K			QW	154 K diode pumped
Fraunhofer Shi [56]	PbSe/PbSrSe	4.2 / 5 / 7.3	300 μW	282 K / 120 K / 30 K	330 mA / 800 mA		MQW	Pulsed / CW / CW
ICSTM Phillips [59,60]	InAs/InAsSb	3-10 / 4.5 / 7	~1 μW / ~0.2 μW	290 K			SLS LED	CW

DH = double heterostructure; SLS = strained layer superlattice; SL = superlattice; MQW = multiple quantum well.

In contrast pulsed operation very near to room temperature (282 K) was very recently reported at 4.2 μm wavelength with lead salt multiple quantum well lasers (PbSe/Pb$_{0.9785}$Sr$_{0.0215}$Se) [56]. These authors suggest that IV-VI laser systems may be preferable to III-Vs for long wavelengths but that III-Vs are likely to be superior for wavelengths below 3 μm. The longest wavelength results with antimonide-based materials well above liquid nitrogen temperatures are reported in [47] and [48] where 3.9 μm operation is achieved in the pulsed mode at 170 K or at 210 K using optical pumping.

HgCdTe/ZnCdTe lasers have been made to operate pulsed at a wavelength of 5.3 μm up to 60 K and up to 154 K at 3.2 μm wavelength using diode pumping [55].

A different approach, not involving band-to-band radiation, is to use transitions between the subbands in a multiple quantum well structure [51-53]. A sophisticated (In,Ga)As/(Al,In)As superlattice structure is employed to obtain laser action where the lower state of the laser transition is separated by an optical phonon energy from the ground state to ensure population inversion. The electrons are recycled through twenty five stages and digital grading is designed to act as a Bragg reflector for the electrons in the excited laser state. The system is described as a 'quantum cascade' laser. A simpler version of this structure is used for LED devices at longer wavelength [54]. Operation at room temperature has yet to be achieved but laser emission at a wavelength as long as 8.4 μm was recently reported at temperatures up to 130 K [53].

In(As,Sb) has been used as a component in the active region of the laser structure at wavelengths between 3.4 and 4.5 μm in [40-43]. Intense luminescence is seen from 'strained layer superlattices (SLSs)' formed by InAs/InAs$_{1-x}$Sb$_x$ or In$_{1-x}$Ga$_x$As/InAs$_{1-x}$Sb$_x$. With these SLSs it is thought that band structure effects can be used to quench non-radiative Auger recombination either by the strain splitting of the valence band [57,58] or by the type II band structure [59,60]. Significant emission at a wavelength as long as 9.8 μm is seen from an uncooled LED made from these SLSs [59,60].

A direct demonstration of the suppression of the Auger processes [57-62] by band structure engineering has been performed by saturation and pump-probe measurements using a free electron laser in InAs/InAs$_{1-x}$Sb$_x$ SLSs [63]. In comparison with the Auger lifetime obtained from similar measurements on InSb (room temperature bandgap about 7 μm), the lifetime deduced from the alloy, whose measured bandgap was 11 μm, was some 50 times longer.

The reverse biassing of the p-n junction can reduce the emission of photons below the density expected from a black body of the same temperature. This negative luminescence [64] can be used to cool the surroundings or as part of LED sources where the enhanced dynamic range of modulation of the emission can be used to good effect in modulated-source gas sensing.

C   INFRARED DETECTORS

The technology for InSb infrared detectors is extremely well developed with two-dimensional photovoltaic arrays consisting of over 65,000 pixels being available. These detectors cooled to between 3 and 30 K have revolutionised infrared astronomy - see for example [65].

The performance of mid-infrared detectors generally decreases rapidly as the temperature is raised above 80 K because of noise associated with the rapidly increasing thermal generation rate. However a large proportion of the electrons and holes can be removed from the active region of devices by means of minority carrier extraction and exclusion with a substantial reduction in noise [64].

D   ELECTRONIC DEVICES

Narrow gap systems also give advantages for the low power/high speed operation of devices such as field effect transistors (FETs) and resonant tunnelling diodes (RTDs). The high conduction band offset between InAs and AlSb (1.6 eV) makes this combination particularly attractive for RTDs. The combination of InAs with GaSb/InAs$_{1-x}$Sb$_x$ gives rise to the possibility of interband tunnelling through the crossed gap alignment. Reference [66] discusses the many different types of tunnel device which become possible through the combination of InAs, AlSb and GaSb in the same structure. The highest frequency operation yet achieved with any microelectronic device has been with an RTD InAs/AlSb device operating at 712 GHz [67,68].

These structures were grown on GaAs substrates as it had been demonstrated that the high density of threading dislocations caused by the large lattice mismatch had little or no effect on the leakage currents. The structure started with a short-period (5 x 2 nm/2 nm) In$_{0.7}$Ga$_{0.3}$As/InAs smoothing superlattice grown on to the semi-insulating substrate followed by an InAs contact region which started with 1000 nm of n$^+$ InAs doped with 5 x 10^{18} cm^{-3} Si donors, followed by 200 nm doped at 2 x 10^{18} cm^{-3} and then by 75 nm doped at 2 x 10^{17} cm^{-3}. The two AlSb barriers were of thickness between 1.5 and 1.8 nm and the central InAs well was of thickness 6.4 nm and undoped. The top contact region started with 10 or 20 nm of InAs lightly doped with Si (2 x 10^{16} cm^{-3} and designed to be fully depleted under operating bias) followed by 100 nm at 2 x 10^{18} cm^{-3} and then by an n$^+$ region of thickness 100 nm doped at 5 x 10^{18} cm^{-3}. The metallisation of the top contact consisted of 20 nm of In followed by 400 nm of gold. Sidewall metallisation was used to make the bottom contact. The peak to valley current ratio was up to 3.4 at 300 K with a peak current density of 3 x 10^5 A cm^{-2}.

Better high frequency response compared with InGaAs/AlInAs RTDs has been demonstrated through free electron laser studies at near terahertz frequencies [69]. Switching times as short as 1.7 ps have also been measured directly by electro-optic sampling [70].

The type II alignment of the InAs/AlSb/GaSb system allows many possibilities for interband tunnel structures. Generally higher peak-to-valley ratios are found with interband devices but the peak current densities are lower. There can however be problems associated with parasitic hole conduction which contributes to the valley current. In [71] this hole current was suppressed by the addition of two monolayers of AlAs to the double barrier structures consisting of a single InAs well surrounded by AlSb barriers. The inclusion of AlAs improved the peak-to-valley ratio from 15 to 30 at room temperature.

The same group also report a lateral three terminal resonant interband tunnelling field effect transistor with tunnel characteristic giving a peak-to-valley ratio of 8 at room temperature [72].

With FETs the attractive features of the InAs/AlSb combination are the very high saturation drift velocity for InAs and the high mobility together with the high room temperature resistance of AlSb which can be used for internal gates. Luo et al [73] first realised an FET employing an InAs channel but the device had poor transconductance because of leakage through the (Al,Ga)Sb barriers employed.

The leakage current can be much reduced by employing AlSb as the barrier because of the much greater conduction band offset. However the type II band alignment means that there is no barrier for the holes and accumulation of holes in the barriers can lead to an undesirable 'kink' effect in the device characteristics. Bolognisi et al [74] used an $AlSb_{0.9}As_{0.1}$ layer to separate the AlSb from a surface passivation layer of $Al_{0.8}Ga_{0.2}Sb$ and to provide a barrier for holes. Microwave FET devices with a gate length of 0.5 μm have shown a peak unity current cut-off frequency (f_t) of 93 GHz [74]. An f_t of 600 GHz has been predicted for a gate length of 0.25 μm [75].

Another effect which limits the performance of FETs with InAs channels is the occurrence of impact ionisation at high electric fields on account of the small bandgap of InAs. Impact ionisation causes the source-drain I-V to turn up instead of saturating. However the impact ionisation occurs at higher electric fields than anticipated from measurements with bulk InAs. Li et al [75] have shown that devices can be made to operate at channel electric fields (20 kV/cm) well above the predicted threshold for impact ionisation. A number of possible reasons have been put forward to explain this postponement of impact ionisation: (i) the increase in the effective bandgap by size quantisation in the InAs channel; (ii) spatial transfer of the holes formed by impact ionisation out of the InAs channel because of the type II band alignment and the lack of a barrier for holes at the InAs/AlSb interfaces; (iii) ballistic motion of the electrons across the narrow gate region; (iv) accumulation of holes which escape from the InAs in the region between the channel and the substrate. Brar et al [76] showed that the impact ionisation was suppressed by using an epitaxial back gate biassed to the source potential. An asymmetric structure was used with a back gate consisting of a conducting p-type GaSb layer. It was shown that, unless the gate was appropriately biassed, most of the rise in drain current near impact ionisation is caused by a feedback mechanism where the holes escaping to the substrate act as a positively charged parasitic gate leading to a much larger increase in the channel current than do the electrons directly generated by the impact ionisation itself.

Structures with gate lengths as small as 0.2 μm have been fabricated [76]. AlSb used as a gate insulator can withstand electric fields of $\sim 10^6$ V/cm (5 V across 50 nm of AlSb) before breakdown occurs even with devices with such small gates [77].

E   OTHER DEVICES

Magnetic field sensors using Hall or magnetoresistance devices are currently being developed for automotive applications where temperatures up to 250°C may be experienced. A high field sensitivity is expected because of the high mobilities expected with narrow gap materials. However the intrinsic excitation of carriers across the bandgap can provide a problem as this conflicts with the other design criterion for a magnetic field sensor for such applications which must have characteristics which change little with temperature. The use of delta doping or heterostructures involving InAs can reduce the change in carrier concentration with temperature to acceptable limits and can provide excellent performance [78].

A mass market for Hall elements is emerging for control of the commutator switching in brushless DC motors for personal computers and other situations where electromagnetic noise has to be minimised. In such applications InAs/GaAlAsSb quantum wells [79] are displacing InSb homoepitaxial films [80] because of the better temperature stability of the heterostructures. The current annual production of the InSb devices is 700 million.

Two clear advantages of narrow gap materials over other semiconductor systems have been identified for mesoscopic studies. The increased energy separation of the 1- or 0-D subbands arising from the low effective mass of the conduction band means that mesoscopic effects can be followed to substantially higher temperatures (it should be possible to follow effects to about three times higher temperature for a particular feature size if InAs/(AlGa)Sb rather than GaAs/(AlGa)As is employed for a given structure). Thus Inoue et al [81] were able to observe the quantised conductance of an InAs/(AlGa)Sb split-gate quantum wire device up to 80 K. The second advantage identified by Koester et al [82] is that ballistic effects can be observed to longer channel lengths; e.g. fully ballistic quantised conductance was observed with a channel length of 1 μm. It was thought that this enhanced length performance arose because the subband separation was greater than the amplitude of the potential fluctuations in the wire.

Looking further forward to the future, hybrid superconductor-semiconductor devices are becoming of increasing interest [83].

A full review of the device-related properties of GaSb-based structures is given in [84].

REFERENCES

[1]   E.R. Brown, S.J. Eglash, G.W. Turner, C.D. Parker, J.V. Pantano, D.R. Calawa [*IEEE Trans. Electron Devices (USA)* vol.41 (1994) p.879]

[2]   W. Walukiewicz [*Mater. Res. Soc. Symp. Proc. (USA)* vol.104 (1988) p.483; also *J. Vac. Sci Technol. (USA)* vol.37 (1988) p.4760]

[3]   C. Nguyen, B. Brar, H. Kroemer, J.H. English [*J. Vac. Sci Technol. B (USA)* vol.10 (1992) p.898]

[4]   R.G. Egdell, S.D. Evans, Y.B. Li, R.A. Stradling, S.D. Parker [*Surf. Sci. (Netherlands)* vol.262 (1992) p.444]

[5]   S.M. Newstead, T.M. Kerr, C.E.C. Wood [*J. Appl. Phys. (USA)* vol.66 (1988) p.4184]

[6]   S. Subbana, G. Tuttle, H. Kroemer [*J. Electron. Mater. (USA)* vol.17 (1988) p.297]

[7]   G. Tuttle, H. Kroemer, J.H. English [*J. Appl. Phys. (USA)* vol.65 (1989) p.5239]

[8]   C. Nguyen, B. Brar, H. Kroemer, J.H. English [*Appl. Phys. Lett. (USA)* vol.60 (1992) p.1854]

[9]   T. Malik, S. Chung, J.J. Harris, A.G. Norman, R.A. Stradling, W.T. Yuen [*Int. Conf. on Narrow Gap Semiconductors* Santa Fe (IoP Conference Series - in press)]

[10]   K. Nakashima [*Jpn. J. Appl. Phys. (Japan)* vol.20 (1981) p.1085]

[11]   I. Poole, M.E. Lee, I.R. Cleeverly, A.R. Peaker, K. Singer [*Appl. Phys. Lett. (USA)* vol.57 (1990) p.1645]

[12]   G. Tuttle, J. Kavanaugh, S. McCalmont [*IEEE Photonics Technol. Lett. (USA)* vol.5 (1993) p.1376]

[13] C.A. Wang, M.C. Finn, S. Salim, K.F. Jensen, A.C. Jones [*Appl. Phys. Lett. (USA)* vol.67 (1995) p.1384]

[14] I.T. Ferguson et al [*Appl. Phys. Lett. (USA)* vol.59 (1991) p.3324]

[15] S.R. Kurtz, R.M. Biefeld, L.R. Dawson, K.C. Baucom, A.J. Howard [*Appl. Phys. Lett. (USA)* vol.64 (1994) p.812]

[16] S.R. Kurtz, R.M. Biefeld [*Appl. Phys. Lett. (USA)* vol.66 (1995) p.364]

[17] Y.B. Li, R.A. Stradling, A.G. Norman, P.J.P. Tang, S.J. Chung, C.C. Phillips [*Proc. 22nd Int. Conf. on Physics of Semiconductors* Vancouver (World Scientific, 1995) p.1496]

[18] S.-H. Wei, A. Zunger [*Phys. Rev. B (USA)* vol.52 (1995) p.12039]

[19] Y.H. Zhang, R.H. Miles, D.H. Chow [*IEEE J. Selected Topics Quantum Electron. (USA)* vol.1 (1995) p.749]

[20] D.M. Symons et al [*Phys. Rev. B (USA)* vol.51 (1995) p.1729]

[21] D.M. Symons et al [*Semicond. Sci. Technol. (UK)* vol.9 (1994) p.118; also *Phys. Rev. B (USA)* vol.49 (1994) p.16614]

[22] G. Tuttle, H. Kroemer, J.H. English [*J. Appl. Phys. (USA)* vol.67 (1990) p.3032]

[23] J.P. Cheng, J. Kono, B.D. McCombe, I. Lo, W.C. Mitchel, C.E. Stutz [*Phys. Rev. Lett. (USA)* vol.74 (1995) p.450; also *Proc. Int. Conf. on Physics of Semiconductors* Vancouver (World Scientific, 1995) p.751; also *Phys. Rev. Lett. (USA)* vol.74 (1995) p.450]

[24] L.L. Chang, N. Kwai, G.A. Sai-Halasz, R. Ludeke, L. Esaki [*Appl. Phys. Lett. (USA)* vol.35 (1979) p.939]

[25] G.R. Booker et al [*J. Cryst. Growth (Netherlands)* vol.146 (1995) p.495]

[26] R.J. Nicholas et al [*Physica B (Netherlands)* vol.201 (1994) p.271]

[27] J. Shen, H. Goronkin, J.D. Dow, S.Y. Ren [*J. Vac. Sci. Technol. B (USA)* vol.13 (1995) p.1736]

[28] C. Nguyen, B. Brar, H. Kroemer [*J. Vac. Sci. Technol. B (USA)* vol.11 (1993) p.1706]

[29] R.M. Feenstra, D.A. Collins, D.Z.-Y. Ting, M.W. Wang, T.C. McGill [*Phys. Rev. Lett. (USA)* vol.72 (1994) p.2749]

[30] J. Spitzer et al [*J. Appl. Phys. (USA)* vol.77 (1995) p.811]

[31] S.J. Chung, A.G. Norman, W.T. Yuen, T. Malik, R.A. Stradling [to be published in Proc. 22nd Int. Symposium on Compound Semiconductors]

[32] C.R. Bolognisi, H. Kroemer, J.H. English [*J. Vac. Sci. Technol. B (USA)* vol.10 (1992) p.877]

[33] C.R. Bolognisi, H. Kroemer, J.H. English [*Appl. Phys. Lett. (USA)* vol.61 (1992) p.213]

[34] C. Nguyen, B. Brar, H. Kroemer, J.H. English [*J. Vac. Sci. Technol. B (USA)* vol.10 (1992) p.898]

[35] T.D. Golding et al [*Appl. Phys. Lett. (USA)* vol.63 (1993) p.1098]

[36] J.R. Meyer, C.A. Hoffman, T.D. Golding, J.T. Zborowski, A. Vigliante [*Inst. Phys. Conf. Ser. (UK)* no.144 (1995) p.1612]

[37] I. Mengalis, R.J. Phelan, R.H. Rediker [*Appl. Phys. Lett. (USA)* vol.5 (1964) p.99]

[38] H.K. Choi, S.J. Eglash [*Appl. Phys. Lett. (USA)* vol.59 (1991) p.1165]

[39] H.K. Choi, S.J. Eglash [*Appl. Phys. Lett. (USA)* vol.61 (1991) p.1154]

[40] H.Q. Le, G.W. Turner, S.J. Eglash, H.K. Choi, D.A. Coppeta [*Appl. Phys. Lett. (USA)* vol.64 (1994) p.152]

[41] S.J. Eglash, H.K. Choi [*Appl. Phys. Lett. (USA)* vol.64 (1994) p.833]

[42] H.K. Choi, G.W. Turner [*Appl. Phys. Lett. (USA)* vol.67 (1995) p.332]

[43] S.R. Kurtz, R.M. Biefeld, L.R. Dawson, K.C. Baucom, A.J. Howard [*Appl. Phys. Lett. (USA)* vol.64 (1994) p.812]

[44] A.N. Baranov, A.N. Imentkov, V.V. Sherstnev, Y.P. Yakovlev [*Appl. Phys. Lett. (USA)* vol.64 (1994) p.2480]

[45] R.H. Miles, D.K. Chow, Y.H. Zhang, P.D. Brewer, R.G. Wilson [*Appl. Phys. Lett. (USA)* vol.66 (1995) p.1921]

[46] D.K. Chow et al [*Appl. Phys. Lett. (USA)* vol.67 (1995) p.3700]

[47] Y.H. Zhang [*Appl. Phys. Lett. (USA)* vol.66 (1995) p.118: also *Int. Conf. on Narrow Gap Semiconductors* Santa Fe (IoP Conference Series - in press)]

[48] Y.H. Zhang, R.H. Miles, D.H. Chow [*IEEE J. Selected Topics Quantum Electron. (USA)* vol.1 (1995) p.749]

[49] H. Lee et al [*Appl. Phys. Lett. (USA)* vol.66 (1995) p.1942]

[50] T. Ashley, M. Carroll, C.T. Elliott, R. Jeffereis, A.D. Johnson, G.J. Pryce [to be published]

[51] C. Sitori, F. Capasso, J. Faist, D.L. Sivco, A.L. Hutchison, A.Y. Cho [*Appl. Phys. Lett. (USA)* vol.66 (1995) p.4]

[52] J. Faist, F. Capasso, C. Sitori, D.L. Sivco, A.L. Hutchison, A.Y. Cho [*Appl. Phys. Lett. (USA)* vol.66 (1995) p.538]

[53] C. Sitori, F. Capasso, J. Faist, D.L. Sivco, A.L. Hutchison, A.Y. Cho [*Appl. Phys. Lett. (USA)* vol.66 (1995) p.3242]

[54] J. Faist, F. Capasso, C. Sitori, D.L. Sivco, A.L. Hutchison, A.Y. Cho [*Appl. Phys. Lett. (USA)* vol.67 (1995) p.3057]

[55] H.Q. Le, A. Sanchez, J.M. Arias, M. Zandian, R.R. Zucca, Y.Z. Liu [*Int. Conf. on Narrow Gap Semiconductors* Santa Fe (IoP Conference Series - in press)]

[56] Z. Shi, M. Tacke, A. Lambrecht, H. Bottner [*Appl. Phys. Lett. (USA)* vol.66 (1995) p.2537]

[57] A.R. Adams [*Electron. Lett. (UK)* vol.22 (1986) p.249]

[58] S.R. Kurtz, R.M. Biefeld, L.R. Dawson [*Phys. Rev. (USA)* vol.51 (1995) p.7310]

[59] P.J.P. Tang et al [*Semicond. Sci. Technol. (UK)* vol.10 (1995) p.1177]

[60] M.J. Pullin, P.J.P. Tang, C.C. Phillips, Y.B. Li, A.G. Norman, R.A. Stradling [*Inst. Phys. Conf. Ser. (UK)* no.144 (1995) p.8-12]

[61] J.R. Lindle, J.R. Meyer, C.A. Hoffman, F.J. Bartoli, G.W. Turner, H.K. Choi [*Appl. Phys. Lett. (USA)* vol.67 (1995) p.3153]

[62] C.H. Grein, P.M. Young, M.E. Flatte, H. Ehrenreich [*J. Appl. Phys. (USA)* vol.78 (1995) p.7143]

[63] B.N. Murdin et al [*Inst. Phys. Conf. Ser. (UK)* no.144 (1995) p.267-71]

[64] T. Ashley [*Inst. Phys. Conf. Ser. (UK)* no.144 (1995) p.345]

[65] G.G. Fazio [*Infrared Phys. Technol. (UK)* vol.35 (1994) p.107]

[66] D.Z.-Y. Ting, E.T. Yu, D.A. Collins, D.H. Chow, T.C. McGill [*Appl. Phys. Lett. (USA)* vol.57 (1990) p.2675]

[67] E.R. Brown, J.R. Soderstrom, C.D. Parker, L.J. Mahoney, K.M. Molvar, T.C. McGill [*Appl. Phys. Lett. (USA)* vol.58 (1991) p.2291]

[68] E.R. Brown, S.J. Eglash, G.W. Turner, C.D. Parker, J.V. Pantano, D.R. Calawa [*IEEE Trans. Electron Devices (USA)* vol.41 (1994) p.879]

[69] J.S. Scott, J.P. Kaminski, S.J. Allen, D.H. Chow, M. Lui, T.Y. Lui [*Surf. Sci. (Netherlands)* vol.305 (1994) p.389]

[70] E. Ozbay, D.M. Bloom, D.H. Chow, J.N. Shulman [*IEEE Electron Device Lett. (USA)* vol.14 (1993) p.400]

[71] S. Tehrani, J. Shen, H. Goronkin, G. Kramer, M. Hoogstra, T.X. Zhu [*Proc. 20th Int. Symp. on Gallium Arsenide and Related Compounds* Freiburg, Germany, 29 Aug.-2 Sept. 1993 (IoP Publishing, Bristol, England, 1994) p.209-14]

[72] S. Tehrani, J. Shen, H. Goronkin, G. Kramer, R. Tsui, T.X. Zhu [*IEEE Electron Device Lett. (USA)* vol.16 no.12 (1995) p.557-9]

[73] L.F. Luo, R. Beresford, K.F. Logenbach, W.I. Wang [*J. Appl. Phys. (USA)* vol.68 (1990) p.2854]

[74] C.R. Bolognisi, E.J. Kane, H. Kroemer [*IEEE Electron Device Lett. (USA)* vol.15 (1994) p.16]

[75] X. Li, K.F. Logenbach, Y. Wang, W.I. Wang [*IEEE Electron Device Lett. (USA)* vol.13 (1992) p.192]

[76] B. Brar, H. Kroemer [*IEEE Electron Device Lett. (USA)* vol.16 (1995) p.548]

[77]   J.R. Boos, W. Krupps, D. Park, B.V. Shanabrook, R. Bennett [*Electron. Lett. (UK)* vol.30 (1994) p.1983 and earlier references]

[78]   J. Heremans, D.L. Partin, C.M. Thrush, L. Green [*Semicond. Sci. Technol. (UK)* vol.8 (1993) p.S424]

[79]   N. Kuze et al [*J. Cryst. Growth (Netherlands)* vol.150 (1995) p.1307]

[80]   I. Shibasaki [*Tech. Digest of 8th Sensor Symposium* (1989) p.211]

[81]   M. Inoue, K. Yoh, A. Nishida [*Semicond. Sci. Technol. (UK)* vol.9 (1994) p.966]; also K. Yoh, H. Taniguchi, K. Kiyomi, M. Inoue, R. Sakamoto [*Tech. Dig. (USA)* (IEDM 91 31.5.1) (1991)]

[82]   S.J. Koester, C.R. Bolognisi, E.L. Hu, H. Kroemer, M.J. Rooks [*Phys. Rev. B (USA)* vol.49 (1994) p.8514]

[83]   C. Nguyen, J. Werking, H. Kroemer, E.L. Hu [*Appl. Phys. Lett. (USA)* vol.57 (1990) p.87]

[84]   A.G. Milnes, A.Y. Polyakov [*Solid-State Electron. (UK)* vol.36 (1993) p.803]